Analog-Digital Converters for Industrial Applications Including an Introduction to Digital-Analog Converters

Frank Ohnhäuser

Analog-Digital Converters for Industrial Applications Including an Introduction to Digital-Analog Converters

Springer

Frank Ohnhäuser
eesy-ic GmbH
Nürnberg
Germany

ISBN 978-3-662-47019-0 ISBN 978-3-662-47020-6 (eBook)
DOI 10.1007/978-3-662-47020-6

Library of Congress Control Number: 2015942903

Springer Heidelberg New York Dordrecht London

Printed on acid-free paper

Springer Vieweg is a brand of Springer
Springer-Verlag GmbH Berlin Heidelberg is part of Springer Science+Business Media (www.springer.com)

Preface

With this book, I want to share the experiences which I gained during my long professional work on analog-to-digital converters (ADCs) based on successive approximation (SAR ADCs) and delta-sigma architectures (ΔΣ ADC). I had the luck to start my career at Burr-Brown, which was a specialized company for highest performance analog integrated circuits such as linear products and data converters. Our team concentrated on improving the performance of ADCs, in successive steps, in terms of accuracy, speed, and power consumption, which for me was a lot of fun as the products were getting better and better. I was able to continue my research studies on ADCs even after Burr-Brown merged with Texas Instruments in 2001 and after I joined eesy-ic in 2012. In particular, the invention of the dynamic error correction (DEC) for ADCs, based on successive approximation, opened the door for all kinds of algorithms that made the design of SAR ADCs particularly interesting. Later, new semiconductor processes with dual-gate oxides enabled further new and exciting ideas and achievements.

In 2007, I started teaching students about the topic of ADCs at the University of Erlangen (Germany) and, in between, also at the University of Klagenfurt (Austria). Teaching and coaching young talents is another area that provides me a lot of fun. Therefore, this book should engage students and engineers, who are new to the topic of ADCs, with explanations of the basic parameters and architectures of ADCs in Chap. 1. I tried to illustrate and explain the topics in a way that is easy to understand. The book then continues with state-of-the-art design methodologies for SAR ADCs in Chap. 2 and for ΔΣ ADCs in Chap. 4. Here design details and schematic examples are given, which should be useful for engineers starting with the design of ADCs and for experienced IC design engineers as well. These chapters further supply background information for development engineers of the end products that include ADCs in their application. Here, the circuits that generate the load on the analog input and the reference and the distortion on the power supply are particularly important. These topics are therefore again covered in Chap. 6. Prior to that, Chaps. 3 and 5 make an excursion into current research topics on SAR ADCs and, respectively, on ΔΣ ADCs. For SAR ADCs, Chap. 3 treats the dynamic error correction and introduces new algorithms using DEC that could achieve new levels in performance. The topic of continuous-time $\Delta\Sigma$ ADCs is chosen for Chap. 5.

During my work on ADCs, I also learned to appreciate the topics of verification and characterization, production test, and quality control. I spent many late hours on the test floor and in the lab to achieve yet another dB in SNR or THD performance or another fraction of an LSB in integral or differential nonlinearity. Actually, I gained the most valuable experience during the debugging of prototypes and test solutions. The full performance of an ADC can only be achieved and measured when the complete environment is optimized for it. Measuring accurately and reliably in the minimum time in mass production is particularly difficult. Young engineers should therefore consider, in particular, a career in the field of production test development. This book provides an introduction to this topic in Chap. 6.

During my work on ADCs, I was often confronted with auxiliary circuitry such as a voltage source to program the trigger level of window comparators that is using digital-to-analog converters (DACs), so Chap. 7 offers a helpful introduction to DACs and particularly explains the different architectures and the performance they are able to achieve. While detailed insights are given, however, with respect to my colleagues working on the design of DACs in general, this book does not claim to be a compendium on highest performance DAC design.

Finally, I want to thank all current and former colleagues and friends for the many interesting years developing ADCs, for the discussions we had, and for the joint developments we made. I am still getting excited if an interesting ADC project crosses my desk. Unfortunately, there are too many people to thank for me to name them individually; however, I want to list them at least as teams. First of all, I want to mention my team at eesy-ic, who still has to deal with me, including many team members, who already had to work together with me at Texas Instruments in Erlangen. I also particularly want to thank my former colleagues and friends at Texas Instruments in Erlangen (Germany), Tucson (Arizona, USA), and Bangalore (India). My special thanks however goes to my family, who had to suffer my absence during the many hours of my spare time that I spent writing this book.

Nürnberg, Germany Frank Ohnhäuser
April 2015

Contents

1 Introduction 1
1.1 Definition of Performance Parameters 3
1.1.1 DC Specifications 5
1.1.2 AC Specifications 13
1.1.3 Digital Interface 20
1.1.4 Power Specifications 24
1.1.5 Jitter 24
1.2 Overview of ADCs 25
1.2.1 Pipeline ADC 27
1.2.2 SAR ADC 29
1.2.3 Delta-Sigma ADC 31
1.3 ADCs Used in Industrial Applications 38
1.3.1 Electrical Motor Drives 38
1.3.2 Power Monitoring Applications 42
References 48

2 ADCs Based on Successive Approximation 51
2.1 Sample and Hold (S&H) 52
2.2 Capacitive DAC 56
2.2.1 Basic CDAC Architectures 59
2.2.2 Trim Solutions 61
2.2.3 Implementation of a Pseudo Differential 16 bit CDAC . . . 70
2.3 Comparator 84
2.3.1 Basic Comparator Topology 84
2.3.2 Layout Considerations 90
2.3.3 Noise Considerations 93
2.3.4 Dynamic Errors in the Comparator 96
2.4 Dynamic Error Correction 100
2.5 Reference Voltage 104
2.5.1 Internal Reference Buffer 104
2.5.2 Reference with External Capacitor 105

2.5.3 An Improved Reference Scheme 107
2.5.4 Noise of the Reference 108
2.6 Noise Calculation 109
2.6.1 A Novel Oversampling Method 111
2.6.2 Noise and Distortion Caused by the Power Supply 113
2.7 Limitations and Next Steps 115
References 117

3 Advanced SAR ADC Design 119
3.1 Floating CDAC Architecture 119
3.2 Dual CDAC Architecture 130
3.3 Digital Trim 141
3.4 INL Trim 145
3.4.1 Calculation of the INL 146
3.4.2 INL Compensation Circuit 149
3.4.3 Results and Measurements 154
3.5 Novel Method to Measure Capacitive Matching 155
3.6 Zero-Power Sampling 165
3.7 Highest Performance SAR ADC 174
3.7.1 Architecture of the Highest Performance ADC 174
3.7.2 Process Selection 179
3.7.3 Results 181
3.8 Lowest Power SAR ADC 192
3.8.1 Process Selection 193
3.8.2 Implementation and Design Challenges 195
3.8.3 Simulation Results 201
References 204

4 Basics on Delta-Sigma Converters 207
4.1 Delta-Sigma ADCs with Higher Order 211
4.2 Multi-bit Architectures 213
4.2.1 Calibration Techniques 216
4.2.2 Dynamic Element Matching 217
4.3 Dithering 219
4.4 Feed-Forward Architectures 220
4.5 Switched-Capacitor Approach 221
4.6 Differential Approach 223
4.7 Bandwidth and Noise 226
4.8 Offset Cancellation 229
4.8.1 Correlated Double Sampling 230
4.8.2 Chopper Techniques 231
4.9 Layout 233
References 234

5 Continuous-Time Delta-Sigma ADCs 237
5.1 Clock Jitter 239
5.1.1 Resistive Feedback 241
5.1.2 Return-to-Zero Architecture 244
5.1.3 Capacitive Feedback 245
5.1.4 Current Feedback 247
5.1.5 Multi-bit Feedback 249
5.2 Excess-Loop Delay 251
5.3 Integrator Requirements 255
5.4 CM Feedback 263
References 264

6 External Driver Circuitry and Test of ADCs 267
6.1 External Connections 267
6.1.1 Driving Amplifier 267
6.1.2 Reference Input 275
6.1.3 Grounding and Supply 278
6.1.4 Digital Interface 279
6.2 Test Methods and Production Test 281
6.2.1 DC Specifications 281
6.2.2 AC Tests 292
6.2.3 Power Consumption 294
6.2.4 Functionality and Continuity Testing 294
6.3 Quality Control in the Mass Production 296
6.3.1 Characterization 296
6.3.2 Capability Study 299
6.3.3 Lab-to-Final Correlation 299
6.3.4 Correlation Lockout 301
6.3.5 Quality Control Run 301
6.3.6 Quality Flow 302
References 303

7 Digital-to-Analog Converters 305
7.1 DAC Specifications 305
7.2 Current Steering DAC 308
7.3 String DAC 311
7.4 R-2R DAC 317
7.5 Delta-Sigma DAC 322
References 327

Index 329

Chapter 1
Introduction

Several statements such as 'We live in a digital world.' or 'Everything is digital.' exist since at least the 1990s that claim the end of analog circuits. The interesting fact is that the analog business is growing faster than the digital.

One reason surely might be the stability of the selling price, as the competition is significantly lower for analog. Why is this the case? Programming is something nearly every engineering student is performing in some way. The design of digital circuitry with VHDL or Verilog is very similar, so that the step for students from programming into the hardware design is small. Analog circuits on the contrary are done on transistor level, which is something that students are not often in contact with. A hesitance exists towards analog design. This book should give an understandable introduction to the analog/mixed signal world and particularly in converter design.

The analog market however can also show a fast growing number of units that are sold every year. The digital applications require feedback from the real analog world, which is given by sensors. Sensors output analog voltages or currents, which need to be conditioned and converted into the digital world. The signal conditioning includes the amplification of the signal, noise filtering and often linearization.

The most famous companies for high performance analog products are Texas Instruments and Analog Devices. Particular interesting is that in 2000, Texas Instruments was basically a DSP company with strong focus on the mobile phone business. Most important customers were Nokia or Ericsson. Within less than 10 years, TI did basically exit the mobile market and invested into the analog market by strategic acquisitions such as Burr-Brown, Unitrode, Powertrends, ChipCon or National Semiconductor. In 2011, nearly 50 % of the revenue came from analog products.

Coming back to financial growth, the numbers of TI do not reflect the growth of the total analog market, as the revenue growth was driven by acquisitions. Figure 1.1 therefore illustrates the revenue from Analog Devices (ADI) as an example.

When looking closer at analog applications, they can be separated into two different fields, the consumer area and the business market. The consumer market

F. Ohnhäuser, *Analog-Digital Converters for Industrial Applications Including an Introduction to Digital-Analog Converters*, DOI 10.1007/978-3-662-47020-6_1

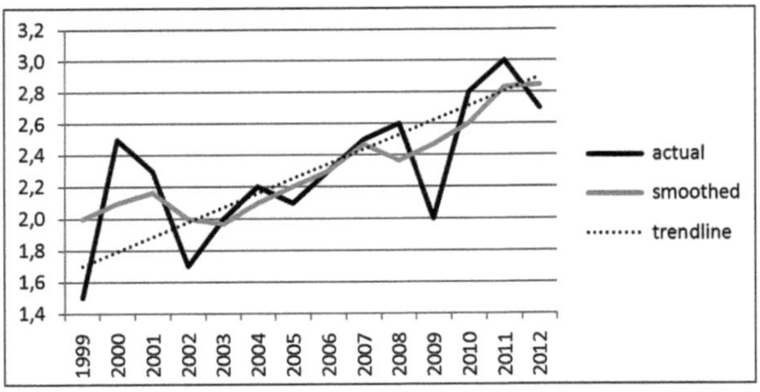

Fig. 1.1 Analog market growth illustrated by the yearly revenue of ADI

is driven by applications such as mobile phones (RX, touch), networks, wireless, play stations, TVs, cameras, audio systems or white goods.

The business side also has a huge variety of applications. There are medical applications such as EVC (EKG), X-ray, CT scanners, which require high resolution at lower speed and particularly a high channel count. Similar to X-ray are security systems such as luggage scanners. Flow meters and motor drives are required in the industrial process environment. A booming new market is the smart grid, where smart-meters are required as well as protection circuits or power inverter. These applications should in the following be called industrial applications.

If being very general, then these two markets differ in the following form. While consumer products need to be cheap, they will be processed on cost efficient processes with small dimensions. If possible the analog circuitry will be integrated together with digital circuits to lower the price on the cost of performance.

In the business area, the costs are still an important parameter, but the required performance has the higher focus. Such performance parameters are the resolution, the linearity, the noise, the power consumption or a robust input voltage range.

This book concentrates on data converters, which are the interface between the analog and the digital world. Analog-to-digital converters (ADC) provide analog inputs for voltages or currents and convert the analog input into a digital word. ADCs are often used to sample and digitize analog sensor outputs, so that the sensor information can be used in digital control algorithms. These control algorithms might further have digital outputs, which are setting parameters in the analog circuitry. In this case, digital-to-analog converters are required. They generate an analog output from a digital input word. Audio applications are a famous example, where DACs are in use. The digital outputs of CDs, DVDs or MP3 streams need to be converted to an analog signal before being applied to a loudspeaker.

As a starting point, the basics of analog-to-digital converters will be presented, which are the definition of the various performance parameters and the different converter types. The book will then continue with some applications together with their technical requirements, which should help to understand the performance parameters. Industrial applications mostly utilize converters based on successive approximation (SAR) and delta-sigma converters. Chapter 2 will describe state-of-the-art techniques and circuit blocks for SAR ADCs. It will discuss design and layout topics. Newest research topics for SAR ADCs are presented in Chap. 3. Circuit architectures for delta-sigma converters, which are state-of-the-art, follow in Chap. 4 and the actual research topic of 'continuous-time delta-sigma ADCs' in Chap. 5.

Converters are mostly implemented in switched-capacitor designs and have a capacitive kick-back to the analog input signals. A proper driver circuitry is required in an application or on test boards for the analog input signal, the reference signal and for supply voltages, which are discussed in Chap. 6.

The development of integrated circuits does not stop at the design. Each IC needs to be tested in production for each specified parameter. Converter noise and parasitic effects, which are also discussed in Chap. 6, make the test of high-performance analog-to-digital converter a real challenge. Costs for the test are dominating the production. Unfortunately, the very interesting and challenging topic of test is rarely taught at universities, so that engineers for production test development are rare as well. Converter test is also described in Chap. 6.

To complement the topic of converters, an excursion to digital-to-analog converters (DAC) is provided in Chap. 7. Please note that the Sects. 1.1, 1.2, Chaps. 2 and 6 are based on [1].

1.1 Definition of Performance Parameters

Users of ADCs will define a converter by different performance parameters. For some applications, it is most important that no codes are missing in the transfer function. Other applications require low noise. Therefore, it is important to know the definitions of the different specified parameters, which are defined by IEEE Std 1241-2000 [2]. This is especially true as the resolution only describes the width of the digital output word. The resolution does not specify the accuracy of the ADCs. Products might claim a 20 bit resolution but are so noisy that only 12 of the 20 bits are stable. The performance parameters can be separated in the DC and the AC parameters and are discussed in the Sects. 1.1.1 and 1.1.2.

Many parameters are expressed in least-significant bits (LSB). An LSB is the total input voltage range (see Sect. 1.1.1.1) divided by the total number of codes. If a 12 bit converter (4096 codes) has an input range of 4.096 V, then an LSB is exactly 1 mV.

An ideal transfer function for a 3 bit ADC is shown in Fig. 1.2. It starts half an LSB below the lowest code transition (negative full-scale or NFS) and ends 1.5 LSBs above the highest code transition (positive full-scale or PFS). A code transition is defined as the analog input voltage, where the ADC changes its output code from a previous value to the next. The DC specifications describe, if the code transitions are at the right place. In the following, they will be explained with the example of a parallel converter.

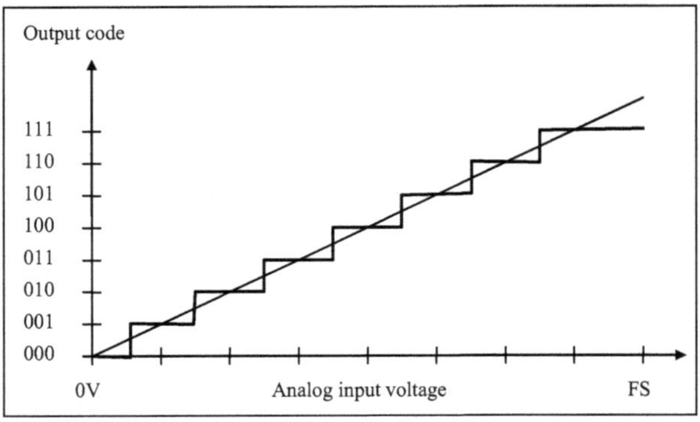

Fig. 1.2 Ideal transfer function of a 3 bit ADC plotting the output code versus the input voltage

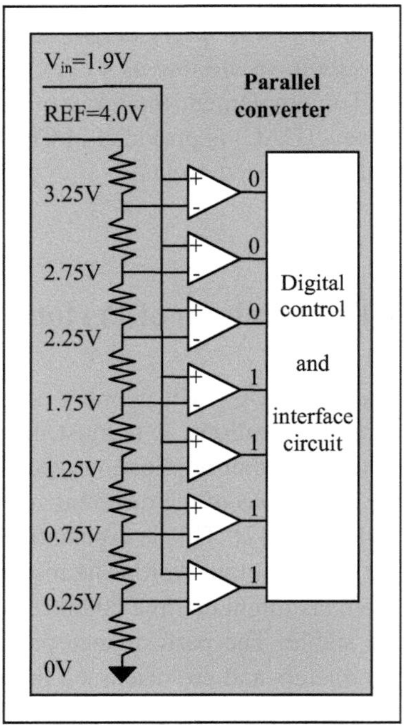

Fig. 1.3 Three bit parallel ADC

The parallel converter or also called flash converter is generating the voltage of each code transition with a resistive divider and is comparing the input voltage to each transition voltage in parallel.

Figure 1.3 shows an example of a 3 bit parallel converter. A resistive divider is typically used to generate the required transition voltages from a reference voltage *REF*, which are then connected to the inverting input of comparators. The analog input voltage V_{in} is connected to the non-inverting inputs.

In this example the analog input voltage is 1.9 V, so that the lower four comparators output ones and the higher three comparators zeros. This coding is called thermometer code and has to be transferred into a binary format by the digital circuitry. In the example of Fig. 1.3, the binary output is 100.

It can be seen that $2^n - 1$ comparators are required, where n is the resolution of the converter. The complexity is increasing exponentially with the resolution. Pure parallel converters do typically not exceed 10 bits.

A complete conversion is performed within every clock cycle. This makes the parallel converter very fast. Sample rates typically exceed 1 GSPS.

There are a few architectures that do some analog coding to reduce the complexity of the converter. Most famous are the folding converters [3].

1.1.1 DC Specifications

The DC specifications describe the accuracy of the transfer function of the converter without dynamic effects. Dynamic effects such as input frequency dependent non-linearity caused by the sample-and-hold stage are covered in the AC specifications that follow in Sect. 1.1.2.

1.1.1.1 Gain and Offset

The transfer function of every converter has defined end points. Unipolar converters start at 0 V and end at the reference voltage or a multiple of the reference voltage. Bipolar converters have a typical input range of minus the reference voltage to plus the reference voltage or multiples of the reference (see Fig. 1.4).

Normally, signals utilize a single wire and are referred to ground. Such signals are called single-ended signals. Such signals work particularly well, when applications use ±15 V supplies and the ground is free of return currents.

Moving to smaller semiconductor process reduced the possible supply voltage range and 5 V single supply applications got standard. Here, the supply current returns to the source through the ground, which also is the reference potential for single-ended signals. Any DC supply current can build up DC voltages across a

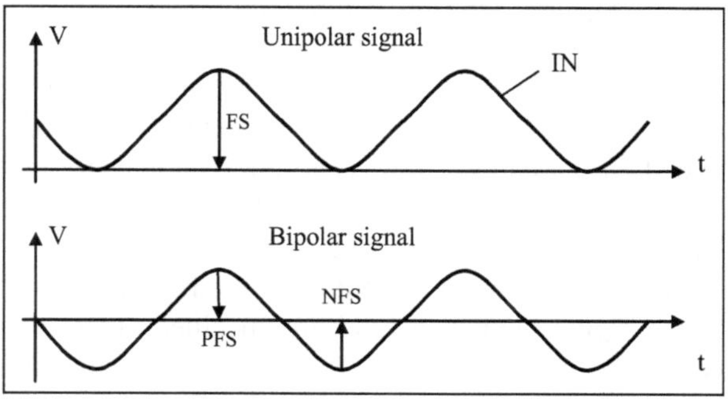

Fig. 1.4 Illustration of unipolar and bipolar input voltage range

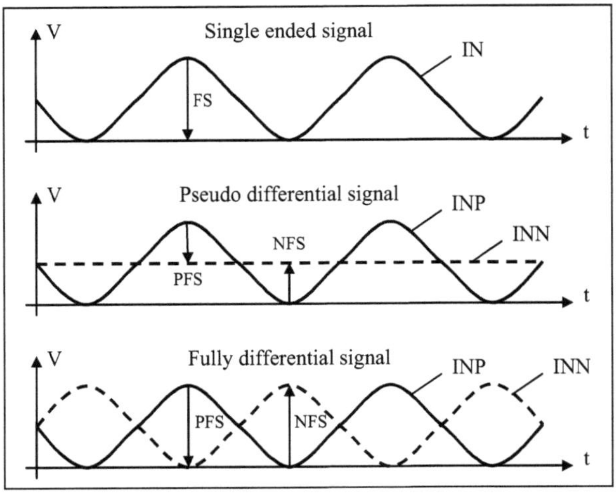

Fig. 1.5 Illustration of the different signal types used in electrical circuits

ground resistance. Solid ground planes are required. Worse are clocked systems that also add a pulsed current through the ground, where unwanted voltages build up through parasitic ground resistances and inductances. Clocked systems are for example digital circuits or DC-DC converters. If the ground in the application, which is used as reference for the single-ended signals, is different than 0 V, then an error is added to signal.

The charge-injection of switches is another error source to single-ended signals. Switches are particularly required in sample-and-hold circuits.

Therefore, modern single supply ADCs are typically differential. This means that the difference in voltage between a positive input *INP* and a negative input *INN* is converted (see Fig. 1.5). Noise and distortions, which are common to both inputs, are suppressed by the differential input structure. Pseudo differential inputs require one input to remain at a constant voltage, which typically is 0 V, but can be 2.5 V for a bipolar ADC. If the ground or 2.5 V connection is tapped in the application at the signal source and routed in parallel to the signal, then the application is getting insensitive from distortion in the ground.

Finally, the negative input of fully differential signals is anti-parallel to the positive input. This allows doubling the input range without leaving the supply range. This way, the signal amplitude can be doubled, which might be important to increase the so called signal-to-noise ratio.

A cutout for the analog input parameters of a typical datasheet is given in Table 1.1. In this example, the ADS12D1800RF from Texas Instruments [4] with folded flash architecture was chosen. Examples for other converter types are given in Sect. 1.2. It can be seen, that a fully differential signal is specified as a 'Full-Scale Differential Input Range'. The common-mode input voltage is limited under 'Voltage Range'. Often, not the common-mode voltage, but the maximum voltage range of the signals INP and INN are specified. This is done, as the common-mode voltage normally is a function of the differential signal amplitude.

Table 1.1 Analog input section of the ADS12D1800RF datasheet [4]

Parameter	Min	Typ	Max	Unit
Analog input				
Voltage range	−0.4		2.4	V
Full-scale differential input range	740	800	860	mVpp
Differential input resistor	91	100	109	Ω
Input capacitance (differential)		0.02		pF
Input capacitance (pin to ground)		1.6		pF

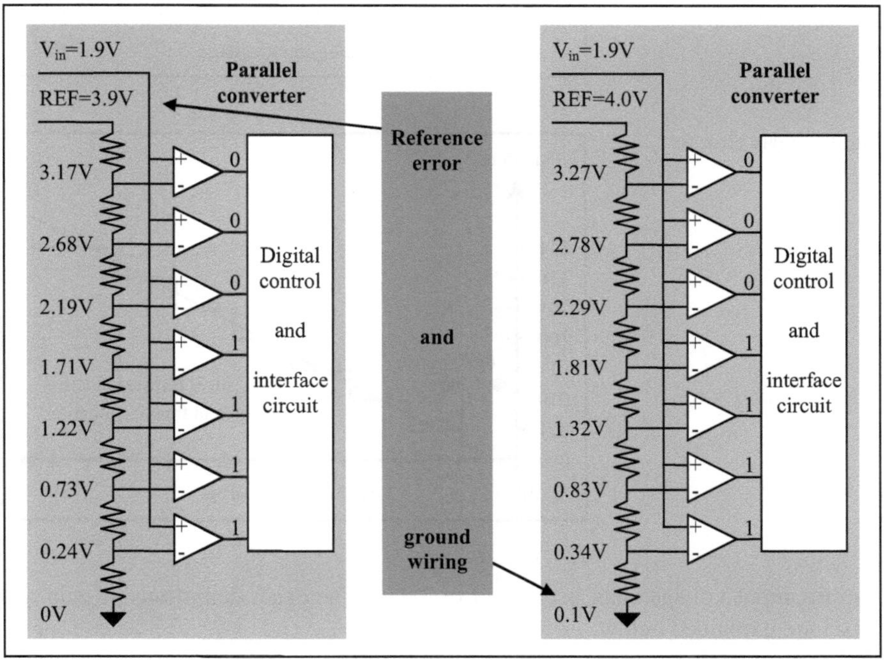

Fig. 1.6 Parallel converter with faulty reference or ground potentials causing end point errors

Beside the input voltage range, the input resistance and input capacitance are typically described as this is important information for the analog driver circuitry.

The end-point errors are defined by the gain and the offset. They are illustrated in the example of the 3 bit parallel converter, which was previously introduced. The ideal first transition for the flash converter in Fig. 1.3 would occur at 0.25 V and the last transition at 3.25 V. In Fig. 1.6, the reference voltage is too low on the left converter. The Negative Full-scale (NFS) is still at 0 V, but the input voltage span is only 3.9 V instead of 4 V. The negative full-scale of the right converter is shifted due to a voltage drop in the ground resistance. This resistance can be caused by a chip-internal wiring from the ground pad to the first resistor or by the ESD (Electro-Static-Discharge) protection cell. Not only that the negative full-scale is shifted, but

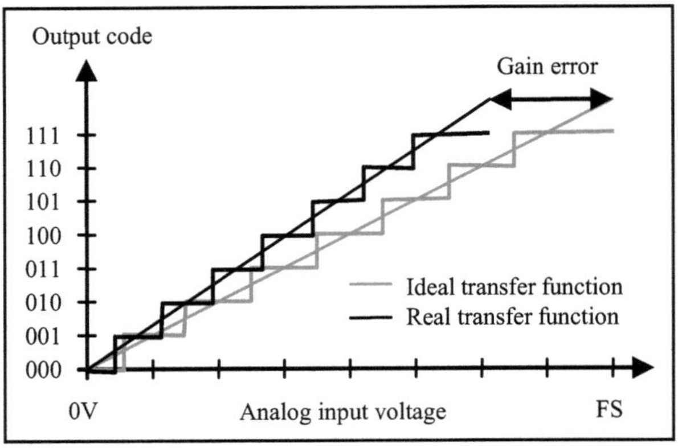

Fig. 1.7 Gain error for unipolar ADCs

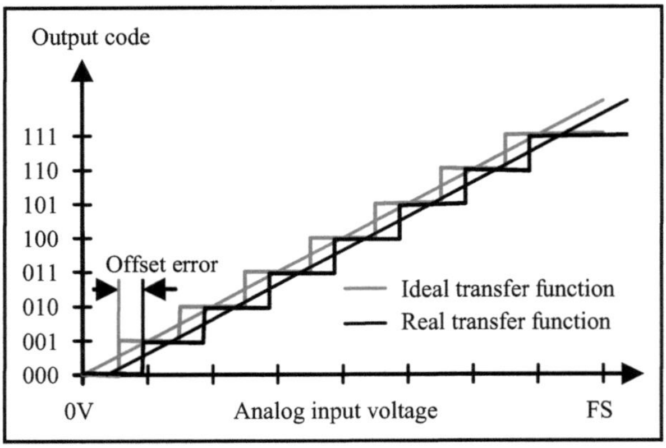

Fig. 1.8 Gain error for bipolar ADCs

also the input voltage span is reduced to 3.9 V. The exact definitions of gain and offset are explained below.

The gain error describes the difference between the ideal input voltage range (full-scale range or FSR) and its real value. The input voltage range is measured to the voltage of the highest code transition minus the input voltage of the first code transition plus two LSBs. The gain error is typically expressed in percent of the full-scale range (also see Figs. 1.7 and 1.8).

$$GainError = \frac{realFSR - idealFSR}{idealFSR} \tag{1.1}$$

The ideal FSR might be identical to the reference voltage *REF*. In this case, the gain error equals

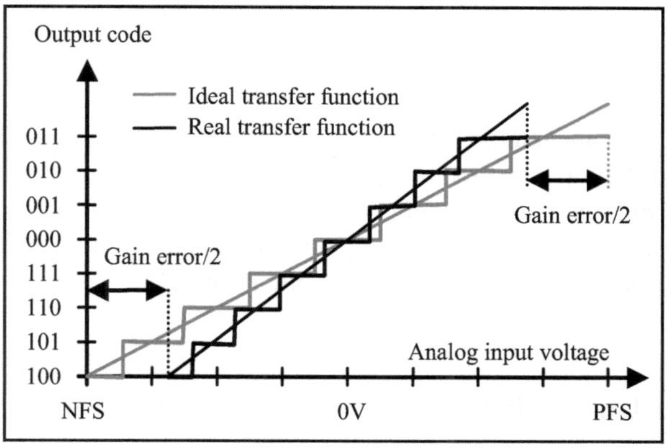

Fig. 1.9 Offset error for unipolar ADCs

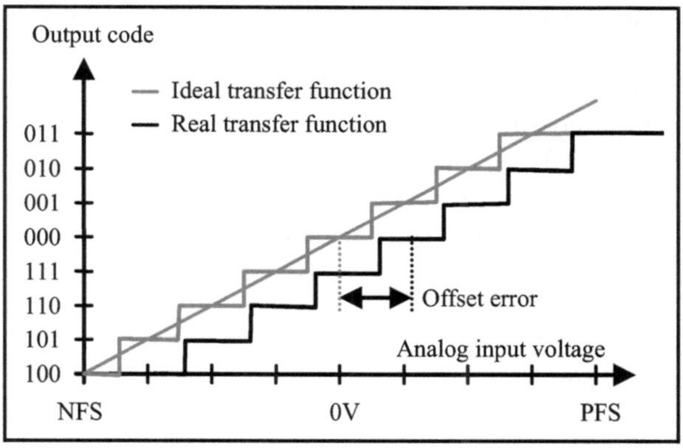

Fig. 1.10 Offset error for bipolar ADC

$$GainError = \frac{(PFS - NFS)}{REF} - 1$$

Caution Some products define a positive and a negative gain error, which should be the PFS minus the mid-scale (MS) and the mid-scale minus the NFS, but they are still referenced to the total FSR. This divides the number in the datasheet by a factor of two and is only used to make a product look better.

The offset error is the shift in the transfer function at a 0 V input voltage. It is therefore also called unipolar zero error or bipolar zero error. The unipolar zero error is identical with the NFS. The first code transition ideally has to occur half an LSB above 0 V, so that the offset error can be measured to the voltage of the first code transition minus half an LSB (see Fig. 1.9). If 0 V are applied to an ideal bipolar converter, then its output should show mid-scale (MS, code 000 in Fig. 1.10), which is half an LSB above the transition between code 111 and 000.

$$UnipolarOffsetError = NFS \tag{1.2a}$$

$$BipolarOffsetError = MS \tag{1.2b}$$

The gain and the offset are often affected in the application by amplification circuits utilizing resistors in the signal path. Therefore, gain and offset are typically calibrated in an application, so that their absolute errors are mostly uncritical. However, only a few applications are calibrated over temperature, so that the gain drift and offset drift are often critical.

1.1.1.2 Differential and Integral Non-Linearity

Every code is supposed to be one LSB wide. If a code width in the transfer function of a converter is different than one LSB, then the code has a differential non-linearity (DNL), which is expressed in Eq. (1.3). Such a non-linearity can have several reasons, like resistor or capacitor mismatch, comparator offsets or voltage coefficients. These error sources are illustrated in Fig. 1.11 for a parallel converter. Here, the comparator offset is included in an effective transition voltage, so that the transitions five and six occur both at an input voltage of 2.5 V.

The DNL is negative for short codes and positive for long codes. If a code does not exist (DNL = −1 LSB), then the converter has 'missing codes'.

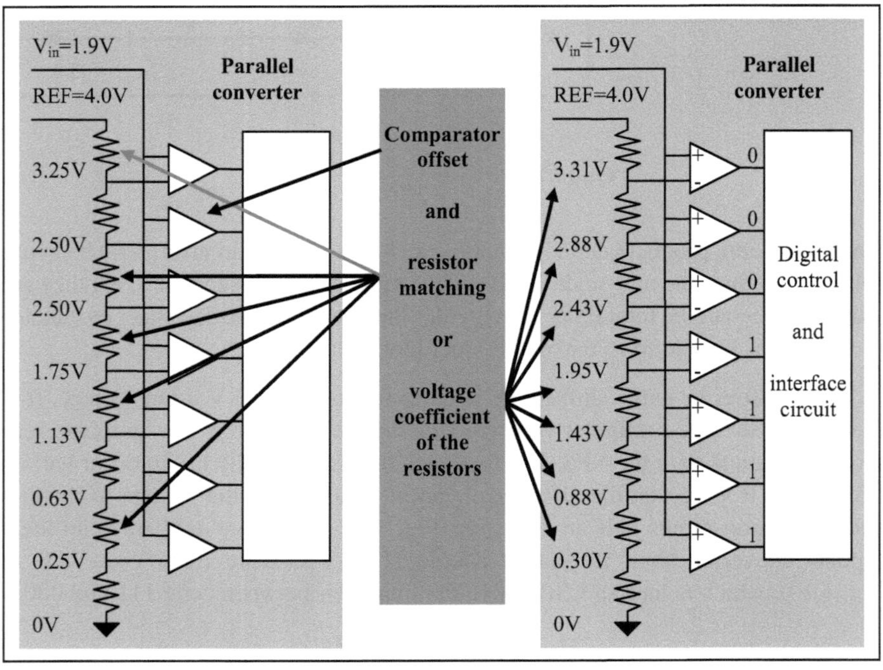

Fig. 1.11 Parallel converter with linearity errors caused by limited resistor accuracy

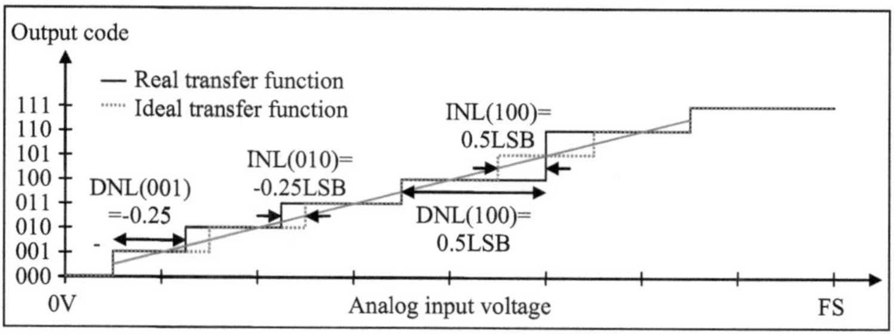

Fig. 1.12 Illustration of linearity errors in an ADC's transfer function

Table 1.2 Calculation of the linearity error of the transfer function in Fig. 1.12

Code	001	010	011	100	101	110
Code width (LSB)	0.75	1	1.25	1.5	0	1.5
DNL (LSB)	−0.25	0	0.25	0.5	−1	0.5
INL	−0.25	−0.25	0	0.5	−0.5	0

Mathematically, the integral non-linearity (INL) sums the differential non-linearity from code 1 to the code of interest. If a straight line is drawn between the NFS and the PFS, then the INL expresses how far a transition is away from that line. Figure 1.12 shows an example with the linearity errors from the left parallel converter of Fig. 1.11. Table 1.2 further explains these linearity errors.

$$DNL(code) = code\ width - 1\ \text{LSB} \tag{1.3}$$

$$INL(code) = \sum_{x=1}^{code} DNL(x) \tag{1.4}$$

Caution: The INL is referred to the line between the endpoints of the transfer function. Several datasheets on the market refer to a best-fit-curve. Here, the reference line is placed, so that the INL is minimized. This makes the INL look better by a factor of two. This fact is demonstrated in Figs. 1.13 and 1.14.

Offset, gain and the two linearity errors are known as DC specifications, because they are measured with DC input voltages. The particular section of the ADS12D1800RF datasheet is shown in Table 1.3. This example includes a best-fit specification for the INL and a separation of the gain error in positive and negative full-scale error.

The DC specifications do not tell how noisy the ADC is, or how well AC input signals are converted. This is expressed by the AC specifications.

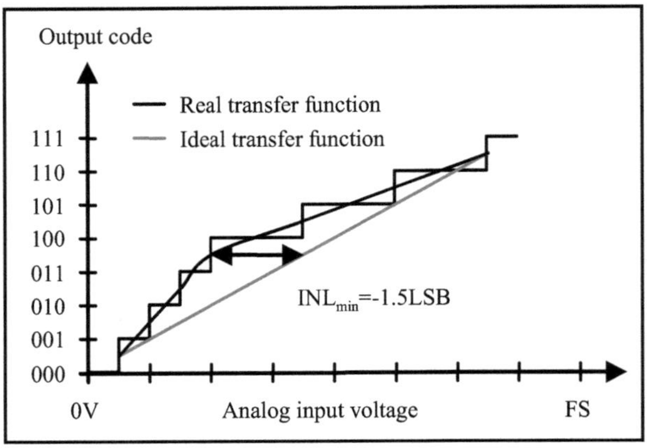

Fig. 1.13 INL measured through the endpoints

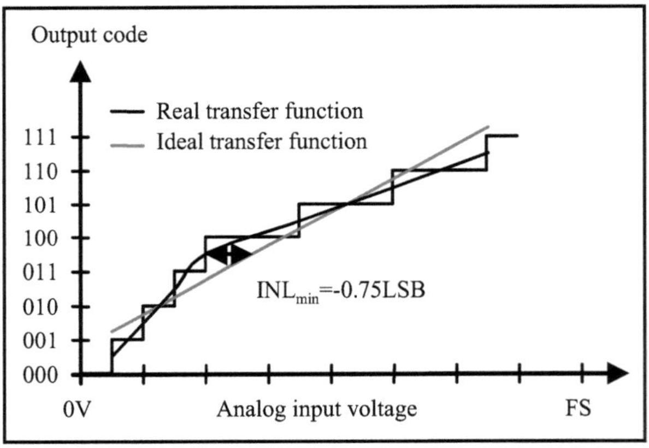

Fig. 1.14 INL measured with a best-fit-curve

Table 1.3 Exemplary DC accuracy section of the ADS12D1800RF datasheet [4]

Parameter	Min	Typ	Max	Unit
Accuracy				
No missing codes	12			Bits
Differential linearity error		±0.4		LSB
Integral linearity error (best-fit)		±2.5		LSB
Positive full-scale error			±25	mV
Negative full-scale error			±25	mV
Offset error		5		LSB

1.1.2 AC Specifications

A digital output code from the converter is expressing a very particular voltage of $V_{code} = code \cdot LSB$. On the other side a complete input voltage range of one LSB is reflected with this code. The delta between the actual input voltage V_{in} and the expressed voltage V_{code} is called 'quantization noise'. The converter is adding further noise to the signal. The total noise is evaluated in the frequency domain.

Base is the Fourier transformation, which is expressed with the Eqs. 1.5 and 1.6

$$F(f) = \int_{-\infty}^{+\infty} f(t) \cdot e^{-j2\pi f t} dt \tag{1.5}$$

$$f(t) = \int_{-\infty}^{+\infty} F(f) \cdot e^{+j2\pi t f} df \tag{1.6}$$

The ADC is converting a time-continuous signal $f(t)$ into a time-discrete signal $f_n(t)$, which is illustrated in Fig. 1.15. Mathematically, the time-continuous signal is multiplied with equidistant pulses $i(t)$

$$i(t) = T_s \cdot \sum_{n=-\infty}^{+\infty} \delta(t - nT_s)$$

The sampled signal $f_n(t)$ can be expressed with

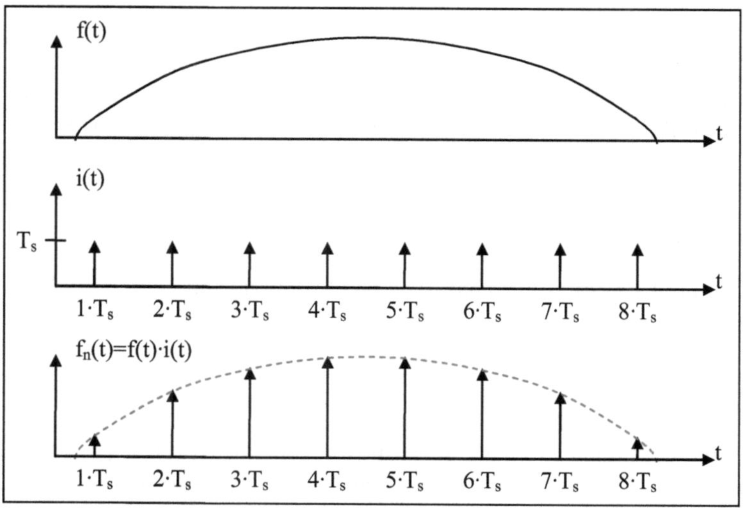

Fig. 1.15 System theoretical generation of a sampled signal with equidistant impulses

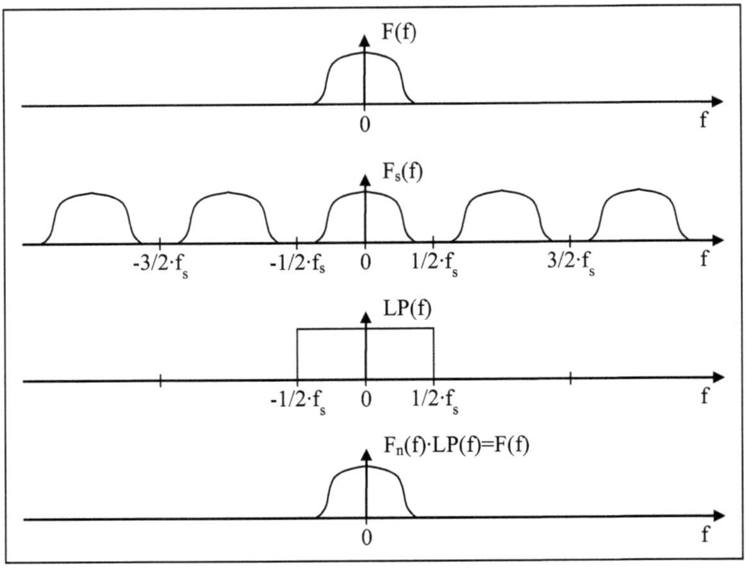

Fig. 1.16 Visualization of the Nyquist theorem

$$f_n(t) = f(t) \cdot \left[T_s \cdot \sum_{n=-\infty}^{+\infty} \delta(t - nT_s) \right] \tag{1.7}$$

Equation (1.8) expresses the Fourier transformed function of Eq. (1.7). Details can be found in the literature [5, 6].

$$F_s(f) = \sum_{\nu=-\infty}^{+\infty} F(f - \nu \cdot f_s) \tag{1.8}$$

The equation shows that a sampled signal $f_n(t)$ generates a periodic spectrum, where the width of the period is the sampling frequency f_s (see Fig. 1.16). The sampling frequency or also called sample rate f_s equals $\frac{1}{T_s}$.

The time-continuous signal can be fully regenerated as shown, if the following two conditions are fulfilled:

(a) The initial spectrum $F(f)$ has to be band-limited from $-\frac{f_s}{2}$ to $+\frac{f_s}{2}$. In this case, $F(f)$ and the base spectrum $(-\frac{f_s}{2} < f < \frac{f_s}{2})$ of $F_s(f)$ are identical.
(b) The sampled signal has to be low-pass filtered $LP(f)$ with a corner frequency of $\frac{f_s}{2}$.

The time-continuous signal can be fully regenerated by low-pass filtering the sampled signal. The maximum signal frequency f_{max} has to be less than half the sample rate $f_s = \frac{1}{T_s}$

$$f_{\max} < \frac{f_s}{2} \tag{1.9}$$

Equation (1.9) is called the Nyquist theorem.

The theory has some limitations. First of all, an ideal low-pass filter is required, which does not exist. Furthermore, sampled signals are expressed as a chain of weighted pulses, while a digital signal would be regenerated with a DAC that outputs a rectangular signal. Furthermore, equidistant values are required, so that a jitter of the sampling times $n \cdot T_s$ generates another error. Finally, the ADC adds the quantization error to the signal.

The spectrum $F_s(f)$ is discrete and periodic, if $f_n(t)$ is not only time-discrete, but also periodic. The signal can be represented with N equidistant discrete values $f[n \cdot T_s]$ in both the time and frequency domain with

$$F\left[\frac{\mu}{N} \cdot f_s\right] = T_s \cdot \sum_{n=0}^{N-1} f[n \cdot T_s] \cdot e^{-j2\pi\mu n/N} \tag{1.10}$$

$$f[n \cdot T_s] = \frac{f_s}{N} \cdot \sum_{\mu=0}^{N-1} F\left[\mu \cdot \frac{f_s}{N}\right] \cdot e^{j2\pi n\mu/N} \tag{1.11}$$

$F\left[\frac{\mu}{N} \cdot f_s\right]$ is expressing the amplitude of the time-continuous signal with the frequency $\frac{\mu}{N} \cdot f_s$. It is also called the amplitude density and has the unit $\frac{\mathrm{V}}{\mathrm{Hz}}$. In the following, such a frequency range is called bin. In the following, the square of the amplitude density of bin μ will be called the power P of the bin μ

$$P[\mu] = F^2[\mu]$$

With Eq. (1.10), the spectrum can be calculated with the N ADC results. This transformation is called the Discrete Fourier Transformation (DFT). Unfortunately, $(2 \cdot N)^2$ multiplications are required for the transformation. If N is a power of 2 ($N = 2^k$), then the algorithm can be modified, which is described by the Fast Fourier Transformation (FFT). The algorithm is explained in the literature [5, 6] and will not further be discussed.

Summarized, the spectrum can be calculated with the output data of the ADC using the FFT algorithm, if one or several periods of a time-continuous signal are sampled equidistantly N-times, with $N = 2^k$. This is also called coherent sampling.

If a sine wave signal is converted by an ADC, then the signal amplitude will fall into one bin, while the other bins contain noise and harmonics. Harmonics are amplitudes that occur with frequencies, which are a multiple of the frequency of the input signal. They are generated, if the transfer function of the ADC is non-linear.

Figure 1.17 shows the result of an FFT with 4096 samples, where a 10 kHz sine wave signal was converted with a 16 bit ADC at a data rate of 1.024 MSPS. N samples generate $\frac{N}{2}$ positive frequency bins. Based on the Nyquist theorem the

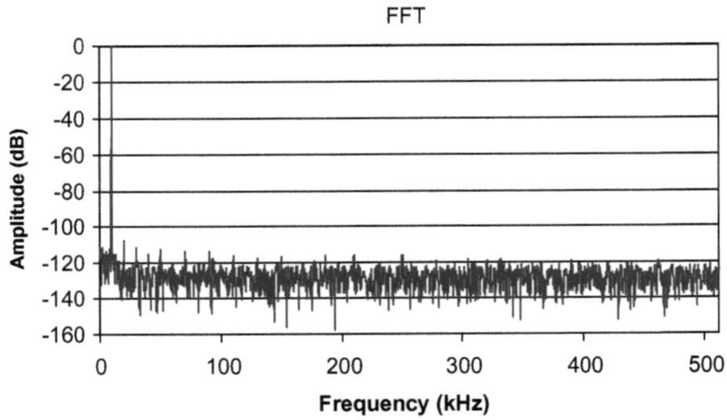

Fig. 1.17 Exemplary FFT with 4096 samples of a 16 bit, 1 MSPS converter

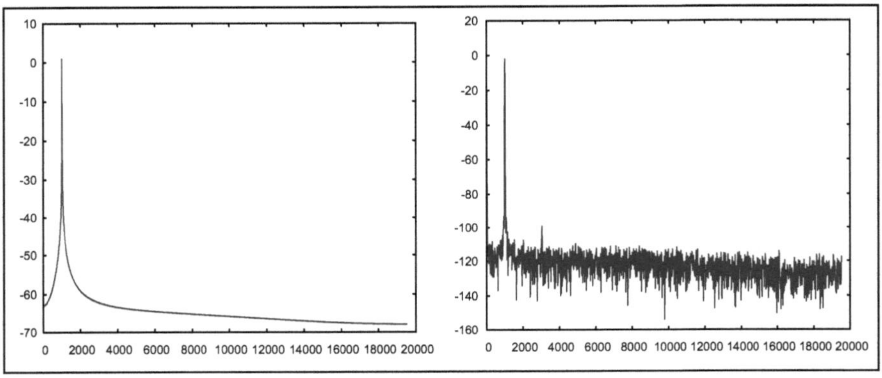

Fig. 1.18 Non coherent FFT result and FFT after applying a Blackman-Harris window

highest frequency bin is equal to half the data rate (512 kHz). Each bin is a sample of the spectrum and the bins are 1024 kHz/4096 = 250 Hz apart. The lowest bin carries the amplitude of the DC input voltage. The amplitude of the 10 kHz input sine wave can be found in bin 40.

The input signal is called the fundamental or the first harmonic. The signal at twice the input frequency would be the second harmonic and so on. For the example above, the harmonics can be found in bins 80, 120, 160 and so on. All other bins carry noise.

Figure 1.18a shows the result of a FFT, where the input signal was not sampled coherently. Coherent sampling for example is not possible if the converter runs from an internal clock signal. In such a case, the input signal can mathematically be conditioned with a window function, which forces the beginning and the end to be at zero. A variety of window functions exists [7], but should not be discussed within this book. The FFT result with the same input data from Fig. 1.18a but after applying a Blackman-Harris window is drawn in Fig. 1.18b.

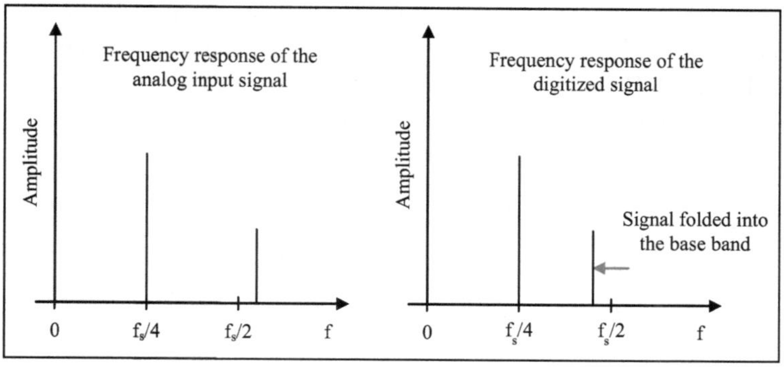

Fig. 1.19 Signal theoretical folding process during the digitizing process

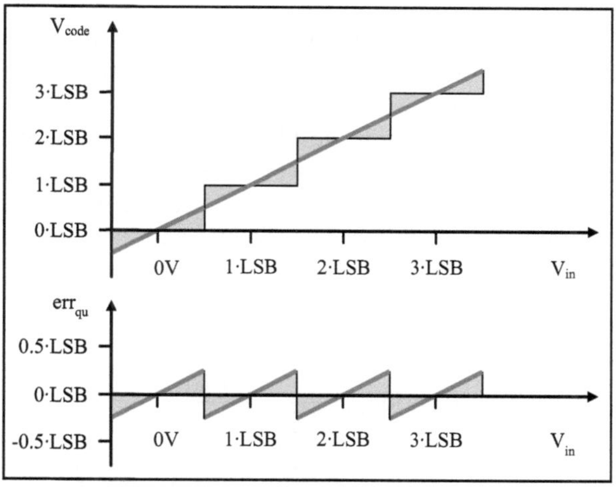

Fig. 1.20 Transfer function of an ADC illustrating the quantization noise err_{qu}

Signals with higher frequencies than half the data rate are folded into the baseband as shown in Fig. 1.19. If for example a fundamental is in bin 1000 of 2048 total bins, then the second harmonic will fall into bin 2000, the third harmonic into bin 1096 (2048-(3·1000-2048)), the fourth harmonic into bin 96, the fifth harmonic into bin 904, the sixth harmonic into bin 1904 and so on.

The digital output of an *n*-bit converter is expressing a voltage corresponding to an integer multiple of one LSB.

$$V_{code} = code \times LSB$$

The difference between the input voltage V_{in} and the voltage V_{code}, which is represented by the digital output *code* is called quantization noise err_{qu} (see Fig. 1.20).

The average power of the quantization noise N_{qu} is calculated with Eq. (1.12). Equation (1.13) is expressing the signal power S, if a sine wave is applied to the input of the ADC. Equations (1.12) and (1.13) can then be used to determine the ideal signal-to-noise ratio (SNR), which is illustrated in Eq. (1.14).

$$N_{qu} = \frac{1}{LSB} \int_{-0.5\,\mathrm{LSB}}^{0.5\,\mathrm{LSB}} err_{qu}^2 dV_{in} = \frac{1}{LSB} \int_{-0.5\,\mathrm{LSB}}^{0.5\,\mathrm{LSB}} V_{in}^2 dV_{in} = \frac{LSB^2}{12} \tag{1.12}$$

$$S = \left(\frac{FS}{2\sqrt{2}}\right)^2 = \frac{FS^2}{8} = \frac{(2^n \cdot LSB)^2}{8} = 2^{2n-3} \cdot LSB^2 \tag{1.13}$$

$$SNR = 10\log_{10}\left(\frac{S}{N_{qu}}\right) = 10\log_{10}\left(\frac{3}{2} \cdot 2^{2n}\right) = 6.02 \cdot n + 1.76 \tag{1.14}$$

Based on Eq. (1.14), an ideal 16 bit ADC will achieve 98 dB in SNR. Unfortunately, the quantization is not the only noise source. Sampling noise, thermal noise and flicker noise from several sources will be added.

The SNR can be calculated with the FFT data. The power of the signal bin s needs to be divided by the sum of the power of all other bins, beside the DC bin P [1] and the nine harmonics (P[k · s] with k = [2, 3, ..., 9]), which are used to calculate the total-harmonic-distortion (THD) with respect to the IEEE specifications.

$$SNR = 10 \times \log_{10}\left[\frac{P[s]}{\sum_{n=2}^{N/2} P[n] - \sum_{k=1}^{9} P[k \cdot s]}\right] \tag{1.15}$$

Equation (1.14) can now be taken to calculate the 'effective number of bits' (ENOB)

$$ENOB = \frac{SINAD - 1.76}{6.02} \tag{1.16}$$

Equation (1.16) is using the SINAD (signal to noise and distortion), which does not only include the noise bins, but also the bins of the harmonics, which is expressed in Eq. (1.17). The harmonics are typically small at low signal frequencies, so that the SNR and the SINAD have similar results.

$$SINAD = 10 \cdot \log_{10}\left[\frac{P[s]}{\sum_{n=2}^{N/2} P[n] - P[s]}\right] \tag{1.17}$$

The harmonics express the non-linearity of a converter and not the noise. Therefore, they are summarized in the total-harmonic-distortion (THD), which takes the sum of the power of the harmonics $P[k \cdot s]$ into ratio to the power of the signal $P[s]$. In typical specifications, the THD includes the first nine harmonics. Some companies might specify the THD with three harmonics only.

$$THD = 10 \cdot \log_{10}\left[\frac{\sum_{k=2}^{9} P[k \cdot s]}{P[s]}\right] \tag{1.18}$$

The spurious-free-dynamic-range (SFDR) expresses the difference between the signal power and the maximum power of the remaining frequency bins.

AC measurements above 40 kHz and with a performance of more than 85 dB in SNR are very difficult to perform, due to the limited performance of available signal generators. In many applications, the SNR is therefore estimated with a DC measurement. A well filtered DC input voltage is applied to the ADC and converted several times. The digital data is collected in a histogram like shown in Fig. 1.21 for a 16 bit SAR ADC.

In the estimation, the rms-noise equals the standard deviation of the output code, which is calculated to 0.50 LSB for the example of Fig. 1.21. The full-scale rms signal of a 16 bit converter is $\frac{2^{16}\text{LSB}}{2\sqrt{2}} = 23170$ LSB, so that the SNR can be estimated to

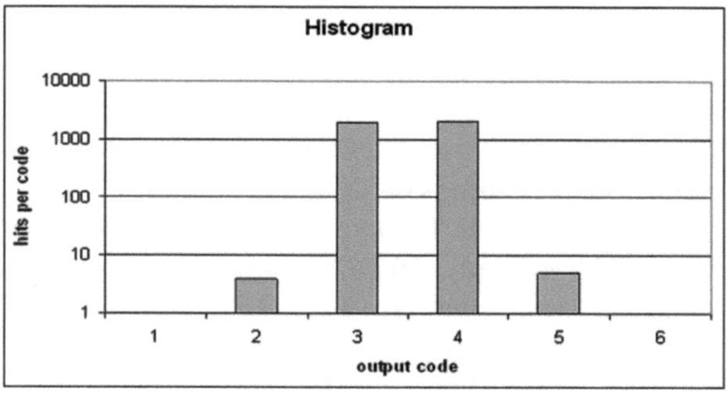

Fig. 1.21 Histogram of an ADC's digital output with a total of 4096 samples

Table 1.4 Exemplary AC accuracy section of the ADS12D1800RF datasheet [4]

Parameter	Min	Typ	Max	Unit
AC accuracy				
SNR	52.9	58.1		dB
ENOBS	8.4	9.3		bits
SFDR	54	71.7		dB
THD		−67	−60	dB
SINAD	52.1	57.7		dB

$$SNR \approx 20 \cdot \log\left(\frac{23170}{0.5}\right) = 93.3\ \text{dB}$$

For the same converter, an SNR of 93.0 dB was measured with the AC measurement at a signal frequency of 10 kHz.

As an example, a cutout of the ADS12D1800RF's AC accuracy parameters is shown in Table 1.4. The SNR, which for an ideal 12 bit ADC would be 74 dB, is clearly limited by the wide bandwidth of 2.7 GHz.

1.1.3 Digital Interface

The digital output of the converter has to communicate with other ICs such as microcontroller, digital signal processors or FPGAs (field programmable gate array). This is done via the digital interface. The digital interfaces of the communicating ICs must be conform in their voltage level and in their protocol.

The voltage levels are defined by JEDEC (Joint Electron Tube Engineering Council—www.jedec.org), which defines standards for the microelectronic industry. There are standards such as ECL and TTL, which do not play a significant role today. For high-speed applications, a standard called LVDS is common, which transfers data with currents. For industrial applications, which are discussed in this book, the CMOS standard is widely spread. It is summarized in Tables 1.5 and 1.6. The voltage levels are typically rated as a function of the supply voltage, which is often named DVDD or VIO.

Particularly important is the digital input voltage range, which is defining a high level V_{IH} and the low level V_{IL}. Similarly, the digital output voltage has to remain in a certain range for a high level output V_{OH} and a low level output V_{OL}. The output voltage surely depends on the load current I_{OH} and I_{OL}.

At the end of the 1990 years, the voltage levels quickly moved from 5 V to a 3.0 V or 3.3 V supply (*DVDD*). Many manufacturers therefore specify the interface voltage in a wide range (see Table 1.7). As an interesting fact, the levels of standard products were not further lowered to 1.8 V. The lower the logic levels are, the more sensitive the signals get for distortion on the printed circuit board (PCB) and an error-free data transfer gets difficult.

Table 1.5 5V CMOS logic levels defined by JESD12-6 [8]

Parameter	Condition	Symbol	Min	Max	Unit
High-level input voltage		VIH	$0.7 \cdot \text{DVDD}$		V
Low-level input voltage		VIL		$0.3 \cdot \text{DVDD}$	V
High-level output voltage	IOH = 1 % rated IOH	VOH	DVDD − 0.1		V
	IOH = rated IOH	VOH	DVDD − 0.8		V
Low-level output voltage	IOL = 1 % rated IOL			0.1	V
	IOL = rated IOL			0.5	V
Input leakage current		II		±1	μA
Three-state output Hi-Z current		IOZ		±10	μA

Table 1.6 3.3 V CMOS logic levels defined by JESD8C.01 [9]

Parameter	Condition	Symbol	Min	Max	Unit
High-level input voltage		VIH	2	DVDD + 0.3	V
Low-level input voltage		VIL	−0.3	0.8	V
High-level output voltage	IOH = −100 μA	VOH	DVDD − 0.2		V
Low-level output voltage	IOL = 100 μA	VOL		0.2	V
Supply voltage	DVDD		3.0	3.6	V

Table 1.7 Digital input and output section of the ADS1282 datasheet from Texas Instruments [10]

Parameter	Min	Typ	Max	Unit
Digital inputs				
VIL	DGND		$0.2 \cdot \text{DVDD}$	V
VIH	$0.8 \cdot \text{DVDD}$		DVDD	V
Digital outputs				
VOL (IOL = 1 mA)			$0.2 \cdot \text{DVDD}$	V
VOH (IOH = 1 mA)	$0.8 \cdot \text{DVDD}$			V
Power supplies				
DVDD	1.65		3.6	V

In addition to the right voltage levels, two ICs must have an identical protocol, before they can communicate with each other. Let's assume the digital output word of the ADC is 16 bit wide, then these 16 bits can be transferred on an interface with 16 wires in parallel. Some additional address and synchronization lines are required. Such an interface is called parallel interface. An example is illustrated in Figs. 1.22 and 1.23. A variety of parallel interfaces exist.

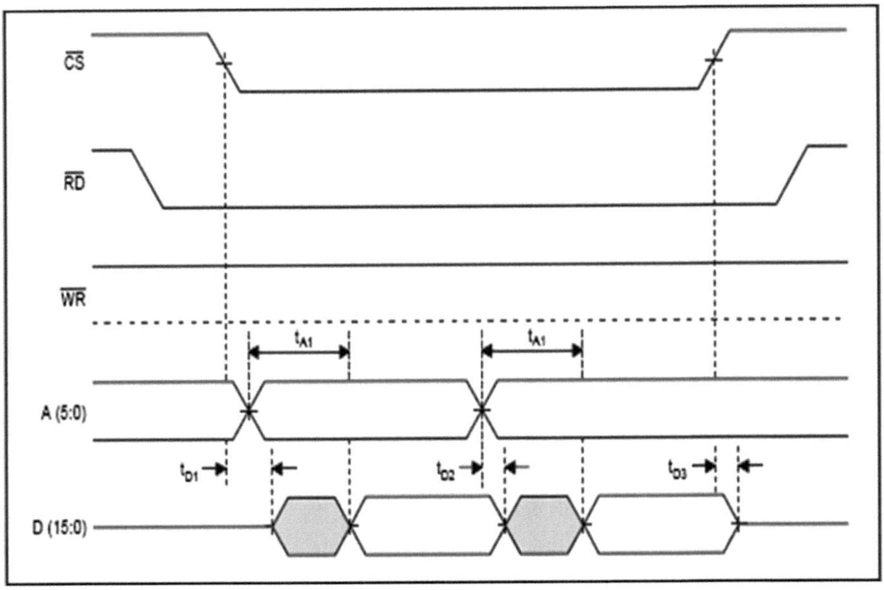

Fig. 1.22 Process of reading data from an IC using a parallel interface

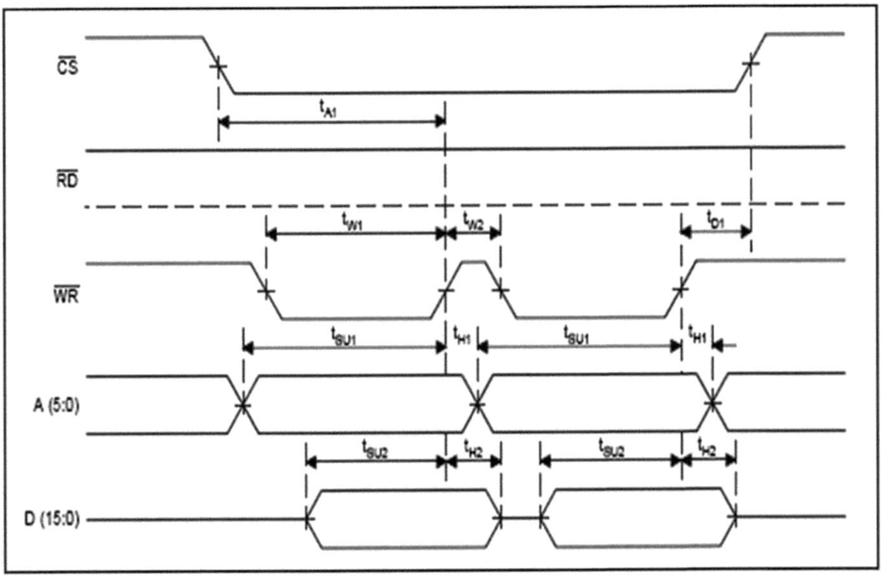

Fig. 1.23 Process of writing data to an IC using a parallel interface

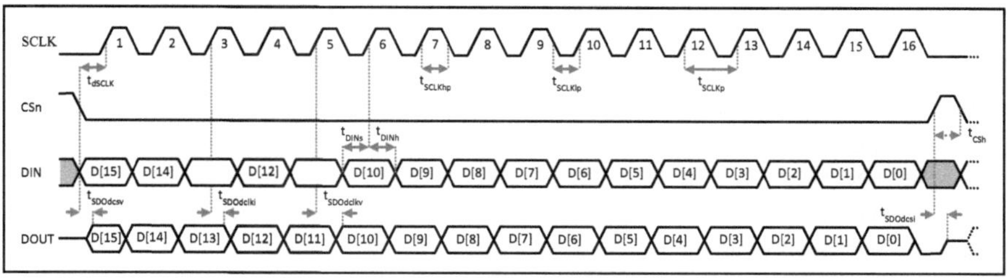

Fig. 1.24 Illustration of a data transfer using a serial port interface

When data is read from a particular IC, the device must first be selected with the Chip Select signal *CS* going low. Note that several devices might connect to the same interface. The read operation is indicated with the Read signal *RD* also getting active low. The selected device will then output the addressed data. If the address changes, then a defined wait time needs to be waited, before the master IC can latch the data in.

Writing data to a device works similar. The device gets selected with *CS* going active low. Optionally, an address might be chosen. The write operation is indicated with the Write signal *WR* going active low. Data is written from the master to the interface. The selected device will latch the data on the interface to the addressed register on the rising edge of *WR*.

The parallel interface can quickly transfer data and is commonly used for fast converters or for converters with a high channel count. As a disadvantage, three synchronization lines, 16 data lines and optional address lines are required. The routing on the PCB is area consuming and ICs with parallel interface need a high pin count.

Therefore, data is often transferred sequentially. They are known as serial port interface (SPI). An example is given in Fig. 1.24.

Again, a device is selected with *CS* getting active low. Data including addresses can be submitted in both directions, from the master to the selected device with *DIN* and vice versa with *DOUT*. One bit is transferred per clock cycle. Such serial interfaces typically start the data transfer with the MSB.

The maximum clock rate is around 50 MHz. This only allows roughly three 16 bit data transfers per microsecond and limits the conversion rate to 3 MSPS for one channel devices. Converters with six channels simultaneous sampling such as the ADS8556 from Texas Instruments or the AD7656 from Analog Devices would be limited to 500 kSPS. As an advantage, the SPI only requires four wires.

The I2C interface can communicate over only two lines (clock and data). As the data line is used from master and slave, it must be high impedance if no data transfer occurs. To remain valid logic levels, a pull-up resistor will be used. The resistor together with the parasitic capacitance on the data line will define a time constant and therefore the maximum data rate (standard mode: 100 kHz, fast mode: 400 kHz, fast mode plus: 1 MHz, high speed mode: 3.4 MHz) on the line. Master and slave will then only pull the line down. Note that the current consumption will then be

Table 1.8 Exemplary power supply section of the dual ADC chip ADS12D1800RF [4]

Parameter	Min	Typ	Max	Unit
Power supplies				
Analog supply current		1360		mA
Track-and-hold supply current		515		mA
Output driver supply current		275		mA
Digital encoder supply current		110		mA
Total supply current		2260	2481	
Power consumption		4.29	4.7	W

high as the full supply voltage is applied to the pull-up resistor. The I2C is only used rarely, as it is typically to slow for industrial applications.

1.1.4 Power Specifications

In some applications, the power consumption and the current consumption are important considerations. In the industrial process control, information is often transferred with currents between 4 and 20 mA. Sometimes, this current is not only used to pass information, but also to supply the electrical circuitry. The current consumption of the control circuitry is therefore limited to 4 mA. Different power saving modes are required, which are typically called nap or sleep mode and power down mode. They differ in their remaining power consumption and recovery time.

Other applications have to minimize the power consumption. This might be hand-held equipment like mobile phones, which are driven from a battery or applications with a high density of electric devices, where the total power dissipation causes cooling problems. Also high frequency devices such as the ADS12D1800RF generate a significant power dissipation as shown in Table 1.8.

1.1.5 Jitter

The AC parameters expect data to be captured in equidistant time steps. Unfortunately, the edge of the digital signal, which freezes the analog input voltage, shows variation, the so called aperture jitter. As the analog input signal is often captured with the clock signal, this effect is often discussed as clock jitter. Dynamic input signals will change their value during the clock variation, so that an erroneous voltage is captured as illustrated in Fig. 1.25.

The maximum error ε is calculated below assuming that the input signal is a sine wave with the amplitude A and the frequency f. The jitter is expressed in Eq. (1.19) with Δt.

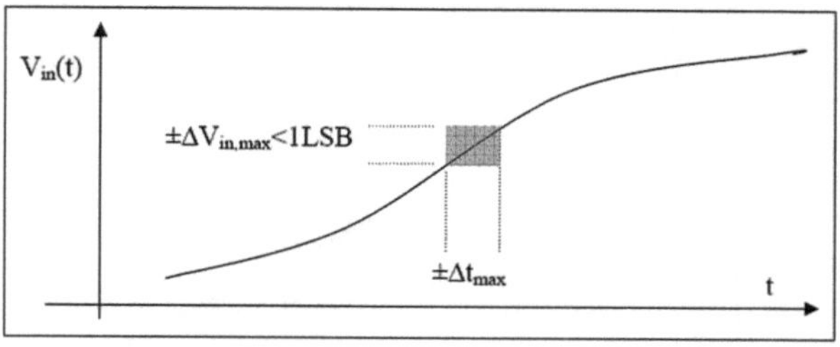

Fig. 1.25 Sampling error of an ADC caused by clock jitter

$$\varepsilon = \frac{dV_{in}}{dt}\Big|_{\max} \cdot \Delta t = \frac{d(A \cdot \sin(2\pi f t))}{dt}\Big|_{\max} \cdot \Delta t = 2\pi f \cdot A \cdot \Delta t < 1\,\text{LSB} \quad (1.19)$$

Equation (1.19) also assumes that the error should ideally be less than 1 LSB. If the analog full-scale range is twice the amplitude, then 2*A* can be replaced by 2^n LSBs. Consequently, the maximum allowed jitter Δt_{max} calculates to

$$\Delta t_{\max} < \frac{1}{2^n \cdot \pi f} \quad (1.20)$$

The jitter requirements increase with resolution and input signal frequency. Let's assume the input signal has a frequency $f = 500$ MHz in a communication application and the resolution of the converter has 12 bits. Then the jitter must be less than 155 fs. In industrial applications, the signal frequency is typically less than 100 kHz. However, the resolution often reaches 16 bit or more. For this case, the aperture jitter must still be better than 50 ps [11].

1.2 Overview of ADCs

Every application has its particular requirements on the ADC. There are different architectures [12–15], which differ in their speed, power consumption, resolution and complexity. The three common standard catalogue ADCs are the pipeline converter, the SAR converter and the delta-sigma converter. An overview is given in Fig. 1.26.

Other converter architectures exist, but they are not commonly used. The flash converter was already introduced in Sect. 1.1. It outputs a conversion result with every clock cycle and therefore achieves very high conversion rates with the drawback on a high component count and low resolution such as 4–8 bit. Flash converters are for example used in some wireless or radar applications as well as in oscilloscopes.

To reduce the component count, the analog input signal can be pre-coded with analog circuitry by basically utilizing simple but fast differential stages and then digitizing the coded analog signals with one comparator per bit in parallel. This

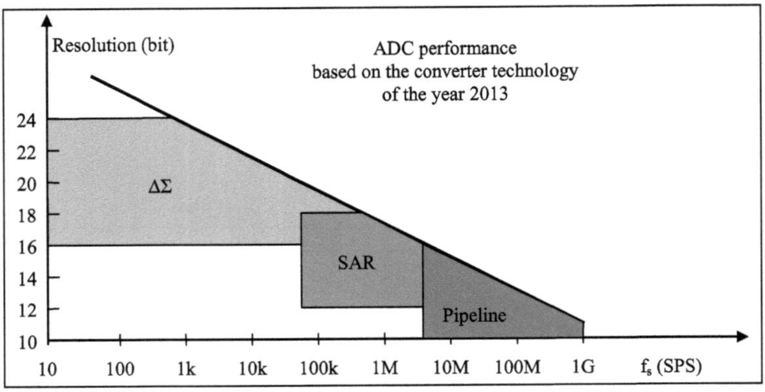

Fig. 1.26 Typical performance range of the various ADC architectures

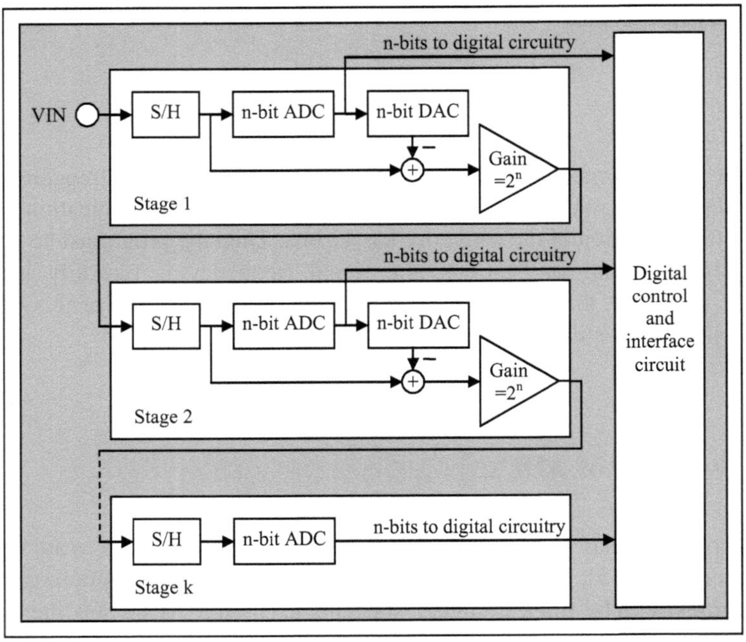

Fig. 1.27 Block diagram of a pipeline ADC with various stages

architecture is called folding flash converter and achieves conversion rates in the GSPS (Giga Samples Per Second) range at 12 bit resolution and nine effective bits. Applications that use the folded converters are basically identical to the applications using flash converters.

Slope converters can be added on the slower end. They ramp the reference signal until it equals to the input signal. Consequently, they are particularly slow without achieving an extraordinary resolution. Due to their simplicity, they were often used in low cost voltmeters.

In the following, the ADC architectures, which are commonly available, are explained in more detail.

1.2.1 Pipeline ADC

Pipeline ADCs sample the input signal with every clock cycle, but they only process a few bits at a time, subtract the digital word from the input voltage to receive the residual voltage, gain it up and pass it further to a next stage, which is evaluating the next bits. Figure 1.27 is showing the general structure of a pipeline converter.

To review the principle method of operation of a pipeline ADC, let's exemplary assume a three bit converter having three stages with one bit resolution each. The input voltage should further be 1.9 V at a reference of 2 V. The full-scale range is twice the reference voltage and therefore 4 V.

The input voltage is first sampled and then compared to the 2 V reference. As the reference is higher, the comparator (1 bit ADC) will output 0. The output of the DAC is also 0 V and V_{in} minus 0 V remains V_{in} (1.9 V). The signal is then gained up by a factor of two (3.8 V) and is passed to the second stage.

With the next clock cycle, the 3.8 V of the output of the first stage is sampled at the input of the second stage. Simultaneously, the first stage will freeze the next sample of the input voltage V_{in}. Now, the sampled voltage of the second stage (3.8 V) is higher than the reference voltage, so that the digital output of the second stage is 1. Therefore, the reference voltage is subtracted from the input voltage and gained up by two, so that the analog output is $(3.8\,\text{V} - 2\,\text{V}) \cdot 2 = 3.6\,\text{V}$.

The third stage is sampling the 3.6 V with the third clock edge and compares it with 2 V again, so that the digital output is 1. The total digital output of the ADC is 011. Be aware, that the described example shows a symmetrical transfer function, where the first code transition is one LSB above 0 V. Regarding to the specifications in Sect. 1.1, the ideal transfer function is asymmetrical with the first transition being half an LSB above 0 V and the last transition 1.5 LSB below full-scale. The illustrated architecture therefore has an offset error of half an LSB.

The example shows that the conversion of the pipeline converter was performed with k comparators/ADCs and $k-1$ amplifiers ($k=3$). The circuit complexity of the pipeline ADC is increasing linearly with the resolution and not exponentially like it would be the case for a flash converter. State-of-the-art 12 bit converters with up to 1 GSPS achieve 65 dB in SNR. Pipeline converters with calibration have a differential-non-linearity (DNL) better than ± 1 LSB. The power dissipation is in the range of 2 W [16]. A section of the ADS5400 data sheet from Texas Instruments is shown in Table 1.9.

Pipeline converters do not achieve the sample rate of the flash converters, because several settling times are in series. While the flash converter only has the delay of the comparator, pipelines have the comparator (or the ADC), the DAC and the buffer delay. The amount of serial circuitry is also adding noise and errors into

Table 1.9 Exemplary section of the ADS5400 datasheet from Texas Instruments [16]

Parameter	Min	Typ	Max	Unit
Resolution	12			
Analog input				
Full-scale differential input range	1.52		2	Vpp
Common-mode input		AVDD/2		V
Differential input impedance	85	100	115	Ω
Input capacitance		0.8		pF
Common-mode rejection ratio		40		dB
Accuracy				
No missing codes	12			Bits
Differential linearity error	−1	±0.7	2	LSB
Integral linearity error	−4	±2	4.5	LSB
Gain error	−5		5	% of FS
Gain error temperature drift		0.03		% FS/°C
Offset error	−2.5	0	2.5	μV
Offset temperature drift		0.02		mV/°C
AC accuracy				
SNR	57	58.5		dB
SFDR	65	75		dB
THD	63	71.7		dB
SINAD	56	58.5		dB
Sampling dynamics				
Conversion rate	0		1	GSPS
Aperture delay		150		ps
Aperture delay jitter		125		fs
−3 dB input bandwidth		2.1		GHz
Power supplies				
IAVDD5 (5 V)		220	234	mA
IAVDD3 (3.3 V)		225	241	mA
IDVDD3 (3.3 V)		136	154	mA
Total power dissipation		2.28	2.45	W
Total power dissipation (sleep)		13	50	mW
Wake-up time from sleep		1.8		ms
Power-supply rejection ratio		50		dB
Temperature range				
Specified performance	−40		+85	°C

the signal path so that the noise performance of pipeline converters is worse than that of other architectures like the SAR converters.

An additional disadvantage is the so called pipeline delay. Even if the conversion rate is high, the signal has to pass k stages, so that the total through-put time is k-times a clock cycle. The typical application for the pipeline converter has high-frequency, continuous running signals like video imaging (camcorders), where the conversion rate is more important than the through-put time. Consequently, the pipeline converter is more used in consumer applications than in industrial applications and will not further be discussed.

1.2.2 SAR ADC

If a snapshot of a signal at a particular time is required in an application, then the converter based on successive approximation (SAR) is the right choice. It has a similar through-put rate as the pipeline ADC, but is only converting one sample at a time, so that the conversion rate is lower. This reduces the complexity to one comparator and one digital-to-analog converter (DAC), which makes the SAR converter much less power-hungry than the pipeline converter. Also the amount of noise and error sources is reduced together with the amount of internal circuitry.

The architecture of a SAR ADC works like a weight scale, which is shown in Fig. 1.28. The input signal is frozen with a sample and hold capacitor. The sampled input voltage is first compared to half of the reference voltage, which is generated by the DAC. If the input signal is higher, then the DAC will increase its output voltage by a quarter of the reference. If the input voltage is still higher, then another eighth of the reference voltage is added. With every clock cycle, the DAC will approximate the input voltage by another bit of accuracy.

The SAR converter is successively evaluating bit for bit. If the SAR converter has an *n*-bit resolution, then it requires *n* clock cycles for the conversion.

Let's exemplary again assume a 3 bit converter. In this case, a 3 bit DAC will be used. For the first approximation, the input voltage of 1.9 V would be compared to half of the 4 V reference (2 V). As $V_{in} < \frac{REF}{2}$, the MSB is 0. The 'weight' is reduced to $\frac{REF}{4} = 1\,\text{V}$ for the second approximation. This time, the input voltage is higher, so that the second bit is 1. The DAC has to add a 'weight' of $\frac{REF}{8} = 0.5\,\text{V}$. The input voltage of 1.9 V is still higher than 1.5 V, so that the third bit is also 1. The ADC output is 011 again.

The performance totally relies on the accuracy of the DAC. The comparator will add noise and offset. State-of-the-art SAR converters achieve an 18 bit resolution at 1.6 MSPS and 18 mW [17] or 16 bits at 10 MSPS and 150 mW [18]. The power is

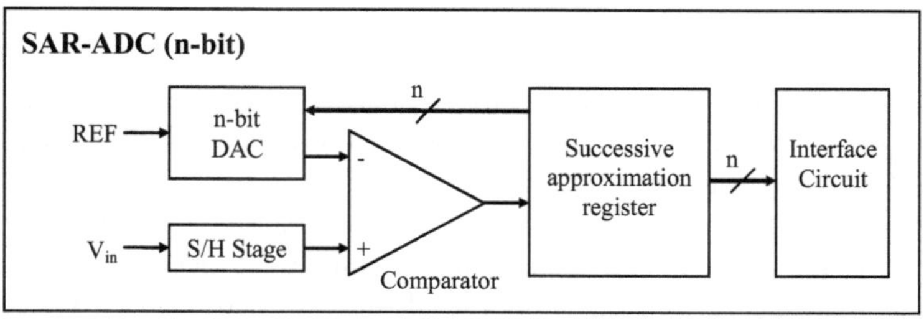

Fig. 1.28 Block diagram of a SAR ADC

Table 1.10 Exemplary section of the AD7982 datasheet from Analog Devices [19]

Parameter	Min	Typ	Max	Unit
Resolution	18			
Analog input				
Voltage range	−REF		+REF	V
Absolute input voltage	−0.1		REF + 0.1	V
Common-mode input range		REF · 0.5		V
Analog input CMRR		67		dB
Leakage current		200		nA
Accuracy				
No missing codes	18			Bits
Differential linearity error	−0.85	±0.5	+1.5	LSB
Integral linearity error	−2	±1	+2	
Gain error	−0.023	+0.004	+0.023	% of FS
Gain error temperature drift		±1		ppm/°C
Zero error		±100	+700	μV
Zero temperature drift		0.5		ppm/°C
Power supply rejection ratio		90		dB
AC accuracy				
SNR	95.5	98		dB
SFDR		−115		dB
THD		−120		dB
SINAD		97		dB
Sampling dynamics				
Conversion rate	0		1	MSPS
Transient response			290	ns
Aperture delay		2		ns
−3 dB input bandwidth		10		MHz
Power supplies				
VDD	2.375	2.5	2.625	V
VIO	2.3		5.5	V
Standby current		0.35		μA
Power dissipation (10 kSPS)		70	86	μW
Power dissipation (1 MSPS)		7.0	8.6	mW
Temperature range				
Specified performance	−40		+85	°C

not only required to achieve the high speed, but also a good noise performance with up to 100 dB in SNR. Table 1.10 lists the specifications of the AD7982 from Analog Devices [19] as an example.

SAR converters do neither have the highest speed nor the highest performance, but they have the widest field of applications. Examples are electrical motor control applications, medical applications, industrial process control, touch-screen products and many more.

1.2.3 Delta-Sigma ADC

The highest resolution is achieved with delta-sigma ($\Delta\Sigma$) modulators [20–22], which are processing the input signal A before digitizing it. The basic modulator structure is shown in Fig. 1.29. The difference between the analog input A and the digital output Y is integrated and passed to an ADC, which in the easiest case is a comparator. The comparator is adding quantization noise N to the integrated signal.

Modulators are mostly implemented in switched capacitor designs, but continuous time modulators are getting more popular again, as they achieve higher speeds. In the following calculation of the transfer function, the continuous time domain is chosen, where the integrator is expressed by $1/s$. s is used for the complex Laplace variable. The structure in Fig. 1.29 can be expressed by the following function

$$Y = (A - Y)/s + N$$
$$Y = A \cdot \frac{1}{1+s} + N \cdot \frac{s}{1+s} \tag{1.21}$$

The transfer function for the input signal is $F_A(s) = \frac{1}{1+s}$, which is expressing a low pass filter. The transfer function for the quantization noise $F_N(s) = \frac{s}{1+s}$ describes a high pass filter. The idea of the delta-sigma modulator is to suppress the quantization noise at low frequencies, where the input signal is expected, and leave it at higher frequencies (noise shaping). The digital filter at the output of the modulator acts as a low pass filter and will cut off all high frequency signals including the quantization noise. The delta-sigma modulator together with the digital filter is called delta-sigma ADC.

Figure 1.29 shows a modulator with one integrator, which is a filter of first order. Higher order filters will produce an even more effective noise shaping, which can be seen in Fig. 1.30. The equation for the ideal signal-to-noise ratio for a modulator with order m, the oversampling ratio OSR and the n-bit ADC can be calculated to [23]

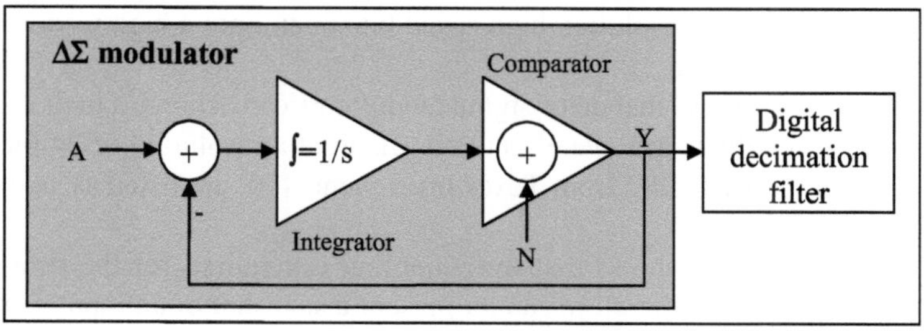

Fig. 1.29 Block diagram of a first order delta-sigma converter

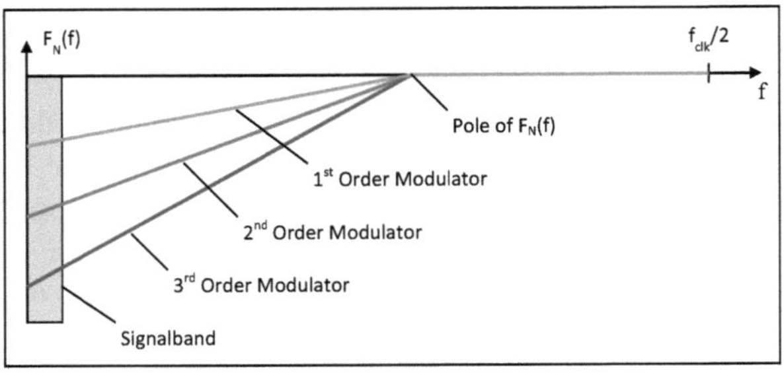

Fig. 1.30 Noise transfer function of a first, second and third order delta-sigma modulators

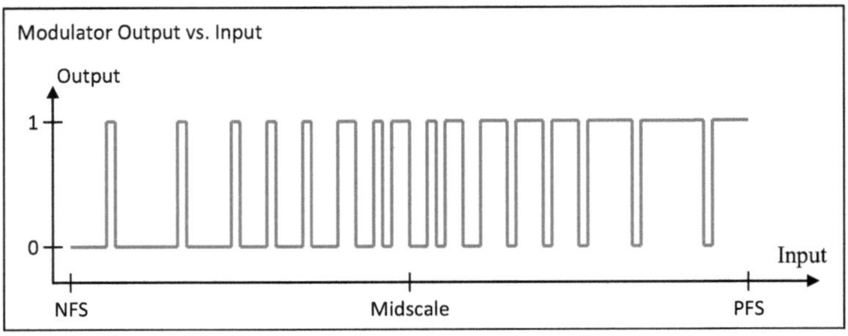

Fig. 1.31 Digital bit stream output of a delta-sigma modulator

$$SNR_m = (20 \cdot m + 10) \cdot \log_{10}(OSR) - 10 \cdot \log_{10}\left(\frac{\pi^{2m}}{2m+1}\right) + 6.02 \cdot n + 1.76 \tag{1.22}$$

The output of the comparator is a stream of ones and zeros (see Fig. 1.31) and not a binary word. It is called a bit stream. It outputs mainly zeros if the input signal is close to negative full-scale, a lot of ones if it is close to positive full-scale and an equal amount of zeros and ones at mid-scale.

The digital filter will not only low-pass filter the bit-stream, but also decimate the input signal from a low resolution high speed bit stream into a high resolution low speed binary digital word.

Equation (1.22) shows that delta-sigma modulators can achieve a high accuracy. State-of-the-art products achieve effectively 21 bit at a resolution of 24 bit. Specifications from the ADS1282 from Texas Instruments [10] are listed as an example in Table 1.11.

Equation (1.22) also shows that oversampling is required for the decimation. Delta-sigma modulators therefore cannot present a snapshot of an input signal, but always an average over a period of time. The digital filter will furthermore delay the signal. As a result, modulators are often used to convert low frequency continuous time signals, like audio signals (f_{in} = 20 Hz–40 kHz) or power supplies for energy

Table 1.11 Exemplary section of the ADS1282 datasheet from Texas Instruments [10]

Parameter	Min	Typ	Max	Unit
Analog input				
Full-scale input voltage		±REF		V
Absolute input voltage	AVSS + 0.7		AVDD-1.25	V
Differential input impedance		1		GΩ
Common-mode input impedance		100		MΩ
Input bias current		1		nA
Crosstalk		−135		dB
DC performance				
No missing codes	31			Bits
Integral linearity error		0.00005	0.0004	% FSR
Gain error	−1.5	−1.0	−0.5	% of FS
Gain error temperature drift		2		ppm/°C
Zero error		50	200	μV
Zero temperature drift		0.02		μV/°C
Power supply rejection ratio	80	90		dB
AC accuracy				
SNR	120	124		dB
SFDR		123		dB
THD		−122	−114	dB
Digital filter response				
Passband ripple			±0.003	dB
Passband (−0.01 dB)		0.375 × fDATA		Hz
Bandwidth (−3 dB)		0.413 × fDATA		Hz
High-pass filter corner	0.1		10	Hz
Stop band attenuation	135			dB
Stop band		0.500 × fDATA		Hz
Group delay		31/fDATA		s
Settling time (latency)		62/fDATA		s
Power supplies				
DVDD	1.65		3.6	V
AVDD current		4.5	6.5	mA
AVDD standby current		25	50	μA
AVDD power-down current		1	15	μA
DVDD current		0.6	0.8	mA
DVDD standby current		25	50	μA
DVDD power-down current		1	15	μA
Power dissipation		25	35	mW
Power dissipation (standby)		90	250	μW
Power dissipation (power-down)		10	125	μW
Temperature range				
Specified performance	−40		+85	°C

meters ($f_{in} = 50$ Hz). They are also very common for discrete measurements, where the conversion rate is not an issue such as the measurement of temperature or weight.

Table 1.11 shows a section regarding the digital filter response, which is a specialty of delta-sigma converters. Particular important are the flatness and the bandwidth of the filter. Furthermore, the filter will add a delay in time, which depends on the data rate.

This book will concentrate on the analog circuit design. There are plenty of papers and books that concentrate on digital filter design [24–27]. This book will only give a short introduction.

As a rule of thumb, the noise of a modulator of order m will raise with m-times 20 dB per decade. The digital filter should be of higher order to efficiently suppress the noise. On the other side, the higher the order of the filter m_f, the more delay in time has the filter and the worse will the pass-band flatness be. The order of the filter is therefore typically chosen to $m_f = m + 1$.

A filter type, which is widely spread due to its efficient size, is the SINC filter. An example with second order is shown in Fig. 1.32. Y_{mod} is representing the bit stream from the delta-sigma modulator such as illustrated in Fig. 1.31. The signal is first applied to a low-pass structure in form of integrators to suppress the noise. This low-pass filter is required to avoid folding of higher frequency noise into the remaining signal-band. Once the high frequencies are suppressed, the data frequency can be reduced to the Nyquist rate by applying the oversampling ratio. In the digital filter design this is basically a reduction of the clock frequency of the following circuitry. Now that the data rate is reduced, it is important to straighten the pass-band. This is done by adding the same amount of differentiators as integrators were used before. The output Y_{out} is then filtered and in binary format.

The term z^{-1} in Fig. 1.32 represents a delay by one clock cycle and is realized on silicon with a register. It can be seen that the integrators continuously sum the bitstream of the modulator. Sooner or later an overflow will occur. It is important that the output Y_{out} remains deterministic, which means that only one overflow must occur during OSR clock cycles. This will require a minimum register width W_R that is dependent on the oversampling ratio OSR and the order of the filter m_f. The register width can be calculated with Eq. (1.23).

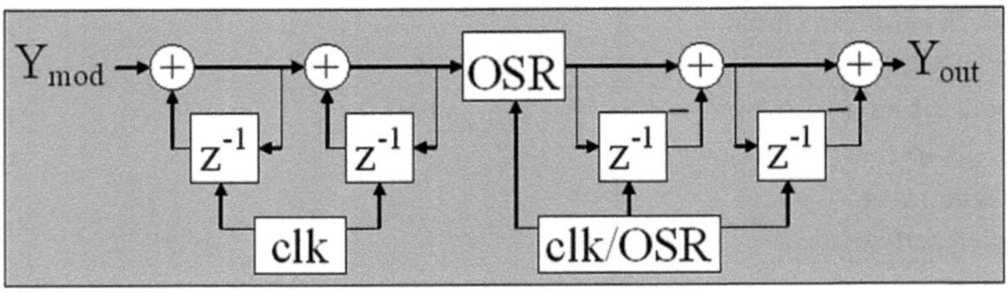

Fig. 1.32 Block diagram of a second order SINC filter

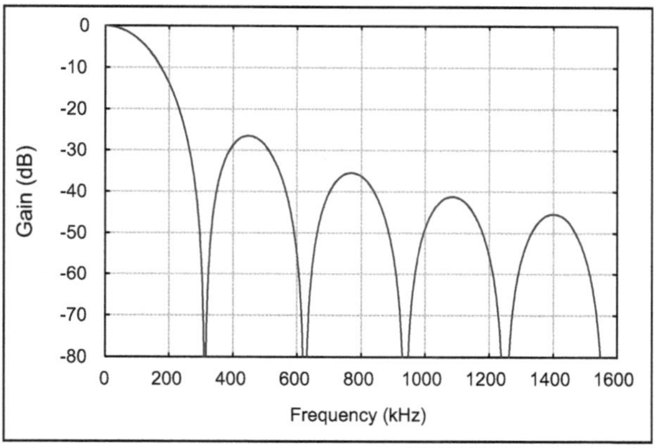

Fig. 1.33 Filter responds of a second order SINC filter (10 MHz bit stream, OSR = 32)

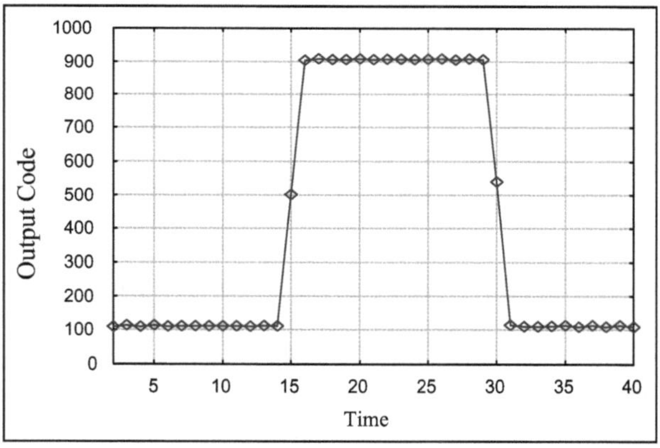

Fig. 1.34 Pulse responds of a second order SINC filter (10 MHz bit stream, OSR = 32)

$$W_R = m_f \cdot \log_2(OSR) \tag{1.23}$$

The filter responds of the second order SINC filter is shown in Fig. 1.33 for a modulator operating at 10 MHz and an OSR of 32. Notches occur at frequencies of multiples of the data rate. The data rate is defined by the modulators clock frequency divided by the oversampling ratio. Section 1.3 shows that these notches can be used to suppress signals such as carrier signals.

Figure 1.34 illustrates the pulse responds of the same filter. The differentiators will cause that the output Y_{out} requires m_f output data rates until it can follow the analog input step. As a consequence there is a difference between group delay and settling time.

The group delay t_{gd} defines the time delay of a filter, when an analog input signal is applied with a frequency smaller than Nyquist. In this case the digital filters are in a steady state. Note that the delay is fix for a given filter with constant clock rate. For analog input steps, the digital filter first has to follow the step. The so called

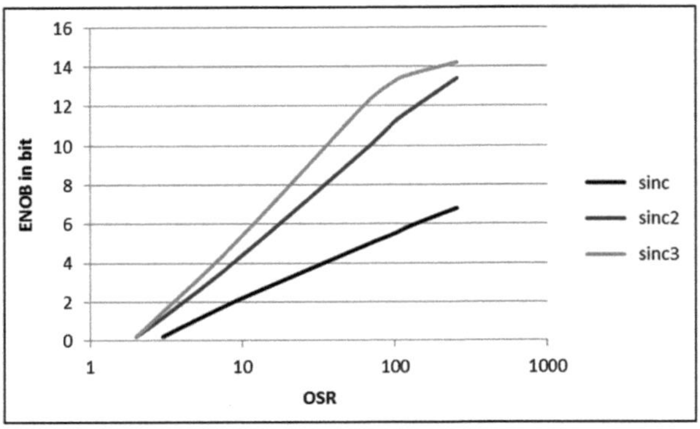

Fig. 1.35 Illustration of ENOBs versus OSR for the AMC1203 from Texas Instruments [28]

settling time t_s is the consequence. T_{clk} in Eqs. 1.24 and 1.25 represents the clock cycle of the modulator clock.

$$t_{gd} = 0.5 \cdot m_f \cdot OSR \cdot T_{clk} \tag{1.24}$$

$$t_s = 1 \cdot m_f \cdot OSR \cdot T_{clk} \tag{1.25}$$

Commercial delta-sigma modulators typically provide graphs in the datasheet that support the design of digital filters, which might for example be implemented in FPGAs. The curves in Fig. 1.35 for example show, which OSR is required for a filter with order 1, 2 or 3 to achieve a certain effective number of bits. If a electrical motor drive for example requires a resolution of 12 bits for the current measurement, then a SINC filter of order 3 and OSR of 64 can be used for this modulator. The used modulator is of second order with a clock rate of 10 MHz. Consequently, the group delay could be calculated to

$$t_{gd} = 0.5 \cdot 3 \cdot 64 \cdot 100\,\text{ns} = 9.6\,\mu\text{s}$$

For the same resolution, a SINC2 filter with an OSR of 128 could be used. The delay could be calculated to 12.8 μs. The filter of third order would therefore be more sufficient.

Figure 1.35 shows a bend for the SINC 3 filter at an OSR of 100. From this point on, the circuit noise will dominate the quantization noise. As a consequence, the noise will only improve by 3 dB per doubling the OSR for OSR > 100.

For the same motor control circuit, the currents need to be observed for failures and need to be turned off within a short period of time such as 3 μs if an overcurrent occurs. The ideal filter can be defined with the help of Fig. 1.36. With a 3 μs settling time, the highest resolution of 5.8 effective bits can be achieved with a SINC filter of second order. Equation (1.25) can be used to calculate the required OSR to

$$OSR = \frac{t_s}{m_f \cdot T_{clk}} = \frac{3\ \mu\text{s}}{2 \cdot 0.1\ \mu\text{s}} = 15$$

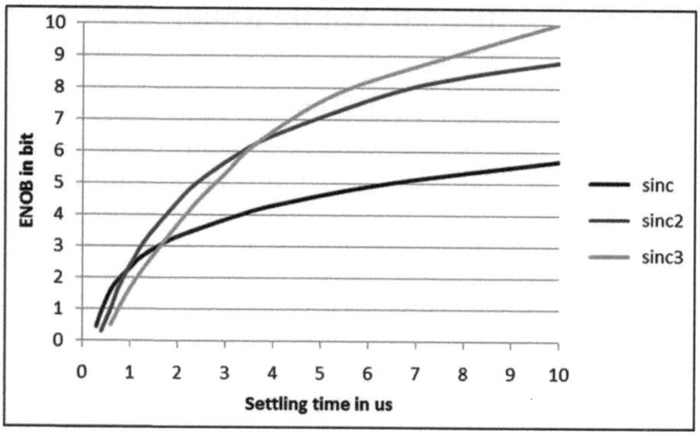

Fig. 1.36 Illustration of ENOBs versus settling time for the AMC1203 from Texas Instruments [28]

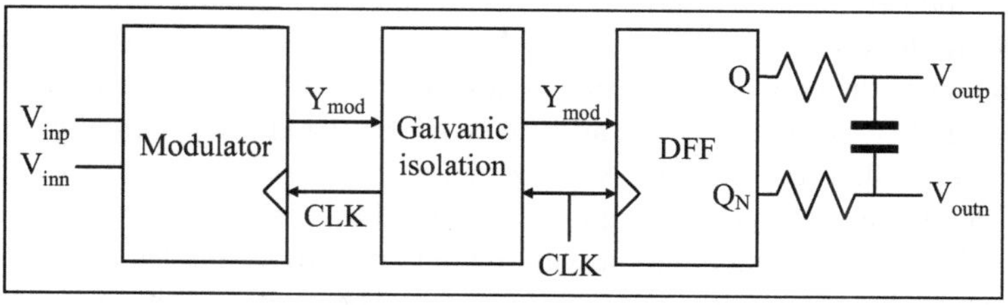

Fig. 1.37 Block diagram of an isolation amplifier

In the mentioned motor control application, two SINC filters would operate in parallel. The third order filter with OSR = 64 would be used for the normal regulation loop. A second order filter with OSR = 15 for overcurrent protection.

Please note that the bitstream can also be filtered with an analog circuitry as shown in Fig. 1.37. An analog output would be the consequence. This fact is actually utilized in some commercial products such as the HCPL-7840 from Avago [29] or the AMC1200 from Texas Instruments [30]. These products transmit a differential voltage through an isolation barrier of several hundreds of volts.

Every converter technology has its unique advantages and is suited for special applications. The next chapter will introduce some industrial applications and will show the usage of the above ADC architectures in the application. It will also discuss the required performance parameters.

1.3 ADCs Used in Industrial Applications

The performance parameters and the ADC architectures that were discussed in the previous chapters should now be applied to two fields of applications, which are electrical motor drives in Sect. 1.3.1 and power monitoring applications in Sect. 1.3.2.

1.3.1 Electrical Motor Drives

The field of electrical motor drives is huge. They are used in machinery such as for milling, turning lathe or robotics. They are used for transportation in production such as in printing machines or assembly lines. Escalators, elevators or ergometers are for personal use. A new boom started with hybrid cars and will continue with e-cars. In industrial applications, AC drives are often used, as they are more reliable than DC motors as they are brushless. Now these motors are normally running synchronous to the power supply. Frequency inverters are used to generate supplies with required frequencies. They can further be regulated for positioning or speed and are then called servo drives. A block diagram is shown in Fig. 1.38.

The three phase AC power supply is converted to a DC link voltage. Using the method of pulse-width modulation with the help of IGBTs, the DC link voltage is converted back into a three phase AC supply. The speed of the new supply is variable in frequency, so that the robust AC motors can be used over a huge frequency span and for very accurate positioning systems (for example tooling machines).

At least two currents to the motor have to be measured. The third current can be calculated, but needs to be monitored for failure. The currents are transferred into

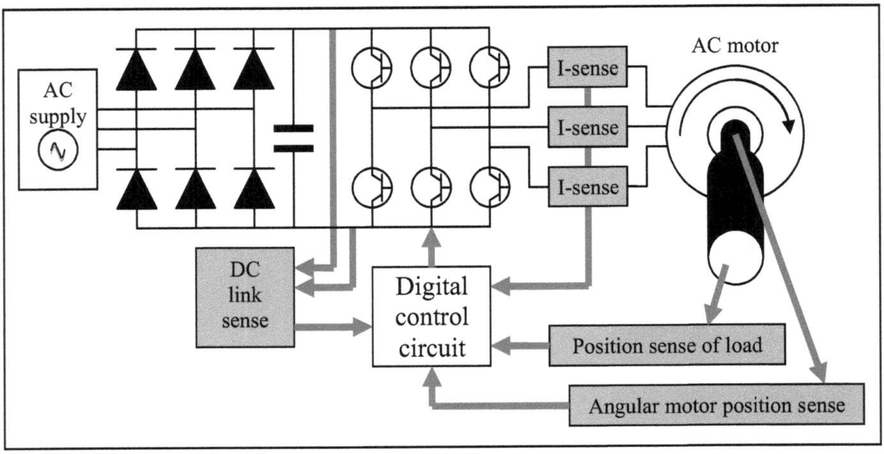

Fig. 1.38 Block diagram of a servo drive for electrical motors with its required sensors

the rotating coordinates of the motor, so that the angular motor position needs to be sensed. This is required for the control of the torque.

More accurate systems also measure the position of a load. Often machines also include a reference position where the coordinates of the load are measured. The motor control system is storing the coordinates of this reference mark and refers all other positions relative to these coordinates. Some additional control variables of interest might be the DC link voltage and the temperature of the motor. Finally, the application might support an analog input port. This analog input is normally used to set the anticipated speed or position.

1.3.1.1 Sensing the Phase Currents

Most important are the currents to the motor. The IGBTs need to switch as soon as the currents are changing their sign. The magnitude is further required for the motor control loop, which is typically running at a switching frequency of the IGBTs of 16 kHz. The currents need to be turned off within a few microseconds, if an over-current is detected. Otherwise, additional damage can be generated inside the power stage or the motor.

Two ways for current measurement are most commonly in use. Motors with high currents use magnetic sensor, which are mostly based on the Hall Effect. These sensors output a voltage between 0 and 5 V that directly connects to the input of a SAR converter. A conversion is triggered, once the motor control loop requires the magnitude of the current. The SAR converter can trigger conversions at any point of time and the conversion rate is sufficiently fast for the system. Sign comparators are used in parallel to monitor the direction of the current. Window comparators monitor failure currents continuously. The required resolution is between 10 and 12 bit.

Shunt resistors are used for current sensing for motors with currents less than 20 A. The voltage across the shunt must remain as small as possible to keep the power dissipation in the shunt reasonable. A full-scale range of $\pm$200 mV or less is normal. The small signal amplitude requires a low noise ADC architecture, so that delta-sigma modulators are suited best. Also note that the currents are measured in the phase, which will float between a positive and negative DC link voltage. Galvanic isolation to the digital control circuit is required. The bitstream of the delta-sigma modulator only requires one digital signal to be isolated (see Sect. 1.2.3), which adds a cost advantage [31].

1.3.1.2 Monitoring the Position with Optical Sensors (Encoders)

The encoder consists a carrier of glass with up to 8192 stripes that is rotating with the load or the motor respectively. Light is shining through the carrier onto a set of photo diodes that are generating a current. If a diode is between two stripes the current is high and if a stripe is on top of a diode, less or no current is generated.

A sine and a cosine wave are generated out of the signals of the diodes with one period per strip.

The sine and cosine are digitized with a set of comparators. Their outputs connect to up-down counters. The angle or angular speed of the motor/load at a certain point of time can be calculated by the number of counted pulses in relation to the total number of stripes on the carrier. The resolution can be increased with two time counters that are used to compare the amount of clock cycles between two stripes and the delay from the last stripe to the time of interest. The additional resolution is dependent on the speed of the motor. The faster the motor, the less clock cycles are between two adjacent stripes. The resolution of the angle can also be increased by converting the sine/cosine wave with an analog to digital converter. Here, usually a 10 Bit ADC is accurate enough.

If a motor rotates with 6000 rpm and the encoder consists of 4096 stripes, then the input frequency will reach 410 kHz. This application is typically expecting input frequencies up to 500 kHz. The Nyquist theorem must not be fulfilled, if the signal is not continuously sampled and is not reconstructed in the digital domain. However, the bandwidth of the sample and hold stage must be high enough. The signal will then be sampled at a particular point of time in the motor control loop. Again, SAR converters have the capability to trigger a conversion at a particular point of time and are most suited here. This is also important, when the machine reaches a reference position.

Some systems try to continuously sample the sine and cosine signals. They reconstruct and filter the signal in the digital domain to improve the signal integrity, which suffers from distortion of the IGBTs and the use of long cables. In this case, the Nyquist theorem must be fulfilled. SAR ADCs with conversion rates higher than 1 MSPS are required.

1.3.1.3 Monitoring the Position with Inductive Sensors (Resolver)

A resolver is based on inductive coupling. Three coils are positioned inside the motor; two of them are positioned in a 90° angle and are in a fixed position. The third coil, which rotates with the motor, is usually stimulated with a sine wave by an external source. This sine wave is induced as carrier signal to the other two stationary coils, so that the amplitude of the induced signal is dependent on the angle between a fixed and the rotating coil. A sine and a cosine wave are generated as the fixed coils are positioned in a 90° angle, which are illustrated in Fig. 1.39. The magnitude of the sine and the cosine are converted with the ADCs and the angle is calculated inside the digital control loop with an arctan operation. The resolver is cheaper and more robust compared to the encoder, but less accurate.

The sine and cosine waves of the resolver have the same frequency as the motor, while the frequency of optical sensors is a multiple of the motor frequency. Using the encoder, a high resolution can already be achieved by counting the pulses and only for the additional resolution the analog to digital converter is used. Using the resolver, the accuracy of the position is totally dependent on the resolution of the

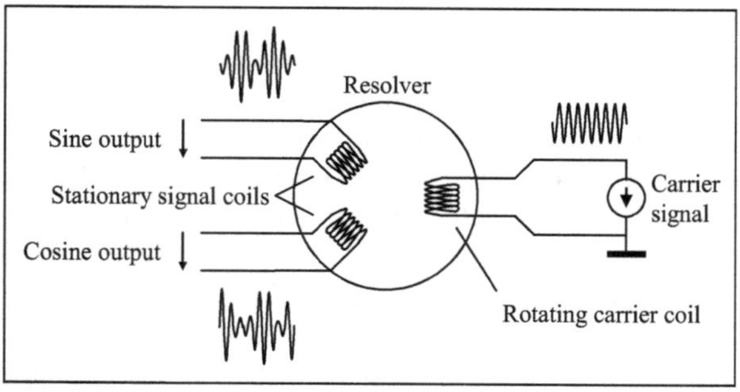

Fig. 1.39 Illustration of a resolver with its according signals

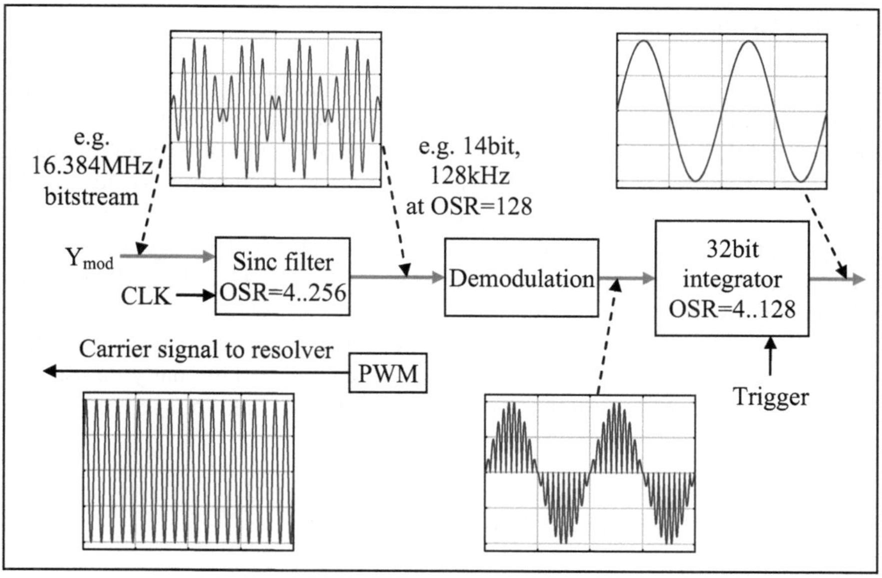

Fig. 1.40 Digitizing a resolver output using a delta-sigma ADC and a dual digital filter [32]

ADC. As the system noise is usually high, a 12 bit converter is typically used together with oversampling to reach a 13 or 14 bit resolution for the angle. Some high-end systems are using 14–16 bit ADCs.

Sine and cosine need to be demodulated. This is typically achieved by sampling the signals at the peak of the carrier signal. Again, it is important to convert the signal at a particular point of time, so that SAR converters are most commonly used.

The carrier signal typically operates at 8 kHz, so that delta-sigma converters would also work in this application. The bitstream from the modulator would be filtered in two steps as illustrated in Fig. 1.40. A first filter generates a digital word with a medium resolution around 12 bit at a medium frequency such as 128 kSPS. Now, the signal needs to be demodulated. This means the 12 bit digital signal needs to be multiplied with +1, if the carrier is positive, and with −1, if it is negative. The time delay of the carrier signal through the resolver, the wiring, the modulator and the first filter needs to be taken into account.

Frequencies, which are an integer multiple of the modulator clock divided by the used over-sampling ratio, are suppressed. These points are called notches. In this concept, the decimation ratio of the integrator is chosen in a way that the carrier frequency is falling into a notch and is canceled by the digital filter. The carrier signal can also be generated digitally. As an example, at a fixed carrier frequency is programmed in bitstream format and is applied to the carrier coil of the resolver. The resolver shows a low-pass behavior and generates a low distortion carrier signal.

1.3.1.4 Auxiliary Signals

Some applications set the anticipated speed with an analog input voltage. This voltage would remain constant for a long period of time, so that a delta-sigma approach would be ideal. A motor control environment is typically rough for analog signals as distances are large and a significant amount of current is switched with the IGBTs generating distortion. A signal range of ±10 V is therefore preferred in this application. As an interesting fact, there are basically no delta-sigma converters with such an input voltage range on the market, so that also here the SAR converter is frequently used.

The SAR converter can easily multiplex between channels. A SAR converter, which is used to monitor an analog input voltage, could additionally be used for other tasks such as measuring the temperature or the DC link voltage.

1.3.2 Power Monitoring Applications [33]

The measurement of the power flow is crucial in the power grid to minimize the loss in the power lines. The renewable energies also cause an energy flow in the grid, which is hard to predict. The power flow therefore must be controlled, so that power lines will not get overloaded. This paper is summarizing the requirements for power measurements.

The smart grid starts in every household. Smart meters should communicate the need of electric power to the grid. There are applications, which do not require the power immediately such as washing machines, heating systems or refrigerators. These applications can request a good start time from the power grid. There are also other applications such as lightning or cooking, which need the power immediately. For these applications, the power needs to be measured accurately for billing, but also to control the power flow.

Several households are bundled in concentrators, which basically have the same requirements as the smart meters. However, the resolution and accuracy of the

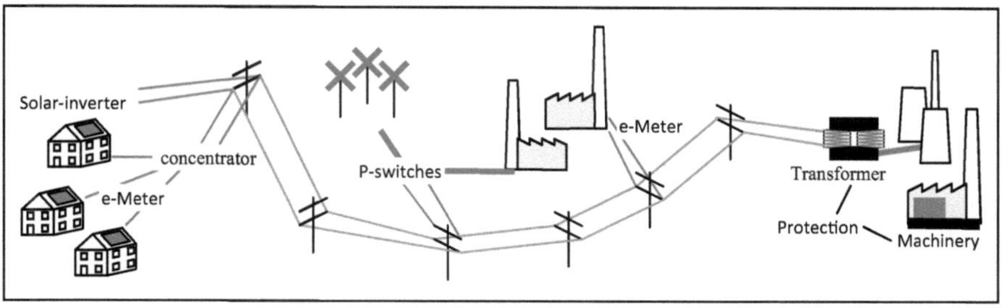

Fig. 1.41 Components in the power grid requiring power measurement

power measurement needs to have higher dynamic range. The concentrators then connect to the power grid.

Renewable energies such as solar energy coming from private households or wind energy cause a varying power flow in the grid. This might quickly cause a local overload of a power line. The power flow must be measured and turned off immediately, if an overload or a malfunction occurs. Power relays are required, which measure the power at a significantly higher speed. These power relays simulate the expected power and if the measured power is outside the expected range, the relays will open.

Additionally, power switches (switchgears) might turn on and off to route the power on the shortest path to the consumer to minimize the energy loss in the grid. Such high power switches, need to disconnect power lines at the right point of time to avoid sparkling in the switch. Similarly they need to connect power lines, when the voltages are in phase. Here an accurate current and voltage measurement is required as well. An overview is shown in Fig. 1.41.

The various applications in the power grid are examined below.

1.3.2.1 Continuous Power Monitoring

The power measurement requirements can be separated in two applications. Smart meter, concentrators or high power switches need to monitor currents and voltage continuously. Delay and conversion rate are less of an issue compared to accuracy and dynamic range.

Nevertheless, the applications differ in the level of accuracy and cost sensitivity. Separate classes were defined such as 0.2 or 0.5, meaning the accuracy has to be 0.2 % of the measured value over a wide power range.

As an example, a class 0.2 meter might have the current ranges from 100 mA to 15 A. Each range needs to be measured with an accuracy of better than 0.1 %, because of additional errors in the system such as the voltage measurement. Furthermore, the error of the current measurement is produced from the current sensor and the digitizer or in other words the analog-to-digital converter. The full-scale range of the current measurement unit above is 15 A. The 100 mA range still

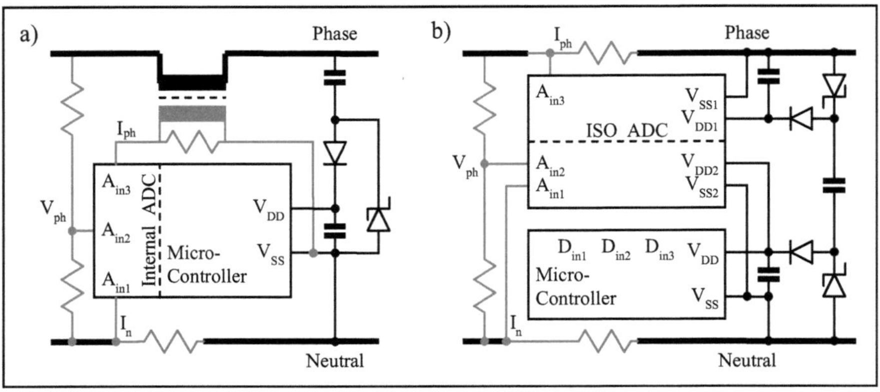

Fig. 1.42 Block diagram of e-meters with an (**a**) typical approach and (**b**) with shunt measurement

requires the 0.1 % accuracy, which results in an accuracy of 100 μA. If this is digitally expressed, then the resolution in number of bits n must be

$$n = \log_2\left(\frac{15\,\text{A}}{100\,\text{mA}} \cdot \frac{1}{0.1\,\%}\right) = 17.2\,\text{bit} \tag{1.26}$$

This resolution must be achieve even if there are a series of errors. Some are caused by the current sensors. The cheapest sensor is the open-loop current transformer (CT). Their linearity is poor as it fully depends on the magnetic core. Disregarding the poor linearity, they are most widely used in e-meter. A typical block diagram of an e-meter is shown in Fig. 1.42.

Microcontrollers calculate the power and perform the measurements with internal ΔΣ ADCs. The neutral line acts as the ground potential. The supply voltage V_{DD} of the controller is generated cost effectively using diodes and capacitors. The current transformer, which senses the phase current, has the required galvanic isolation inherently included. The voltage measurement utilizes a resistive divider. Often, the current through the neutral line is also evaluated. A shunt resistor is typically used as isolation is not required. The bad linearity of the CT is compensated by calibration. This approach is drawn in Fig. 1.42a.

Higher accuracy sensors run in a closed loop, where the magnetic field is compensated. The magnetic core is then always in the same operating point, so that its non-linearity has no influence on the accuracy. The compensation current is typically provided as the sensor output. Such sensors are available from companies such as LEM or VAC. Their accuracy is typically in the sub 1 % range, but their price is significantly higher than the open-loop current transformers.

Therefore, the measurement using shunt resistors in the phase is an interesting alternative. The phase current generates a voltage drop across the shunt resistor, which can then directly be measured by an ADC. The problem is that the ADC must operate at the phase potential. Sufficient galvanic isolation is required and a floating supply needs to be generated, which adds costs to this approach. Isolated ADCs were first provided by HP/Avago [34]. Over the last years, Analog Devices [35] and Texas Instruments [28] entered this market, so that these solutions will get more cost effective. Furthermore, it is to be expected that higher integration can reduce the system costs.

Figure 1.42b proposes such a solution that enables the use of the accurate shunt measurement, but keeps the costs under control. Costs can particularly be reduced, if the calibration of the linearity over temperature is eliminated, which is required for current transformers, and if a microcontroller without ADC can be used. The proposal measures the phase current and includes the required isolation simultaneously. Additionally, it should also measure the neutral current.

The end-point errors of the proposed ADC such as gain and offset affect the full-scale range. Consequently, their accuracy needs to match the full-scale requirements such as 0.1 %. A 1-point calibration at room temperature should therefore be sufficient. The gain drift can be relaxed to 28 ppm/°C. ADCs with internal reference typically achieve such rating.

More critical are the linearity requirements as expressed in Eq. (1.26). As long as the linearity does not show jumps in the transfer function, the linearity can be calibrated. Jumps in the transfer function are expressed by the differential non-linearity (DNL). Small DNL values are typically provided by ADCs based on delta-sigma modulation. For cost effective manufacturing, system calibration should only be performed at room temperature. The linearity has to remain stable over temperature.

ADCs based on delta-sigma modulation need to be used in this application, which have the best DNL performance that is very important in this application as described above.

The effective number of bits and the noise of the ADC are not significant in these metering applications, as the power is integrated over a long period of time. If the current for example is converted with 10 kSPS then 10,000 samples are measured within a 1 s time windows. The noise within this window is reduced by 3 dB with doubling the oversampling ratio. An integration of 10,000 values will therefore improve the noise by 6.5 bits. Based on Eq. (1.26), the ADC noise could be as low as 11 ENOBs. The noise of delta-sigma ADCs is typically significantly better than this value.

1.3.2.2 Relay Protection

Large and expensive machines or applications, which can cause severe damage if defect, are supervised with protection circuits. Standard examples are generator protection, motor protection or transformer protection. Here, the electric power needs to be controlled with higher data rate, so that a fast reaction time is possible. Typical data rates are 10 kSPS. The delta-sigma converters, which are used in metering applications, cannot be used here, because of their lower speed and filter delay.

Due to the speed requirements, SAR ADCs are widely used in this application. They also provide a sweet spot of low power consumption, fast conversion rate and reasonable resolution.

Unfortunately, their DNL and noise performance is limited to roughly 16 bit and cannot meet the performance of delta-sigma modulators. Therefore, products with gain adjustments are used to support the dynamic range that is required for the current measurement. The widest input voltage range is typically chosen to ±10 V. Additional ranges with ±5 and ±2.5 V are provided. The switching of the input ranges could be avoided, if the ADCs could provide an increased accuracy of two additional bits at a reasonable price and power consumption.

The noise requirements of the protection systems are typically higher than in metering application as the higher conversion rate limits a possible oversampling. Assuming that no oversampling is performed, the signal-to-noise ratio should be higher than 90 dB, if input range adjustment is provided.

The biggest challenge is the required channel count. Let's assume that the part is used in a transformer protection, where the currents and voltages of three phases and the neutral line are sensed. These measurements need to be performed on the primary side and the secondary side, so that a total of 16 signals need to be measured simultaneously. Today's products, which are specialized for this application, can provide a maximum of eight channels [36–38]. Obstacles are the required die area and the consumed power consumption. If one channel for example requires a typical supply current of 8 mA at a 5 V supply, then the 16 channels together consume 640 mW. The heat dissipation is getting a sever issue.

Furthermore, transistors, which withstand the range of ±10 V, are large and consume a significant amount of chip area. The eight channel ADCs, which were mentioned before, typically require a die area of 25 mm^2. These products already divide the input voltage to an internal 5 V range and run the conversion on a 5 V semiconductor process. An increase in channel count is not possible with today's techniques as such products would not fit into the anticipated packages such as TQFP or TSSOP. Furthermore, the yield would drop significantly due to the large die area.

Often, the protection equipment needs to support all kind of current sensors. Worst are current transformers, which output a single ended signal. Ground distortion and cross coupling on the printed circuit board (PCB) degrades the performance of the signal. Better would be a differential signal routing. This is another reason, why ±10 V input ranges are strictly required by the protection relay manufacturers [39].

SAR ADCs have the drawback that 5 V input ranges and therefore 5 V semiconductor processes provide the optimum between noise, conversion rate and power dissipation. Moving towards lower process dimensions would reduce the input range and therefore the size of the LSB, while the noise sources like the thermal noise of the comparator or the sampling noise (kT/C) remain constant. The signal-to-noise ratio would therefore drop linear to the input voltage range.

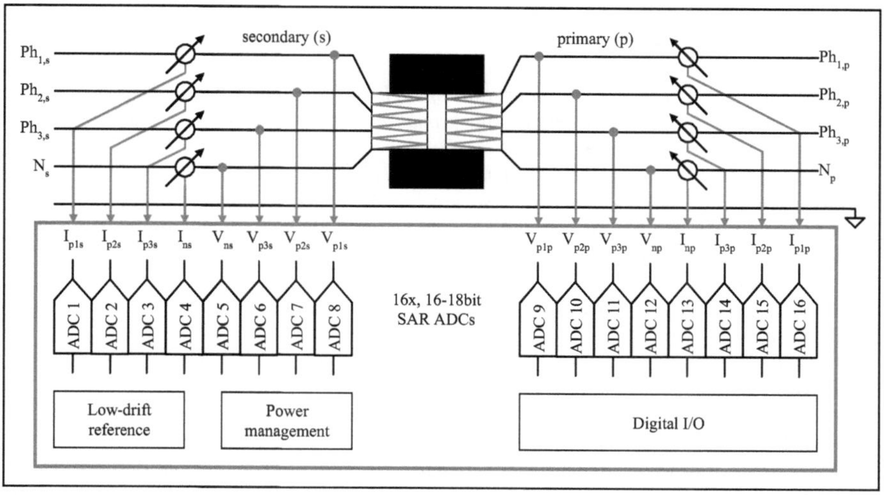

Fig. 1.43 Proposed IC with 16 ADCs required for applications for power protection

Therefore, the ADC manufacturers need to invent new algorithms such as presented in Sect. 3.2, where large input voltage ranges can be converted with low supply voltages and therefore on processes with smaller dimensions, without internally reducing the signal amplitude. Moving to a 1.8 V process might reduce the power consumption by a factor of 10 and the die area by factor of 2. Due to a smart switching scheme, the internal signal amplitude remains undivided and the proposed architecture provides signal-to-noise ratios and linearities such as state-of-the-art products.

This enables the proposed circuit shown in Fig. 1.43. Basically, 16 SAR converters with a preferred resolution of 18 bits should be integrated on silicon. The input voltage range needs to remain at ± 10 V. Research also needs to be performed on production level. The product verification of such circuits consumes a significant calibration and test time and adds most of the manufacturing cost.

Moving to such a large channel count will also cause a challenge on dynamic interferences of the internal supply voltage. The proposed smaller process dimension would also help to integrate supply compensation circuits, which actually increase the circuit stability and can improve the yield even with an increased integration level.

Industrial applications typically require ADCs with low to medium conversion rates at medium to high resolutions. SAR ADCs and Delta-sigma ADCs are commonly used. The following chapters will therefore describe architecture, design and layout requirements for both converter types.

References

1. Frank Ohnhäuser, *Theory and realization of high-end analog-to-digital converters (ADC) based on the principle of successive approximation (SAR)*, PhD Thesis at the University of Erlangen-Nürnberg, Erlangen, 2008.
2. *IEEE standard for terminology and test methods for analog-to-digital converters*, IEEE Instrumentation and Measurement Society, 2000, URL: http://ieeexplore.ieee.org/xpl/articleDetails.jsp?tp=&arnumber=929859&queryText%3DIEEE+Standard+for+Terminology+and+Test+Methods+for+Analog-to-Digital+Converters, status of 3/11/2015.
3. Frank Ohnhäuser, *Entwurf eines Parallel-Analog-Digital-Umsetzers mit analoger Vorkodierung in GaAs Technologie*, student project at the Friedrich-Alexander-University, Erlangen 1995.
4. Product datasheet of the ADC12D1800RF, Texas Instruments, 2013, URL: http://www.ti.com/lit/ds/symlink/adc12d1800rf.pdf, status of 3/23/2015.
5. Rolf Unbehauen, *Systemtheorie – Grundlagen für Ingenieure*, Oldenburg Verlag, München Wien 1993.
6. Dieter Kress, Ralf Irmer, *Angewandte Systemtheorie*, VEB Verlag Technik, Berlin 1989.
7. Martin Allinger, *Development and implementation of a time efficient trim method for SAR analog-to-digital converters*, PhD thesis at the University of Erlangen, Erlangen, 2012.
8. JESD12-6, *Interface Standard for Semicustom Integrated Circuits*, 1991, URL: http://www.jedec.org/sites/default/files/docs/jesd12-6.pdf, status of 02/27/2015
9. JESD8c.01, *Interface Standard for Nominal 3V/3.3V Supply Digital Integrated Circuits*, 2006, URL: http://www.jedec.org/sites/default/files/docs/jesd8c-01.pdf, status of 02/27/2015
10. Product datasheet of the ADS1282, Texas Instruments, 2010, URL: http://focus.ti.com/lit/ds/symlink/ads1282.pdf, status of 3/11/2015.
11. Robert Loehr, *Entwicklung eines Taktgenerators mit Frequenzabgleichmöglichkeit und optimiertem Phasenrauschen in 180nm CMOS Technologie*, Master Thesis at the Friedrich-Alexander-University, Erlangen 2012.
12. Dieter Seitzer, *Elektronische Analog-Digital-Umsetzer, Verfahren, Bauelemente, Beispiele*, Springer Verlag, Berlin/Heidelberg 1977.
13. Rudy J. van de Plassche, *CMOS integrated analog-to-digital and digital-to-analog converters*, Kluwer Academic Publisher, Bosten/Dordrecht/London 2003.
14. Behzad B. Razavi, *Principles of data converter system design*, *Wiley-IEEE Press*, New York/Chichester/Weinheim/Brisbane/Singapore/Toronto 1995.
15. W. Friedrich Oehme, Mario Huemer, Markus Pfaff, *Elektronik und Schaltungstechnik – ein verständlicher Einstieg*, Fachbuchverlag Leipzig im Carl Hanser Verlag, München/Wien 2006.
16. Product datasheet of the ADS5400, Texas Instruments, 2010, URL: http://focus.ti.com/lit/ds/symlink/ads5400.pdf, status of 3/11/2015.
17. Product datasheet of the LTC2379-18, Linear Technology, 2011, URL: www.linear.com/docs/30158, status of 3/11/2015.
18. Product datasheet of the AD7626, Analog Devices, 2009, URL: http://www.analog.com/static/imported-files/data_sheets/AD7626.pdf, status of 3/11/2015.
19. Product datasheet of the AD7982, Analog Devices, 2007, URL: http://www.analog.com/static/imported-files/data_sheets/AD7982.pdf, status of 3/11/2015.
20. David A. Johns and Ken Martin, *Analog Integrated Circuit Design*, John Wiley & Sons Verlag, NewYork/Chichester/Brisbane/Toronto/Weinheim/Singapore, 1997.
21. Steven R. Norsworthy, Richard Schreier, Gabor C. Temes, *Delta-Sigma Data Converters, Theory, Design and Simulation*, Wiley-IEEE Press, New York 1996.
22. Sangil Park, *Principles of Sigma-Delta Modulation for Analog-to-Digital Converters*, Motorola, 1990, URL: http://www.numerix-dsp.com/appsnotes/APR8-sigma-delta.pdf, status of 3/11/2015.

23. Frank Ohnhaeuser, *Untersuchung von Sigma-Delta-Modulatoren höherer Ordnung mit Multibit-Rückführung*, diploma thesis at the Friedrich-Alexander-University, Erlangen 1996.
24. Rüdiger Kusch, *Decimation Lowpass Filters for Sigma-Delta Modulators*, diplom.de, 1998.
25. P.Y.K. Cheung, *A comparison of decimation filter architectures for sigma-delta A/D converters*, in the proceedings of the IEEE International Sympoisum on Circuits and Systems, pp. 1637–1640, 1991.
26. D. V. Morozov, M. M. Pilipko, A. S. Korotkov, *Decimation filter of the delta-sigma analog-to-digital converter with ternary data encoding*, Russian Microelectronics, Volume 40, Issue 5, pp. 352–360, 2011.
27. Rajaram Mohan Roy Koppula, Sakkarapani Balagopal, Vishal Saxena, *Efficient Design and Synthesis of Decimation Filters for Wideband Delta-Sigma ADCs*, Boise State University, Boise, 2011.
28. Product datasheet of the AMC1203, Texas Instruments, 2011, URL: http://focus.ti.com/lit/ds/symlink/amc1203.pdf, status of 3/11/2015.
29. Product datasheet of the HCPL-7840, Avago, 2012, URL: www.avagotech.com/docs/AV02-1289EN, status of 3/11/2015.
30. Product datasheet of the AMC1200, Texas Instruments, 2012, URL: http://focus.ti.com/lit/ds/symlink/amc1200.pdf, status of 3/11/2015.
31. Frank Ohnhaeuser, *Analog isolation*, in the journal Power System Design, pp. 22–23, 2011, URL: http://www.powersystemsdesign.com/print-archives-dir/psde_dec11.pdf, pp. 22–24, status of 3/11/2015.
32. Frank Ohnhaeuser, Michael Reinhold, Mikael Badenius, *Resolver arrangement*, US Patent 7196643, Texas Instruments, 2007, URL: http://www.google.com.ar/patents/US7196643, status of 3/11/2015.
33. Frank Ohnhaeuser, *Integrated IC solutions for power measurement in Smart Grid Applications*, in the proceedings of the International Conference on Smart Grid – Technology, Economics and Policies (SG-TEP2012), 4 pages, 2012.
34. Product datasheet of the HCPL-7860, Avago, 2011, URL: http://www.avagotech.com/docs/AV02-0409EN, status of 3/11/2015.
35. Product datasheet of the AD7400, Analog Devices, 2013, URL: http://www.analog.com/static/imported-files/data_sheets/AD7400.pdf, status of 3/11/2015.
36. Product datasheet of the ADS8568, Texas Instruments, 2011, URL: http://focus.ti.com/lit/ds/symlink/ads8568.pdf, status of 03/11/2015.
37. Product datasheet of the AD7606, Analog Devices, 2006, URL: http://www.analog.com/media/en/technical-documentation/data-sheets/AD7606_7606-6_7606-4.pdf, status of 3/11/2015.
38. Product datasheet of the MAX11049, Maxim Integrated, 2011, URL: http://www.maximintegrated.com/datasheet/index.mvp/id/6429, status of 3/11/2015.
39. Frank Ohnhaeuser, Mario Huemer, *History and outlook for high voltage and high performance SAR A/D converter*, n the *e&i Journal (Journal of the Austrian Electrotechnical Association, ÖVE)*, No. 4, pp. a17-a20, April 2008.

Chapter 2
ADCs Based on Successive Approximation

Chapter 1 discusses the various performance parameters and architectures of ADCs. The SAR ADC is presented as the ADC that is most frequently used in industrial applications, because it provides a high resolution (12–18 bit) at a medium sample rate (around 1 MSPS). This chapter therefore presents design and architectural basics and details regarding the components of a SAR ADC [1]. The principle block diagram is shown again in Fig. 2.1.

The input voltage V_{in} is frozen on a sample and hold stage, which is discussed in Sect. 2.1. Topics are the charge injection from the switches and the voltage coefficient from the sampling capacitor that will cause an offset error as well as linearity errors. The size of the sampling capacitor also determines the sampling noise.

The DAC, which is the heart of the SAR ADC, is explained in Sect. 2.2. Its offset and linearity are directly reflected in the ADC's transfer function. It is normally implemented with capacitors (CDAC) based on charge redistribution, where the capacitors are switched between the reference voltage and ground to set the appropriate output voltage. The switches to the capacitors are controlled via the comparator (Sect. 2.3) and the successive approximation register.

Settling issues and other error sources can be corrected towards the end of the conversion. A dynamic error correction is discussed in Sect. 2.4, which also opens the door for future correction schemes (Chap. 3). Settling effects of the reference are limiting the speed, noise and power consumption of the converter. Section 2.5 describes considerations regarding the reference topology.

Noise is an important specification. Different sources are analyzed and are summarized in Sect. 2.6. Finally Sect. 2.7 briefly discusses the limitations of state-of-the-art ADCs and leads to the modern research topics in Chap. 3.

F. Ohnhäuser, *Analog-Digital Converters for Industrial Applications Including an Introduction to Digital-Analog Converters*, DOI 10.1007/978-3-662-47020-6_2

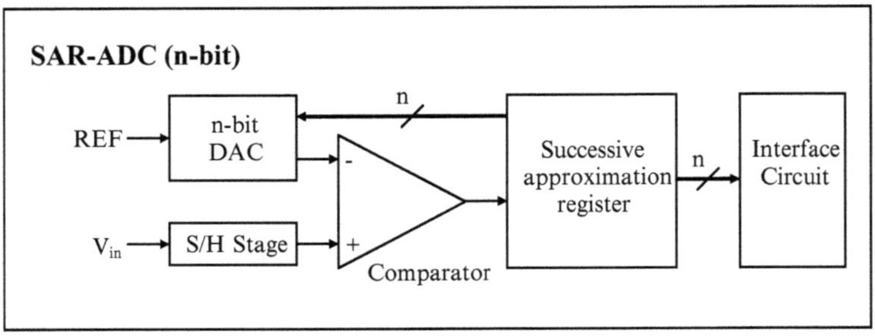

Fig. 2.1 Block diagram of a SAR ADC

2.1 Sample and Hold (S&H)

The S&H circuit is basically a capacitor, which is charged to the input voltage during the sampling time. For the conversion, the input voltage V_{in} is frozen on the capacitor by opening the input switch, which is typically realized as CMOS switch (see Fig. 2.2). Unfortunately, the charge injection of the input switch will generate an offset, which is heavily dependent on the input voltage.

In addition, the on-resistance of the input switch is voltage dependent. The on-resistance is forming a voltage dependent low-pass filter together with the sample capacitor and the parasitic capacitance of the switch. The voltage dependency causes distortion that is very significant at higher signal frequencies. The ratio of the transistor width to length $\frac{W}{L}$ of the input switch has to be large to minimize this effect. The high $\frac{W}{L}$ will on the other side increase the charge injection.

The circuit can therefore be improved with the structure in Fig. 2.3. The charge is frozen with the switch sw_1 to ground. This transistor remains at the same operating point independent of the input voltage, so that the on-resistance is constant and does not affect distortion. The width W of sw_1 can remain small to minimize charge injection. After freezing the charge with the small switch sw_1, the big input switch sw_2 can be opened without adding further charge to the capacitor. Not that sw_1 is called hold switch.

Finally, sw_3 is connecting the capacitor back to ground. This way, the capacitor is connected to a defined potential. As the voltage across the capacitor is frozen, the potential at the node V_{sample} equals $-V_{in}$.

The S&H from Fig. 2.3 will generate a good linearity, as the charge injection of sw_1 is now independent of the input voltage. However, it will still cause a severe offset error, which will change with the supply voltage and temperature.

This problem can be solved, if the sampling is performed fully differential like shown in Fig. 2.4.

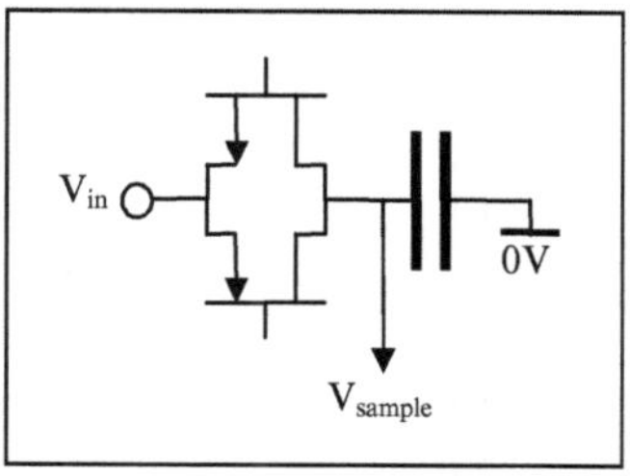

Fig. 2.2 Basic S&H circuitry

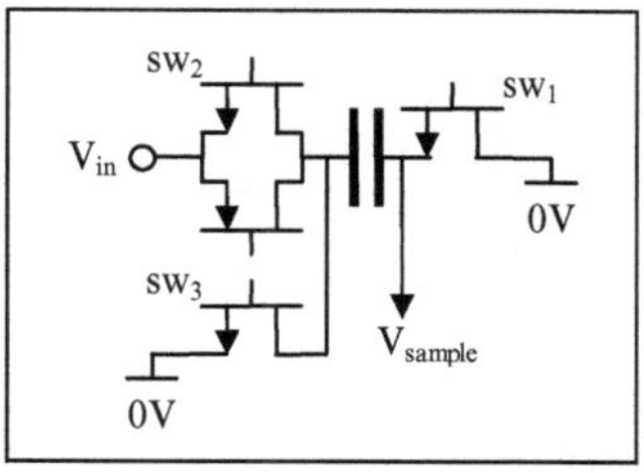

Fig. 2.3 Double switch S&H

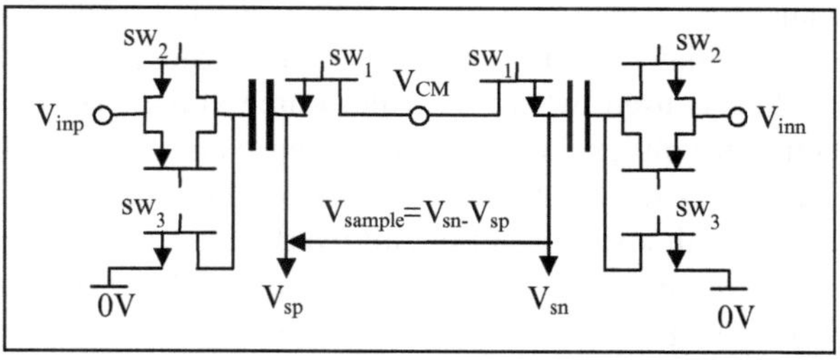

Fig. 2.4 Fully differential S&H circuit

The voltage $V_{sp/n}$ after freezing the input voltage $V_{inp/n}$ can be calculated to

$$V_{sp/n} = V_{CM} - V_{inp/n} + V_{charge}$$

V_{charge} is expressing an offset caused by the charge injection of the hold switches. A fully differential voltage is sampled, which can be calculated to

$$V_{sample} = V_{sn} - V_{sp} = V_{inp} - V_{inn}$$

Identical charge is injected by the two hold switches into V_{sp} and V_{sn}. The charge injection is eliminated in the differential voltage V_{sample}. This structure further allows adding a common-mode voltage V_{CM}, which can be used to select the ideal common-mode operating point of the comparator.

On typical 5 V (or less) CMOS processes, the on-resistor is low enough to realize the input switch sw_2 with a standard CMOS switch. The W ratio of the NCH and the PCH transistor is chosen to receive the most constant on-resistance over the input

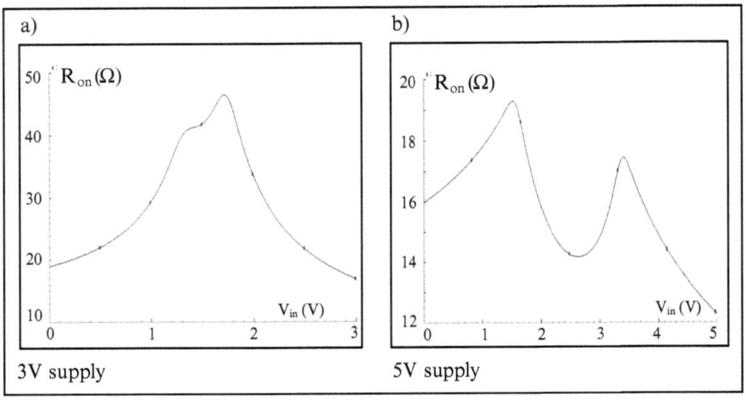

Fig. 2.5 Typical on-resistance of a CMOS sample switch at (**a**) 3 V supply and (**b**) 5 V supply

voltage as shown in Fig. 2.5, where the transistors were optimized for 5 V operation.

ADCs with a high voltage input range (+/−10 V) need to use high voltage input transistors, which show a severe body effect together with a high on-resistance and an essential parasitic capacitance. This causes high distortion. In these cases, a method is used, where the gate-source voltage of the input switch is biased to a constant voltage during the sampling phase. This method is called bootstrapping. The on-resistance of such switches is compared in Fig. 2.6. Note that bootstrapping comes with a significant circuit overhead.

In any case, the sample and hold stage fully relies on the quality of the capacitor. The voltage coefficient of standard Poly-NWell capacitors is so high that the integral linearity suffers. High-end ADCs require processes with special capacitors like poly-poly capacitors, metal-poly capacitor or metal-metal capacitors. Effects of the voltage coefficient are discussed in Sect. 3.4 [2].

A sample and hold circuitry can be considered as an *RC* circuitry, where the *R* is the on-resistance R_{on} of the input switch and the *C* the sample capacitor C_s. Every resistor is adding thermal noise [3] with the density n_{Ron} of

$$n_{Ron} = \sqrt{4kTR_{on}} \tag{2.1}$$

In this equation, *T* is expressing the temperature and *k* the Boltzmann constant.

Thermal noise is equally distributed over the whole frequency span. The bandwidth of the input circuitry however is limited to $f_{-3\mathrm{dB}} = \frac{1}{2\pi R_{on}C}$. Even if the noise is suppressed for frequencies above $f_{-3\mathrm{dB}}$, it is not eliminated as shown in Fig. 2.7.

Integrating the frequency responds of the first order low-pass filter will effectively result in $\Delta f = \frac{\pi}{2} \cdot f_{-3\mathrm{dB}}$ [3], so that the rms-voltage of the low-pass filtered noise n_f during the sampling process $V_{rms,samp}$ can be calculated to

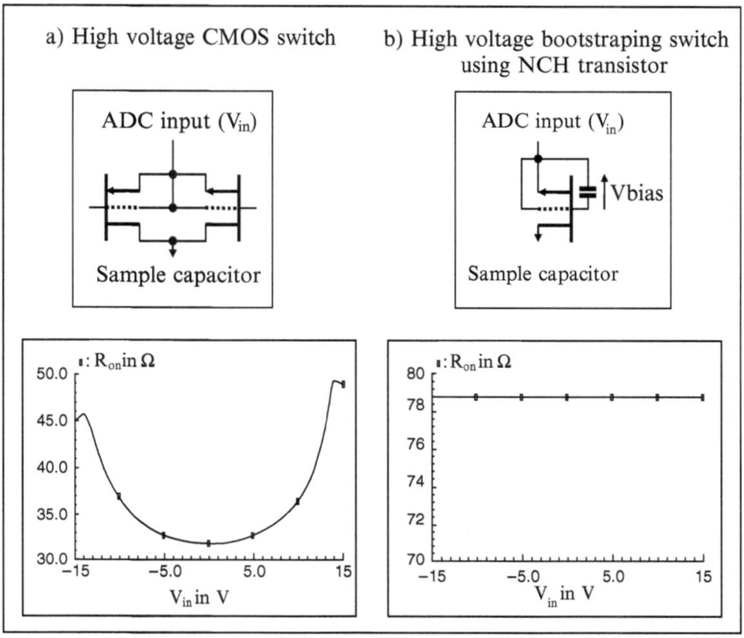

Fig. 2.6 On-resistance of high voltage switches during sampling

$$V_{rms,samp} = \sqrt{\int_0^\infty n_f^2 df} = \sqrt{n_{Ron}^2 \cdot \Delta f} = \sqrt{\frac{kT}{C}} \tag{2.2}$$

The sampling noise can only be influenced with the size of the sampling capacitor. If the input range of the ADC is reduced by a factor of 2, then the size of the LSB is also divided by two and the noise needs to be lowered by a factor of 2. For the sampling noise, this is only possible by increasing the sampling capacitor by a factor of 4.

New CMOS technologies with minimum gate lengths have the disadvantage of small supply voltages, which would limit the input voltage range. Several modern analog processes therefore leave the gate length between 0.35 μm (3.3 V) and 0.6 μm (5 V) and concentrate on other parameters like transistors with low 1/f noise, capacitors with low voltage coefficients or resistors with low temperature coefficient.

Note that the sample and hold circuitry from Fig. 2.4 is sampling on two sample capacitors, so that the noise needs to be considered twice. The rms-voltage needs to be multiplied with the square-root of two. The differential architecture however can be used to increase the signal amplitude by a factor of 2.

In the architecture from Fig. 2.4, the hold switch is connected to a common-mode voltage V_{CM}. Also the common-mode voltage is a potential noise source.

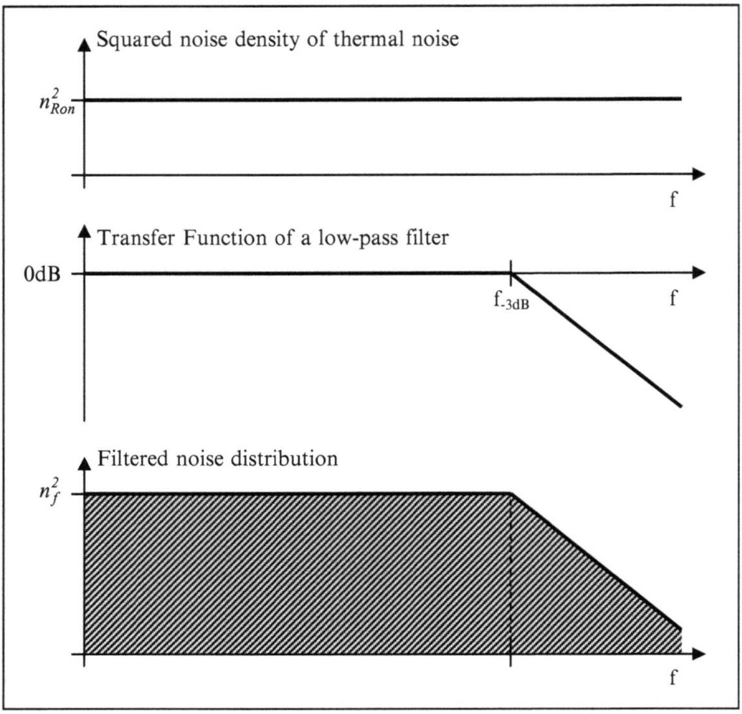

Fig. 2.7 Illustration of thermal noise shaped by a first order filter

However, the bandwidth of the voltage source is typically much smaller than the bandwidth of the sample and hold structure. This will cause the noise to be common to the positive and the negative sample and hold capacitor. The common noise will be suppressed by 60–80 dB by the common-mode rejection of the fully differential DAC and the fully differential comparator.

2.2 Capacitive DAC

The DAC is the heart of the SAR converter. Its differential and integral non-linearity will directly be reflected in the transfer function of the ADC. Typical DAC structures are the string DAC, the R – 2R structure or the current steering DAC (see Chap. 7). All of them show limitations in speed and performance [4].

Ideal is a capacitive DAC (CDAC), which is based on the principle of charge redistribution. An example for 3 bits is shown in Figs. 2.8, 2.9, 2.10 and 2.11.

In Fig. 2.8, all capacitors are connected to the input voltage V_{inp} for sampling. The charge, which is stored across the capacitors, can be calculated to

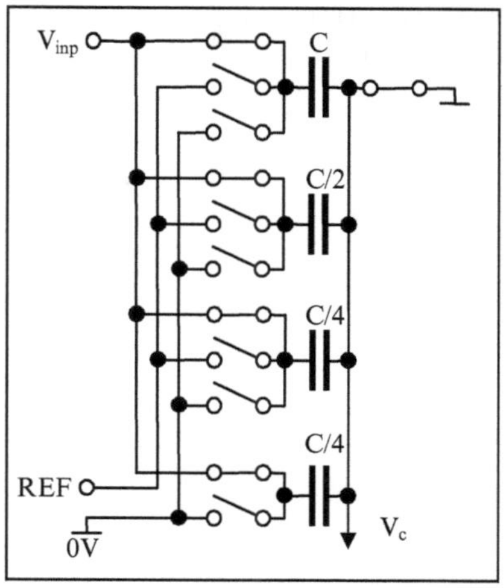

Fig. 2.8 CDAC during sampling process

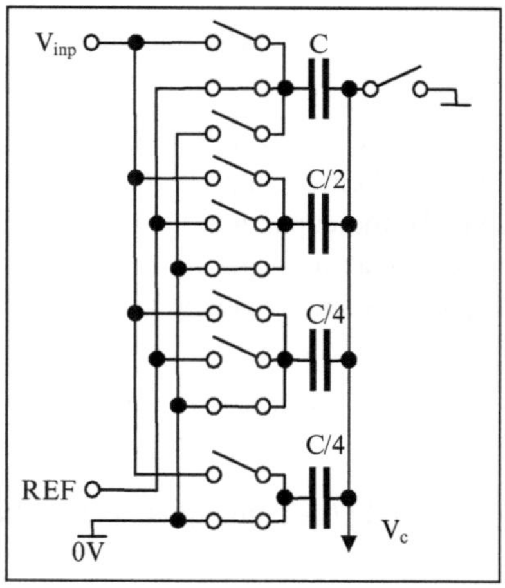

Fig. 2.9 CDAC during MSB decision

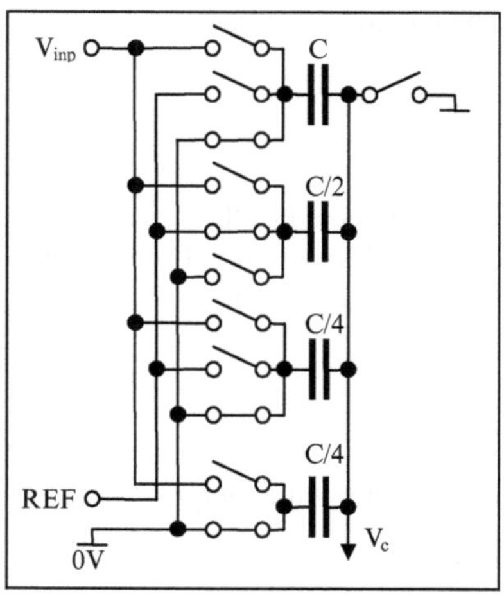

Fig. 2.10 CDAC during MSB-1 decision

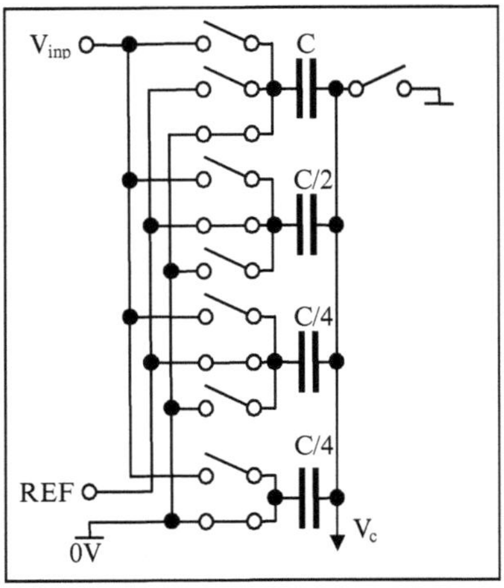

Fig. 2.11 CDAC during MSB-2 decision

$$Q_{samp} = \left(C + \frac{C}{2} + \frac{C}{4} + \frac{C}{4}\right) \cdot V_{inp} = 2C \cdot V_{inp}$$

The charge is frozen by opening the hold switch between V_c and ground. Then, the input switches coupling to V_{inp} open. For the evaluation of most significant bit (MSB), the capacitor of the size C is connected to the reference voltage REF, all other capacitors are connected to ground (see Fig. 2.9). The charge across the capacitors can be calculated to

$$Q_{MSB} = C \cdot (REF - V_c) + \left(\frac{C}{2} + \frac{C}{4} + \frac{C}{4}\right) \cdot (0 \text{ V} - V_c)$$

Q_{samp} and Q_{MSB} are equal, as the charge is frozen. This will lead to

$$2C \cdot V_{inp} = C \cdot (REF - V_c) + C(0 \text{ V} - V_c)$$

The equation can be solved to

$$V_c = \frac{REF}{2} - V_{inp}$$

V_c is connected to the negative comparator input, the positive comparator input is connected to ground. The comparator will therefore compare, if the voltage V_c is less or equal than 0 V

$$V_c = \frac{REF}{2} - V_{inp} \leq 0 \text{ V} \Leftrightarrow \frac{REF}{2} \leq V_{inp}$$

If the comparator output is 1, then the input voltage was higher than half of the reference voltage. For the MSB-1 evaluation, C would remain at REF and $\frac{C}{2}$ would switch from ground to *REF*. In the example of Sect. 1.2.2, the reference was 4 V and the input voltage 1.9 V. V_{in} would be less than half the reference, so that the comparator output and therefore the MSB would be 0. The capacitor C will connect to ground, $\frac{C}{2}$ to the reference voltage.

Figure 2.10 shows the CDAC during the next bit decision. The total charge can now be calculated to

$$Q_{MSB\text{-}1} = \frac{C}{2} \cdot (REF - V_c) + \left(C + \frac{C}{4} + \frac{C}{4}\right) \cdot (0 \text{ V} - V_c)$$

Again, the charge is identical to Q_{samp}, so that

$$V_c = \frac{REF}{4} - V_{inp} \leq 0 \text{ V} \Leftrightarrow \frac{REF}{4} \leq V_{inp}$$

The comparator will now decide, if the input voltage was higher than a quarter of the reference. In the example of $V_{inp} = 1.9$ V, this would be the case, so that the comparator output and therefore the MSB-1 is one. The capacitor $\frac{C}{2}$ remains connected to the reference voltage and the next bit capacitor is switching from ground to the reference as shown in Fig. 2.11, which will lead to

$$Q_{MSB\text{-}2} = \left(\frac{C}{2} + \frac{C}{4}\right) \cdot (REF - V_c) + \left(C + \frac{C}{4}\right) \cdot (0 \text{ V} - V_c) \quad \text{or}$$
$$V_c = \frac{3REF}{8} - V_{inp} \leq 0 \text{ V} \Leftrightarrow \frac{3REF}{8} \leq V_{inp}$$

For the example, this is also true, so that the MSB-2 is also one. The 3 bit digital output of the ADC would be 011.

2.2.1 *Basic CDAC Architectures*

For converters with more than 10 bits, the ratio between the MSB capacitor and the LSB capacitor is getting significant due to the binary scaling (see Fig. 2.12). A serial capacitor can be placed between an MSB and an LSB capacitor array as shown in Fig. 2.13. The serial capacitor is also called scale-down capacitor. The solution with the scale-down capacitor has a larger LSB capacitor and less total capacitance if compared to the straight binary CDAC. Better matching can be achieved at a smaller CDAC size.

An alternative solution is shown in Fig. 2.14, where the LSBs are implemented by adjusting the voltage applied to the smallest capacitor. A resistive divider, also called string DAC, is used to tune the reference voltage, so that V_c changes in LSB steps.

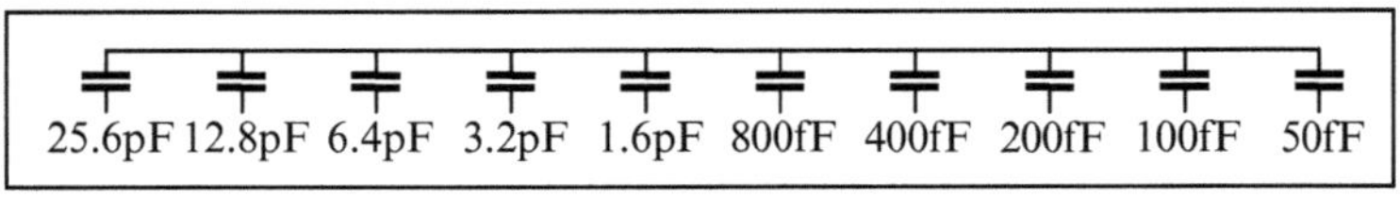

Fig. 2.12 Capacitor array within a standard binary weighted 10 bit CDAC

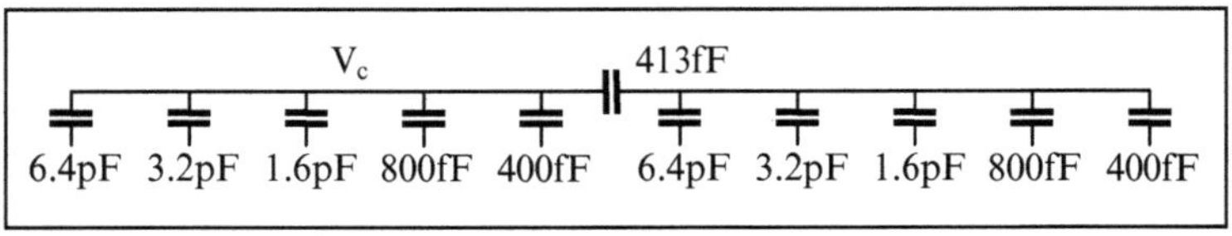

Fig. 2.13 Capacitor array of a 10 bit CDAC with serial scale-down capacitor

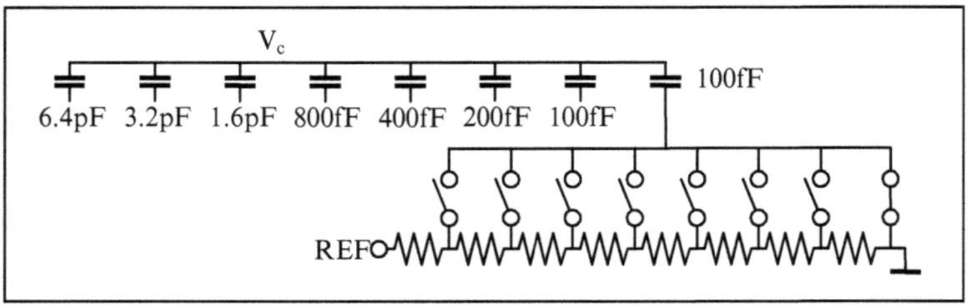

Fig. 2.14 CDAC combined with a resistive string DAC for the LSB evaluation

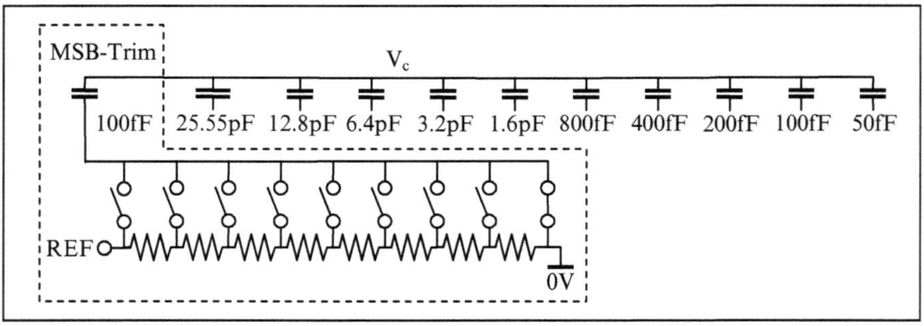

Fig. 2.15 MSB trim circuitry using a resistive string DAC coupling through a trim capacitor

Capacitors typically match up to 0.1 %, so that a 10 bit accuracy can directly be achieved. If the resolution is more than 10 bits, then a calibration or trim solution is required.

Figure 2.15 shows a trim solution for the MSB capacitor. The MSB capacitor is reduced by a value corresponding to one LSB and a trim capacitor of 2 LSBs is switched in parallel. If this trim capacitor is charged to half the reference voltage synchronously with the MSB capacitor switching to *REF*, then the MSB capacitor has the same charge redistribution than before. But now, the weight can be changed by either trimming or calibrating the ADC. The weight of the MSB capacitor can be

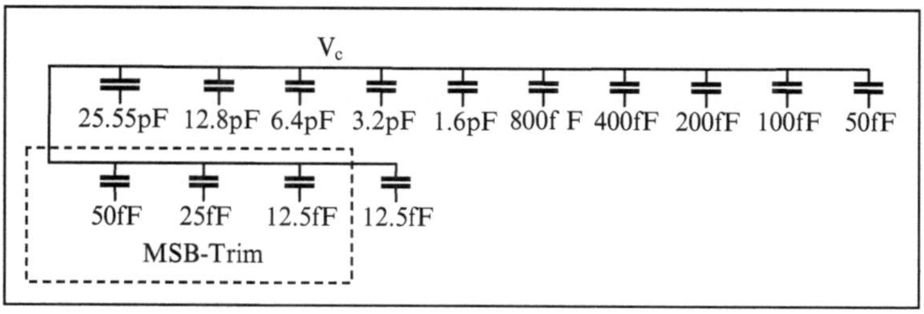

Fig. 2.16 MSB trim circuitry using additional capacitors

increased by 1 LSB, if the trim capacitor is charged to the reference voltage. In the same way, the weight is reduced by one LSB, if the 100 fF trim capacitor remains constantly at ground.

The disadvantage of this structure is that the reference has to drive a DC current through the string DAC, which is adding to the power consumption. This is especially true for 16 bit and 18 bit ADCs, where several bits are trimmed. An alternative is a capacitive trim, like shown in Fig. 2.16.

Programmed trim capacitors are switching in parallel with the MSB. The small size of the trim capacitors is causing an issue, which is discussed later.

2.2.2 Trim Solutions

Trim solutions are in common use since several decades [5–7]. The typical approach was laser cutting of thin film resistors or links, which are made out of poly or metal, during wafer probing. Unfortunately, the assembly process will shift the measured parameters. This is due to the mechanical packaging stress, but also to an inhomogeneous dielectric of the packaging material, which is called molding compound. This will change the parasitic capacitors inside the CDAC and influence the DNL of particular bits up to 6 LSB for 16 bit ADCs. In addition, the packaging shift has a significant variation associated, so that a perfectly trimmed CDAC is nearly impossible. Perfectly trimmed ADCs are therefore selected as high-grades, which are sold at a higher price.

The laser cutting can be done anywhere on the die and does not require additional circuitry. Therefore, the fuse can also be placed in the wiring, and capacitors can be disconnected easily and very space effectively.

In-package trim solutions were designed, where the trim information is either stored by blowing fuses electrically or by adding memory like EPROM (Electrically Programmable ROM) or OTPs (One-Time-Programmable-ROM).

The approach with the electrically blown fuse is very similar to the laser cutting of fuses. The only difference is that the fuse is blown with a high current flow instead of the laser. Unfortunately, this requires the fuse to be connected to a low impedance transistor and a low impedance ground or supply. Due to the parasitic capacitance of the transistor, the electrically blown fuse can mainly be used to generate digital control signals. Note that fuses can be implemented on a standard CMOS process.

The electrically blown fuses and the OTPs generate digital signals. These signals are used to control switches for trim solutions. Both trim solutions from Figs. 2.15 and 2.16 can be realized with this technique.

Typically, only the matching of the binary weighted bit capacitors was trimmed. But much more than that is possible. Offset, gain and CMRR can also be adjusted inside the capacitor array. A general calculation is performed below on a fully differential CDAC, which will explain the various trim effects.

For the calculation, the CDAC in Fig. 2.17 is used. In most cases, the input voltage is directly sampled on the bit capacitors, but here the sampling capacitors $C_{sn,p}$ are separated from the bit capacitors, which are called $C_{1n,p}$ to $C_{Nn,p}$. The resolution of the CDAC is also kept general to N-bit. The input voltage range is bipolar. The fully differential topology will keep the charge injection of the hold switches (sw_{hp} and sw_{hn}) symmetrical. C_{jn} and C_{jp} ($j\epsilon\{1, \ldots, N\}$) are binary weighted ($C_{jn,p} = C \cdot 2^{-(j-1)}$), where C_{1n} expresses the capacitor for the MSB (Most Significant Bit or bit 1) and C_{Nn} the capacitor for the LSB (Least Significant Bit or bit N).

In sample mode, the inputs V_{inp} and V_{inn} are coupled to the sample capacitors C_{sn} and C_{sp}, while the remaining CDAC capacitors $C_{jn,p}$ are switched to 0 V. The nodes V_{cn} and V_{cp} that short the other pin of all capacitors is connected to a common-mode voltage V_{CM}, which can be used to set an anticipated common-mode operating point

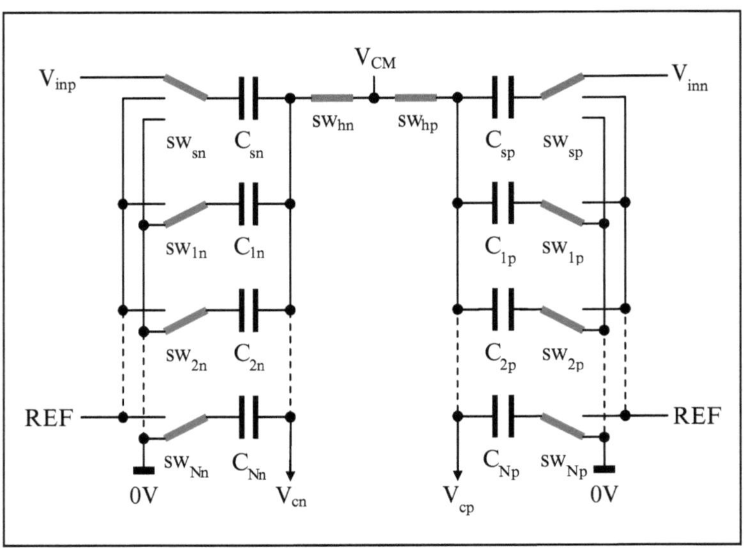

Fig. 2.17 Fully differential CDAC used below to calculate various trim schemes

of the comparator. The sampled charge Q_{sp} and Q_{sn} on the nodes V_{cp} and V_{cn} can be evaluated to

$$Q_{sp,n} = C_{sp,n} \cdot \left(V_{CM} - V_{inn,\,p}\right) + \sum_{j=1}^{N} C_{jp,n} V_{CM} \tag{2.3}$$

The charges Q_{sp} and Q_{sn} are frozen by opening the switches sw_{hn} and sw_{hp}, which in sample mode work at the equal operating point V_{CM} and ideally generate an identical charge injection. Thanks to the equal operating point, the switches further only show a minor influence to the ADC's distortion and can be minimized particular if V_{CM} is chosen to be close to 0 V or VDD, where either the NCH or respectively the PCH transistor are conducting well. The offset error and offset error drift can be optimized. After opening the sample switches, the capacitors C_{sn} and C_{sp} should be disconnected from the input nodes V_{inp} and V_{inn} and should in this example be conducted to the reference voltage *REF* for the complete conversion. The bit capacitors C_{jn} ($j\epsilon\{1, \ldots, N\}$) are now toggling between *REF* and 0 V based on the normal SAR algorithm. For the following calculation, the charge at the bit decision i should be named Q_{ip} and Q_{in}. The state of a particular bit switch $sw_{jn,p}$ ($j\epsilon$ $\{1, \ldots, N\}$) should be expressed by x_{jp} and x_{jn}. $x_{jn,p}=0$ indicates that the switch $sw_{jn,p}$ connects the capacitor $C_{jn,p}$ to 0 V. The capacitor $C_{jn,p}$ shall conduct to the reference *REF* with $x_{jp,n}=1$.

$$Q_{ip,n} = C_{sp,n} \cdot \left(V_{cp,n} - REF\right) + \sum_{j=1}^{N} \left(C_{jp,n} \cdot \left(V_{cp,n} - x_{jp,n} \cdot REF\right)\right) \tag{2.4}$$

The charge after sampling and during a bit decision i equal each other $Q_{sp,n}=Q_{ip,n}$, as V_{cp} and V_{cn} are both in high impedance state during the conversion. Equations (2.3) and (2.4) are set equal and are solved for V_{cp} and V_{cn}. In the next step, the differential comparator input $V_c=V_{cp}-V_{cn}$ is evaluated. The comparator then decides if $V_c>0$. The above steps are expressed in Eq. (2.5):

$$V_{inp}\frac{C_{sn}}{C_{sn}+\sum_{j=1}^{N} C_{jn}} - V_{inn}\frac{C_{sp}}{C_{sp}+\sum_{j=1}^{N} C_{jp}} > REF\left[\frac{C_{sn}+\sum_{j=1}^{N}\left(x_{jn}\cdot C_{jn}\right)}{C_{sn}+\sum_{j=1}^{N} C_{jn}} - \frac{C_{sp}+\sum_{j=1}^{N}\left(x_{jp}\cdot C_{jp}\right)}{C_{sp}+\sum_{j=1}^{N} C_{jp}}\right] \tag{2.5}$$

Ideally, the capacitance of a capacitor on the positive CDAC side equals the capacitance of the respective capacitor on the negative side ($C_{sn}=C_{sp}$ and

$C_{jn} = C_{jp}$). The bit capacitors should further be binary weighted ($C_{jn} = C \cdot 2^{-(j-1)}$ with $j \epsilon \{1, \ldots, N\}$). In this particular example, the sample capacitors should have the capacitance C as well ($C_{sn} = C_{sp} = C$).

2.2.2.1 DNL Trim

The DNL expresses the variance of a particular code from its ideal width of 1 LSB. The following calculation will prove that the DNL is dependent on the matching of the binary weighted bit capacitors. Most critical is the matching of the MSB capacitors (here C_{1n} and C_{1p}), which have to match to the sum of the bit capacitors $C_{jn,p}$ ($j \epsilon \{2, \ldots, N\}$). The code width is defined by two code transitions that can be calculated with Eq. (2.5). For the DNL of the MSB, the upper code transition V_{tr1} is defined with $x_{1n} = 1$ and $x_{jn} = 0$ ($j \epsilon \{2, \ldots, N\}$) and the lower code transition V_{tr2} with $x_{1n} = 0$ and $x_{jn} = 1$. In this example, the ADC should operate in single-ended mode, so that x_{1p} is always 1 and $x_{jp} = 0$ ($j \epsilon \{2, \ldots, N\}$). At a code transition, the ADC transfer function crosses from one ADC output code to the next. This means that at the critical bit decision in the course of the conversion the comparator just changes from output 0 to output 1. This can only be the case if the comparator input $V_C = 0$ V. Consequently, the Eqs. (2.6a) and (2.6b) use the equal sign. The following calculation should further show the effect of a mismatch ΔC of C_{1n}, so that $C_{1n} = C + \Delta C$

$$\begin{aligned} V_{tr1p} \frac{C}{3C + \Delta C - C_N} - V_{tr1n} \frac{C}{3C - C_N} \\ = REF \cdot \left(\frac{2C + \Delta C}{3C + \Delta C - C_N} - \frac{2C}{3C - C_N} \right) \end{aligned} \tag{2.6a}$$

$$\begin{aligned} V_{tr2p} \frac{C}{3C + \Delta C - C_N} - V_{tr2n} \frac{C}{3C - C_N} \\ = REF \cdot \left(\frac{2C - C_N}{3C + \Delta C - C_N} - \frac{2C}{3C - C_N} \right) \end{aligned} \tag{2.6b}$$

If ΔC is neglected in the denominator, then

$$\begin{aligned} V_{tr1} &= V_{tr1p} - V_{tr1n} = REF \cdot \frac{\Delta C}{C} \\ V_{tr2} &= V_{tr2p} - V_{tr2n} = -REF \cdot \frac{C_N}{C} = -1 \text{ LSB} \\ DNL_{MSB} &= V_{tr1} - V_{tr2} - 1 \text{ LSB} = REF \cdot \frac{\Delta C}{C} \end{aligned} \tag{2.7}$$

The capacitor mismatch ΔC is directly proportional to the DNL. It can be corrected by switching additional trim capacitors with the absolute value of ΔC in parallel to the bit of interest if ΔC is negative or anti-parallel to the bit of interest is positive, which here is the MSB. The implementation of the in-package trim method is further explained in Sect. 2.2.3.

Please note that parasitic capacitance at the nodes V_{cp} and V_{cn} will add to the denominator similar as the sampling capacitors. If these capacitors remain at a constant potential all the time, then they will lower the voltage range at the comparator input, but will not influence the linearity, gain or offset. However, the lower voltage range will make the comparator noise more significant, which will reduce the SNR.

2.2.2.2 Gain Trim

Gain and offset can certainly be calibrated and corrected in the digital domain, but this would go on the cost of resolution. If for example a total gain adjustment needs to have a range of 20 % and an offset adjustment of another 5 %, then 25 % of the ADCs input range cannot be used, because they need to be reserved for the signal variations. The ADC retains its total input range, if offset and gain are corrected in the analog domain.

The absolute gain and offset errors of an application are adding up much more than the values of the ADC. Some ADCs on the market [8] therefore have an offset and gain correction implemented that can be controlled by the user through internal registers that are set through the digital interface.

The gain of the ADC is determined by the ratio of the sampling capacitor $C_{sn,p}$ to the sum of bit capacitors $C_{jn,p}$ in the CDAC, which are switching between ground and the reference. This can be proven by calculating the positive full-scale (PFS) and negative full-scale (NFS) with Eq. (2.5). For the PFS, x_{jn} all equal one ($x_{jn}=1$, $j\epsilon\{1,\ldots,N\}$) and x_{jn} are all zero for the NFS ($x_{jn}=0, j\epsilon\{1,\ldots,N\}$). Again the ADC should operate in single-ended mode, so that x_{1p} is always 1 and $x_{jp}=0$ ($j\epsilon\{2,\ldots,N\}$). The calculation assumes that the capacitors on the positive side equal the capacitors on the negative side ($C_{sn}=C_{sp}=C_s$ and $C_{jn}=C_{jp}=C_j$)

$$V_{PFS} = REF \cdot \frac{\sum_{j=2}^{N} C_j}{C_s} \tag{2.8a}$$

$$V_{NFS} = -REF \cdot \frac{C_1}{C_s} \tag{2.8b}$$

$$V_{PFS} - V_{NFS} = REF \cdot \frac{\sum_{j=1}^{N} C_j}{C_s} \tag{2.8c}$$

The input range is proportional to the reference and to the ratio of the sum of the bit capacitors C_j to the sampling capacitor C_s. The gain of a SAR ADC can be adjusted by connecting more or less capacitors to the input pins V_{inp} and V_{inn} during sampling. It is also possible to sample directly on bit capacitors C_{jn} and C_{jp}. Note that the ratio of sample capacitor to the bit capacitors on the negative side has to be identical to the same ratio on the positive side.

$$\frac{C_{sn}}{\sum_{j=1}^{N} C_{jn}} = \frac{C_{sp}}{\sum_{j=1}^{N} C_{jp}}$$

Otherwise, the gain on the positive CDAC array will be different than the gain of the negative array. If the common-mode voltage of the input varies, then it is processed differently on the two sides and will affect the common-mode rejection, which is shown later on.

To increase the input voltage range of the ADC, C_s has to be reduced, which is expressed in Eq. (2.8c). In the previous examples, the sampling was always performed on all CDAC capacitors, so that the input range resulted in 0 V to *REF*. If only the MSB capacitor is used for the sampling, then the input range would be 0 V to 2 *REF*. Several products are on the market, which use this method to switch between a variety of input ranges (±10 V, ±5 V, ±2.5 V, 0–5 V) [9, 10].

Equation (2.8c) can also be used to trim the ADC gain. In the example of Fig. 2.18, capacitors are added to lower the input voltage range by 1 LSB and 2 LSB. Therefore, they need to be connected to the input voltage during sampling and to the reference voltage or ground during the whole conversion.

Note that the scale-down capacitor C_{sd} needs to be readjusted, if additional capacitors are added to the scale-down array.

Originally, the value of the scale-down capacitor was estimated, so that the series capacitance of the scale-down capacitor and the sum of capacitors inside the scale-down array $C_{LSB-array}$ were equal to the smallest capacitor in the MSB array ($C_{MSB\text{-}4} = 400$ fF). In the CDAC from Fig. 2.13, the sum of the capacitors in the LSB array added up to 12.4 pF.

$$\frac{1}{C_{sd}} = \frac{1}{C_{MSB\text{-}4}} - \frac{1}{C_{LSB-array}} \Rightarrow C_{sd} = 413 \quad \text{fF}$$

If the LSB array is extended with the capacitors for the gain correction, then this equation is not valid anymore. If for example the biggest capacitor in the LSB array is switching from ground to the reference, then the charge in the CDAC of Fig. 2.13 is distributed between the other LSB capacitors and the scale-down capacitor. For the CDAC with extended capacitors such as in Fig. 2.18, the charge in the CDAC will also be distributed to the capacitors for the gain adjustment, which will divide the signal and will cause a gain error of the LSB array. The weight of all LSB capacitors together will be lower than the $C_{MSB\text{-}4}$ capacitor, which will generate DNL errors. This can easily be compensated by further increasing the scale-down capacitor.

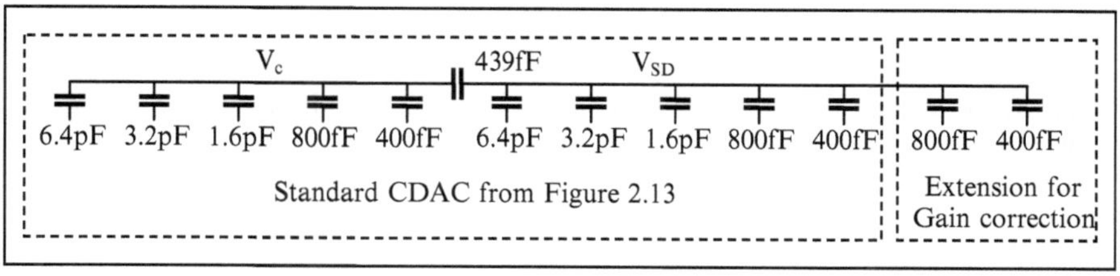

Fig. 2.18 Capacitor array of a 10 bit CDAC with additional capacitors for gain correction

For the correct calculation of the scale-down capacitor, the delta of the scale-down voltage V_{SD} is calculated by taking into consideration that the total charge will not change as it is frozen at the scale-down node. Initially, V_{SD} should be discharged (0 V) and all capacitors in the MSB and the LSB array are coupled to ground. The initial charge Q_1 is therefore zero. Now the biggest capacitor in the LSB array (now called C_{LSBh}) is switched from ground to the reference voltage, which will lead to the following equation

$$Q_2 = C_{LSBh}(V_{SD} - REF) + C_{rest} \cdot V_{SD} = 0 \Leftrightarrow V_{SD} = REF\frac{C_{LSBh}}{C_{LSBh} + C_{rest}}$$

C_{rest} includes the other LSB capacitors, which are equal to C_{LSBh} minus the LSB capacitor C_{LSB}, and it includes the capacitors for the gain adjustment C_{gain}, any parasitic capacitors C_{par} and the series capacitance $C_{SD,MSB}$ of the scale-down capacitor and the capacitance of the MSB array. If the scale-down voltage V_{SD} is applied to the scale-down capacitor C_{SD}, then the change in charge should be half compared to applying the reference voltage to the smallest capacitor in the MSB array C_{MSBl} (here $C_{MSB\text{-}4}$) so that

$$\frac{1}{2}C_{MSBl} \cdot REF = C_{SD} \cdot V_{SD} = C_{SD} \cdot REF \cdot \frac{C_{LSBh}}{2C_{LSBh} - C_{LSB} + C_{par} + C_{gain} + C_{SD,MSB}}$$

If V_{SD} is substituted and if $C_{SD,MSB}$ is estimated to C_{SD} as $C_{MSB} >> C_{SD}$, then

$$C_{SD} = \frac{C_{MSBl} \cdot 2C_{LSBh}}{2C_{LSBh} - C_{MSBl}} \cdot \left(1 + \frac{C_{par}}{2C_{LSBh}} + \frac{C_{gain}}{2C_{LSBh}} - \frac{C_{LSB}}{2C_{LSBh}}\right) \quad (2.9)$$

In the example of Fig. 2.18, parasitic capacitors are not included ($C_{par} = 0$), the gain capacitors are 1.2 pF, the LSB capacitor is 0.4 pF, C_{LSBh} is 6.4 pF and C_{MSBl} is 0.4 pF again, so that the scale-down capacitor can be calculated to 439 fF.

An additional method for a gain adjustment through the reference is also presented in Fig. 2.60 of Sect. 2.5.3. The method with the CDAC capacitors has the advantage that they are based on capacitive matching, so that they work especially accurate and with very low drift.

2.2.2.3 Offset Trim

If an offset voltage V_{off} is added to a signal V_{sig}, then the input voltage V_{in} of the ADC can be expressed by

$$V_{\text{in}} = V_{sig} + V_{off}$$

If the same offset is also added to the normal voltage of the CDAC V_{DAC}, then the CDAC output V_{CDAC} can be expressed to

$$V_{CDAC} = V_{DAC} + V_{off}$$

and the comparator is comparing

$$V_{\text{in}} > V_{CDAC} \Leftrightarrow V_{sig} > V_{DAC}$$

Equation (2.5) shows, how the offset can be compensated through the CDAC. The offset error V_{off} of the bipolar ADC is measured at mid-scale ($x_{1n} = 1$ and $x_{jn} = 0$ for $j \epsilon \{2, \ldots, N\}$), if $V_{inp} = V_{off} + REF$ and $V_{inn} = REF$. If all capacitors are ideal, then Eq. (2.5) leads to

$$V_{off} = \frac{REF}{C_s} \cdot \left(C_1 - \sum_{j=1}^{N} \left(x_{jp} \cdot C_j \right) \right) \tag{2.10}$$

Equation (2.10) assumes that $C_{sp} = C_{sn} = C_s$ and $C_{jp} = C_{jn} = C_j$ ($j \epsilon \{1, \ldots, N\}$).

The generated offset is 0, if C_{1p} is connected to *REF* after sampling and all other capacitors C_{jp} to ground. A positive offset can be generated, if C_{1p} remains at ground. Other capacitors C_{jp} can connect to REF to adjust the offset. Finally, a negative offset is set, if capacitors C_{jp} beside C_{1p} are switched to the reference after sampling [11, 12].

It is also possible to extend the CDAC from Fig. 2.13 with capacitors for offset compensation. In Fig. 2.19, two capacitors are added inside the scale-down array.

Adding capacitors for the offset correction might be required, if the CDAC scheme uses the capacitors C_{jp} for a fully differential conversion process.

Switching the 400 fF capacitor of the offset correction from ground to the reference will have the same effect like switching the 400 fF LSB capacitor to the reference voltage. In this way, an offset of one LSB can be added to V_c.

During the sampling phase, where V_c is connected to a common-mode voltage, the second electrode of the offset capacitors is connected to ground. The selected offset capacitors are switched to the reference as soon as V_c is disconnected from its common-mode source. The selected offset capacitors remain at the reference voltage as long as the conversion is in process and switch back to ground with the start of the next sampling phase.

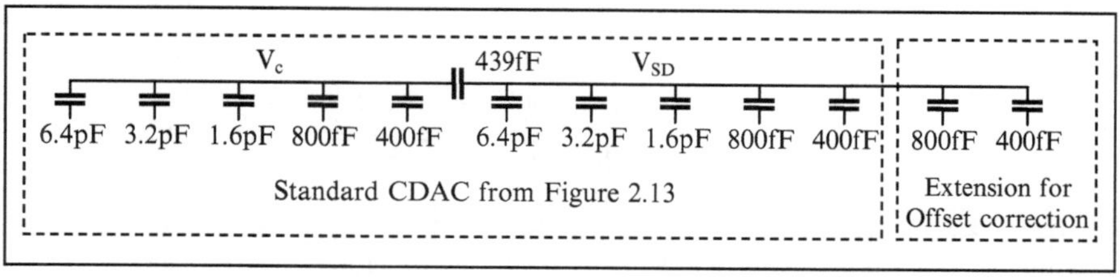

Fig. 2.19 Capacitor array of a 10 bit CDAC with additional capacitors for offset correction

The structure of Fig. 2.19 can also realize negative offset voltages. The only difference is that the capacitors are pre-charged to the reference voltage during the sampling phase and are then connected to ground during the conversion phase.

If capacitors for offset adjustment are added inside the scale-down array, then the scale-down capacitor needs to be adjusted identically to the gain correction. The gain and offset adjustment can both be implemented into one CDAC.

2.2.2.4 CMRR Trim

The common-mode rejection ratio (CMRR) inside the SAR ADC can be considered as the change of the CDAC output voltage $dV_c = dV_{cp} - dV_{cn}$ versus the change of the external common-mode voltage dV_{in}. The differential input voltage during common-mode measurement is zero ($V_{inp} = V_{inn} = V_{in}$). Adding these to Eq. (2.5) leads to

$$CMRR = \frac{dV_c}{dV_{in}} = \frac{C_{sn}}{C_{sn} + \sum_{j=1}^{N} C_{jn}} - \frac{C_{sp}}{C_{sp} + \sum_{j=1}^{N} C_{jp}} = \frac{C_{sn}}{C_{totn}} - \frac{C_{sp}}{C_{totp}} \tag{2.11}$$

To achieve an ideal common-mode rejection ratio $CMRR = 0$, the ratio of the sampling capacitor C_{sp} to the total capacitance C_{totp} on the positive side has to be identical to the same ratio on the negative side. The CMRR can therefore either be trimmed by modifying the sampling capacitance C_s on only one side of the CDAC or by continuously adding or removing capacitance from V_c to ground on only one side of the CDAC.

2.2.2.5 Trim Capacitor Array

Trim schemes regarding Eqs. (2.7) and (2.8a, 2.8b, 2.8c) as well as Eqs. (2.10) and (2.11) require capacitors, which correspond to a fraction of an LSB. These capacitors are generally difficult to generate. Their size can be increased, if they are placed inside the LSB-array as discussed above. However, adding all trim

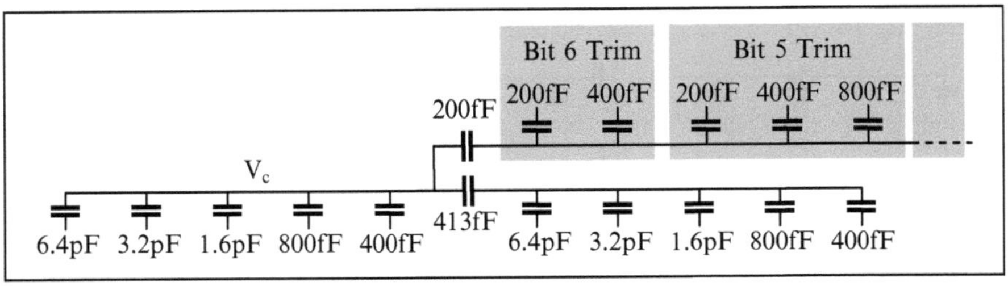

Fig. 2.20 Capacitor array of a 10 bit CDAC with additional trim capacitor array

capacitors to the LSB-array will add too many parasitic capacitors, so that a high variation in the LSB weight might be visible, which would require additional DNL trim range.

An alternative was found by adding a trim capacitor array, which is connected to the MSB array with a trim-scale-down capacitor as shown in Fig. 2.20. The trim capacitors can be realized with a reasonable size, so that they are matching well to the other capacitors. This reduces trim time and production costs. The realization of such a trim solution is shown in the next chapter, where a 16 bit CDAC implementation is discussed.

2.2.3 Implementation of a Pseudo Differential 16 bit CDAC

The CDAC of this example should be based on the CDAC in Fig. 2.13 including a scale-down capacitor. The CDAC is implemented as pseudo differential architecture, which requires a negative side.

Figure 2.21 shows the implementation. C_1 represents the MSB, C_2 the MSB-1 and so on. The MSB capacitors require trimming, which is illustrated with the capacitors C_T. The capacitance is illustrated with a number. The smallest capacitor (C_6) in the MSB array is realized with one unity capacitor. For good matching, the larger capacitors are realized with multiple unity capacitors. C_5 for example is realized with two unity capacitors instead of one capacitor with twice the area of C_6. Reason is that the etching process in production varies slightly from lot to lot depending on the exact temperature or acid concentration. Consequently, the fringe varies from wafer lot to wafer lot causing a change in capacitance, which is dependent on the fringe. Two capacitors that differ in the outline will therefore match worse than two capacitors with the same outline.

The layout of the CDAC and particularly the unity capacitor as illustrated in Fig. 2.22 needs to be performed very carefully:

1. The top plate is mostly responsible for the matching. The corners should have 45° angles, as the edging process will always round off corners. On the other side, a significant amount of capacitor mismatch is generated by the fringe. The

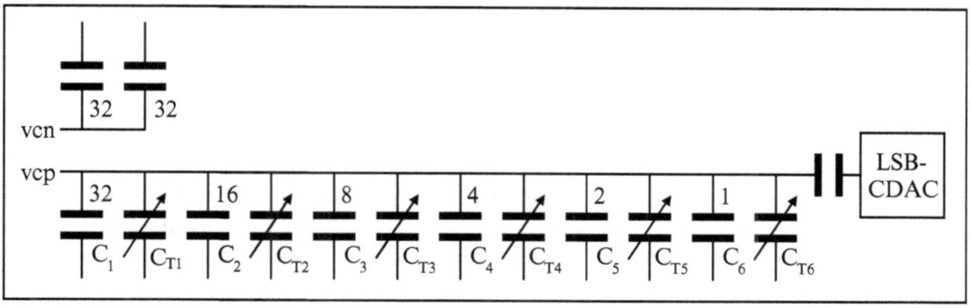

Fig. 2.21 Capacitor array of a pseudo-differential CDAC implementation

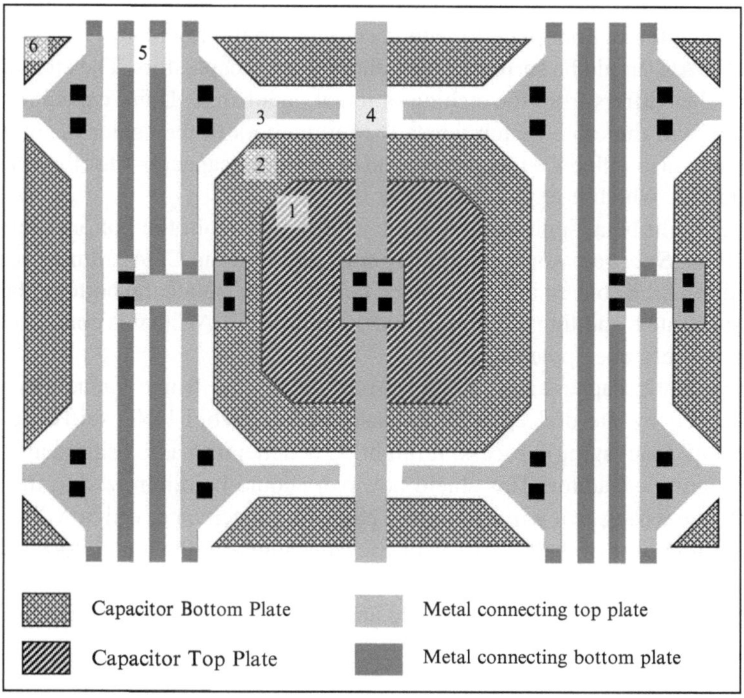

Fig. 2.22 Layout suggestion for a unity capacitor used inside a CDAC

outline to area ratio should therefore be small. Consequently, the chamfer should remain short. Octal structures should be avoided.

2. The bottom plate might have a significant parasitic capacitance to the substrate (up to 10 %), which reduces the settling speed of the capacitors. The corners of the bottom plate might also have 45° corners. The bottom plate typically overlaps the top plate, so that the mismatch caused by the fringe is limited to the top plate. Due to the parasitic capacitance to the substrate, the bottom plate is not used for the high impedance connection V_{cp} and V_{cn} from the CDAC to the

comparator. Consequently the high impedance node is normally utilizing the capacitors top plate as shown in Fig. 2.22.

3. The high impedance node of the capacitor, which connects to the comparator, must be shield. Therefore, the unity capacitor is surrounded by a grounded metal line to shield the capacitor from the connection of the bottom plate. The bottom plate is typically switching between the positive and negative reference potential. Any parasitic capacitance between the wiring and the top plate will influence the capacitance and therefore the DNL. Note that 1 LSB is typically less than 1 fF.
4. The high impedance node should be routed as far away from the bottom plate to minimize the parasitic capacitance.
5. The connection of the bottom plate is wired on the same or on a lower metal layer than the bottom plate to reduce coupling to the top plate.
6. The CDAC should be surrounded by dummy unity capacitors. The etching speed of notches differs from the etching of metal on a fringe. Unity capacitors on the outline will therefore match worse than capacitors in the center of the CDAC. This will also influence the distribution of the unity capacitors for the various bit capacitors, which is explained further below.
7. There might be a significant parasitic capacitance from the bottom plate to the substrate. Substrate noise can therefore couple into the CDAC during sampling and during the conversion. It might be beneficial to isolate the bottom plate from the substrate by adding a NWELL underneath the CDAC that is connected to the negative reference potential.
8. The parasitic capacitance of the wiring and of the fringe capacitors will be influenced by the dielectric of the package. To avoid DNL shifts during the assembly (packaging) process, it might be beneficial to add a shield also on top of the unity capacitor. This shield however should be minimized to the fringes and to the wiring as it will add further parasitic capacitance, which will influence the settling time of the CDAC. Parasitic capacitance to the high impedance node will further act as a capacitive divider for the comparator input signal and will reduce the internal signal amplitude and therefore decrease the signal-to-noise ratio.

For further improvement of the matching, the distribution of the unity capacitors in the layout of the MSB array is important. Figure 2.23 offers some possible solutions. Unity capacitors, which are part to the MSB capacitor, are marked with one. All three approaches keep a point symmetrical distribution. This is crucial to compensate for process gradients. Let's assume the oxide is thicker in the middle of the wafer than on the edge. A gradient of the oxide thickness would then exist in the x- and in the y-dimension. Unity capacitors on the top left side of the MSB array might therefore be thicker than unity capacitors on the bottom right side. A linear gradient would not influence the total matching with a point symmetrical placement of the unity capacitors.

As a second rule, capacitors on the outline might be affected differently by the etching process than capacitors of the inside of the array. The circular approach

Circular Approach							
1	1	1	1	1	1	1	1
1	1	2	2	2	2	1	1
1	2	3	4	3	3	2	1
1	2	4	6	5	3	2	1
1	2	3	5		4	2	1
1	2	3	3	4	3	2	1
1	1	2	2	2	2	1	1
1	1	1	1	1	1	1	1

Diagonal Approach							
1	1	2	3	4	2	1	1
1	1	2	3	4	2	1	1
1	1	2	3	5	2	1	1
1	1	2	3	6	2	1	1
1	1	2		3	2	1	1
1	1	2	5	3	2	1	1
1	1	2	4	3	2	1	1
1	1	2	4	3	2	1	1

Mixed Approach							
2	1	2	1	3	1	4	1
1	3	1	2	1	2	1	2
2	1	3	1	4	1	2	1
1	2	1	6	5	1	1	3
3	1	1	5		1	2	1
1	2	1	4	1	3	1	2
2	1	2	1	2	1	3	1
1	4	1	3	1	2	1	2

Fig. 2.23 Possible distributions of the unity capacitors in the layout of a positive MSB array

from Fig. 2.23 might therefore have a worse MSB matching compared to the diagonal and mixed approach. Note that the other capacitor arrays (LSB-array and capacitors of the negative CDAC side) might attach to the MSB array. Figure 2.23 does not necessarily show the final outline.

The vertical MSB capacitors from the diagonal approach might therefore not be on the outline. Advantageous is also that the number of unity capacitors on the top and bottom are proportional to the capacitor size. In this way, 4 unity capacitors of the MSB (bit 1), 2 of bit 2 and one of bit 3 are positioned on the bottom and the top side in Fig. 2.23 to consider the etching effects well. A matching to 11 bit is possible. Local gradients however are not eliminated.

The best matching is therefore achieved with the mixed approach. A matching to 12–13 bit is possible. This distribution however has the disadvantage of wiring overhead and will show the largest die area.

Also dynamic element matching might be a possible approach. Here, the distribution of the unity capacitors is changing from conversion to conversion. As a drawback, the mismatch of capacitors will then show up as converter noise as the capacitor mismatch will generate a different conversion result for each distribution. The signal-to-noise ratio will suffer. The dynamic element matching would be particular beneficial if it is combined with oversampling. Let's assume that the dynamic element matching is foreseeing four different distributions. An oversampling of four will include all distributions and would actually improve the signal-to-noise ratio due to the oversampling. A matching to 14–15 bits might be possible.

It was mentioned above that the nodes *vcp* and *vcn*, which connect the CDAC with the comparator, are high impedance nodes and are particular sensitive in absorbing noise from the substrate or distortion from digital circuitry or other switching effects. It is therefore further important in the layout to separate the analog signals and particularly the nodes *vcp* and *vcn* from the digital circuitry. Two examples of the top level CDAC floor plan are given in Figs. 2.24 and 2.25. Note that both approaches foresee a fully differential CDAC.

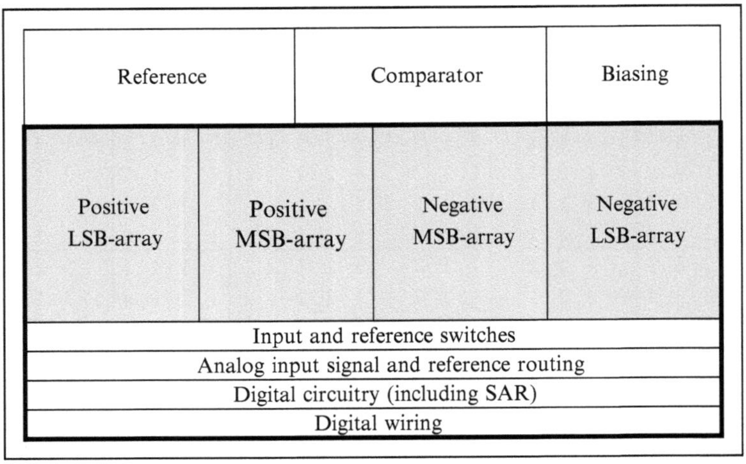

Fig. 2.24 Suggestion of a floor plan of SAR ADCs

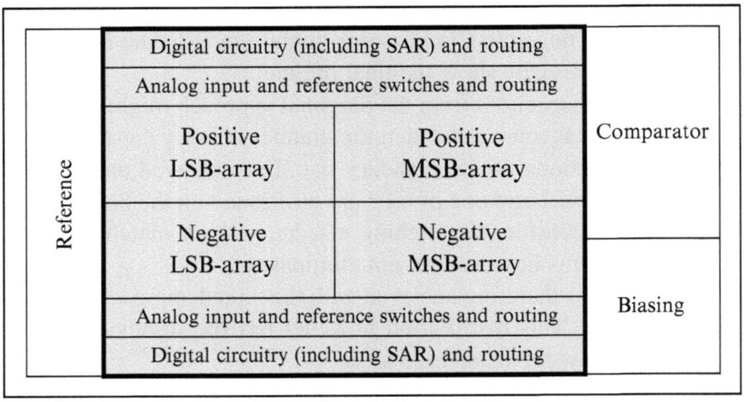

Fig. 2.25 Suggestion of an alternative floor plan of SAR ADCs

If a 5 V process is in use, then the input voltage range is limited to the same. Consequently, an LSB can be calculated to

$$1\ \text{LSB} = \frac{5\ \text{V}}{65536} = 76\ \mu\text{V}$$

The size of the LSB is important to calculate the sample capacitance as the rms-noise $V_{rms,samp}$ is defined by Eq. (2.2). If the peak-to-peak noise is estimated to 6 times the rms noise, then the sample noise has to be less than 12.6 μV or 9 μV if taking into consideration that the noise occurs on the positive side and the negative side and therefore needs to be reduced by the square root of 2. Note that the noise is added geometrically. The required capacitor size can then be calculated to

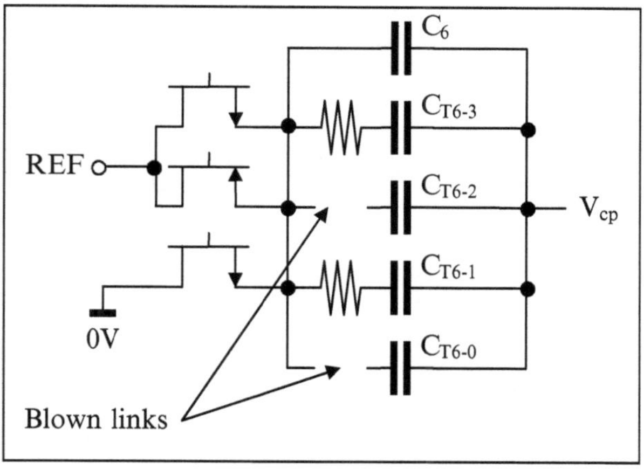

Fig. 2.26 Bit 6 circuit of a CDAC including DNL trim capacitors with laser links illustrated as resistors

$$V_{rms,samp} = \sqrt{\frac{kT}{C}} \Leftrightarrow C = \frac{kT}{V_{rms,samp}^2} = 50 \text{ pF}$$

In the particular implementation, the sample capacitance was chosen to 40 pF. One LSB is therefore represented by 0.61 fF. This number illustrates that trimming of the capacitors at least in the MSB array is required for resolutions higher than 12 bit. This can be done by cutting the wiring to the trim capacitor with a laser. Laser links as shown in Figs. 2.26 and 2.27 are require, so that the schematic for C_6 and C_{T6} in Fig. 2.21 would look as illustrated in Fig. 2.26.

If a trim resolution of half an LSB is anticipated, then the smallest trim capacitor $C_{T6\text{-}0}$ needs to be 0.3 fF. Such a small capacitance cannot be realized with the same capacitor type as the bit capacitor C_6. Typically, metal capacitors of a different type are used. This causes two problems. First, the matching between the two capacitor types is very limited. The trim weight and resolution will vary from lot to lot. Second, the capacitors might age differently, so that the DNL might change over time.

A typical layout of laser links is drawn in Fig. 2.27.

1. The link material is typically poly-silicon or metal.
2. The oxide above the link is removed to avoid laser absorption in the oxide
3. The wiring to the laser link can be of higher resistance as the trim capacitance is small.
4. Laser cutting can generate charge in the substrate, which then can cause latch-up. Biased P- an N-diffusion is surrounding the laser links to absorb the charge.

Laser cutting needs to be done on wafer level before the silicon is packaged. Unfortunately, the molding compound of the packaging is influencing the parasitic capacitors. This will cause a shift of the capacitor matching during the packaging process, so that the matching of the capacitors is limited, even when they are perfectly trimmed on wafer level. An average shift can be adjusted with a trim

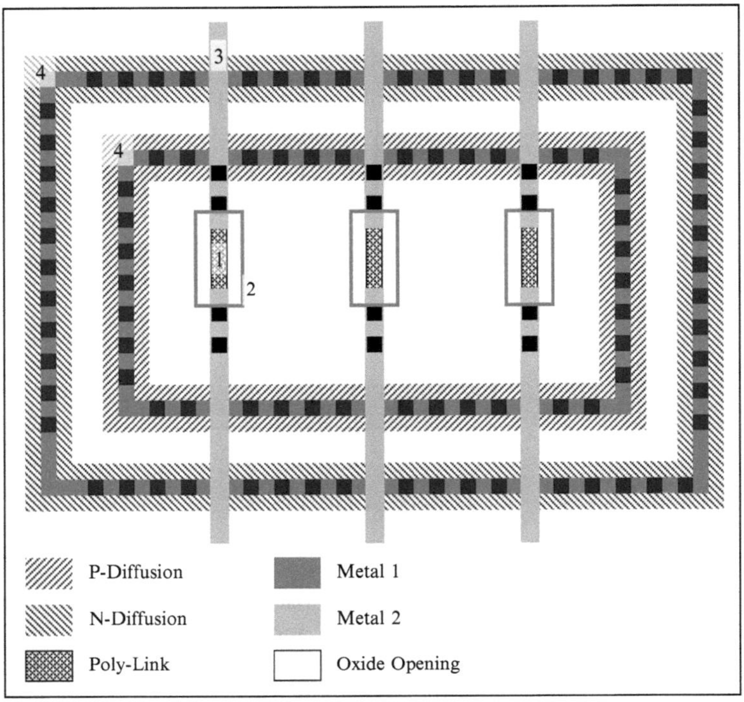

Fig. 2.27 Typical layout for laser links implemented in poly silicon

target, but the variation will cause a capacitor mismatch and therefore DNL. Mostly affected are all trimmed capacitors starting from MSB-5 up to the MSB (*bit* 6 up to 1). These are the ADC output codes 1023, 2047, 4095, 8191, 16383, 32767 and all combinations of these codes. A typical DNL curve is shown in Fig. 2.28.

The DNL curve in Fig. 2.25 is showing spikes in the order of 0.5 LSB to 1 LSB. These are caused by capacitor mismatch due to packaging shift. The DNL is trimmed slightly positive to avoid missing codes. The worst case (1.5 LSB) is at code 49151, which is followed with a short code 49152 (−0.5 LSB). This short and long code combination is caused by a thermal effect of the comparator input stage, which is discussed in the next Sect. 2.3.4.

The packaging shift described above can be avoided with in-package trim, where laser links need to be replaced with switches. Unfortunately, the parasitic switch capacitance has the same magnitude as the trim capacitor. The trim capacitor would therefore be in series with the parasitic capacitance as shown in Fig. 2.29. Its effective trim weight is reduced. The parasitic switch capacitance is furthermore process, supply voltage and temperature dependent. The resulting DNL errors would be more significant than the packaging shift.

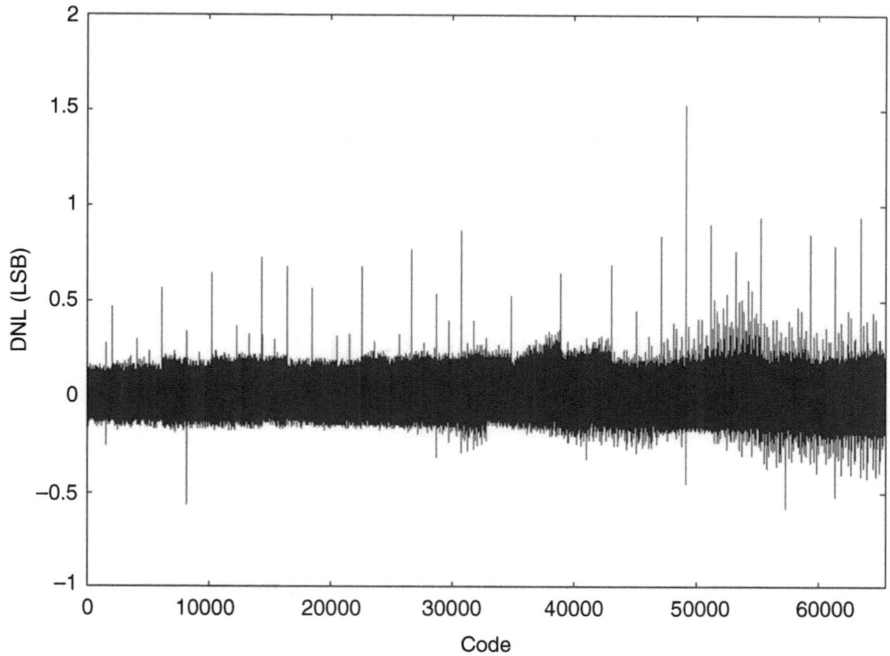

Fig. 2.28 Typical DNL of a SAR ADC trimmed with laser links

Figure 2.29 shows the parasitic source capacitors at the switch to the trim capacitor $C_{T6\text{-}0}$. The sum of the Source-Drain, Source-Bulk and Source-Gate capacitors of the NCH and the PCH transistor are in series to $C_{T6\text{-}0}$.

As a result, the switch side of $C_{T6\text{-}0}$ and all other trim capacitors needs to be connected to a low impedance node, like the reference or ground. This way, the switch side of $C_{T6\text{-}0}$ is at a defined potential, so that the parasitic capacitors only add delay to the switching process, but no linearity errors.

The solution is to switch $C_{T6\text{-}0}$ synchronously with C_6 between the ground and reference, if the C_6 capacitor needs to be increased. If the C_6 capacitor needs to decrease, then $C_{6\text{-}0}$ is switched anti-parallel to C_6. If the capacitor is not selected by the trim register, then the capacitor remains at ground all the time. This is adding parasitic capacitance to the node V_{cp}. C_6 and $C_{T6\text{-}0}$ act here as an example. The same switching scheme would apply to all other trim capacitors as well.

Parasitic capacitors will lower the signal range at the comparator input and lower the SNR. However, the trim capacitors are significantly smaller than the bit capacitors, so that this effect can be neglected.

Note that this switching scheme will cause a gain error, if the activated trim capacitors do not act as sample capacitance as well. This gain error however is small (0.1–0.2 %) compared to the typical reference accuracy (0.5 %), so that it can be neglected especially as it is constant over temperature and supply voltage.

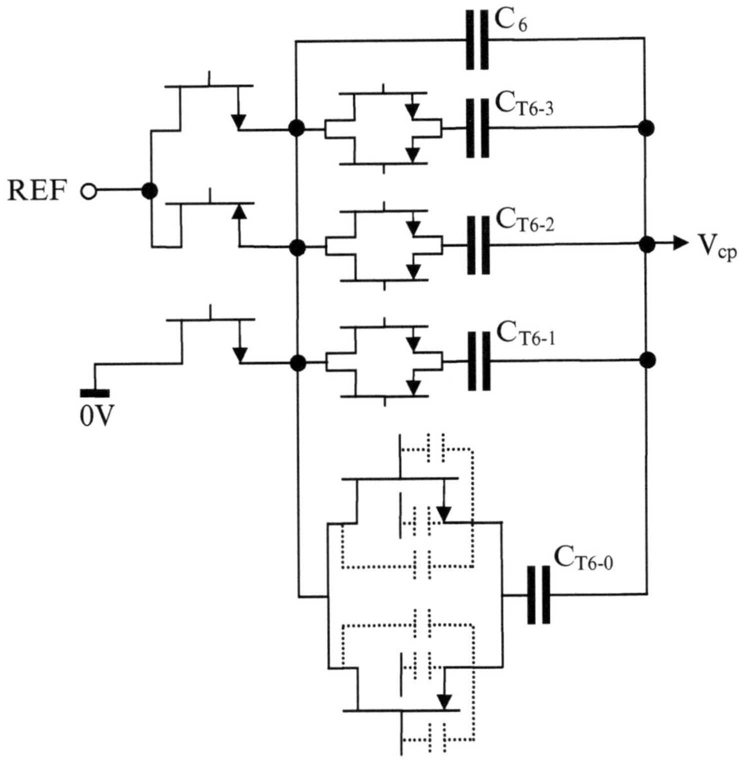

Fig. 2.29 Bit 6 circuitry inside a CDAC with trim switches replacing laser fuses

Another problem is the small size of the trim capacitors. If the DNL should remain in the range of ±0.25 LSB, then the smallest trim weight needs to be 0.5 LSB or $C_{Trim0} = \frac{C_{1p,n}}{2^n}$. This would be too small to implement with the same capacitor type. Using a different material however could generate a long-term capacitor drift, because the different types might age differently over time.

A separate trim array was added with a trim-scale-down capacitor based on the realization in Fig. 2.20, which is connected between the MSB array and the trim array. In this particular design, the size of the trim-scale-down capacitor can be chosen as unity-capacitor.

The trim of the scale-down capacitor would be most advantageous. This is caused as the high impedance node of the LSB DAC has a significant parasitic capacitance to the shielding, which reduces the original signal amplitude of the LSB DAC. The magnitude can reach 20 LSB. This parasitic capacitance will not match with the bit or trim capacitors and will further show higher variance from lot to lot. A trim of the scale-down capacitor is not really required as it can be compensated with an increased trim range of the bit capacitors, which was introduced above.

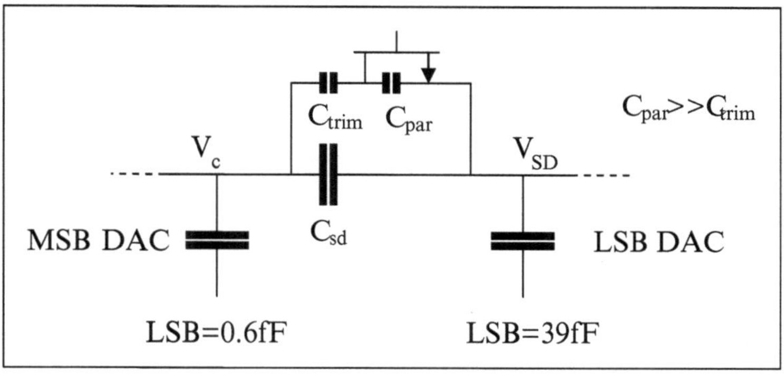

Fig. 2.30 Parasitic switch capacitance C_{par} prohibiting scale-down trim switches

However, the bit capacitors would add an unnecessary gain error. A solution for the scale-down trim would therefore be welcome.

The scale-down capacitor is located between the two high impedance nodes of the MSB- and the LSB-capacitor array, which were previously called V_c and V_{SD}. Switches on either side would add parasitic capacitance to a sensitive node (see Fig. 2.30), so that it is not possible to trim the scale-down capacitor in a similar way with switches as being done for the bit capacitors.

With a total sample capacitance of 40 pF, one LSB is represented by 0.61 fF for a 16 bit ADC. An electronic trim (in-package trim) of such a capacitor would require a transistor in series with the trim capacitor (see Fig. 2.30). Now the parasitic capacitance of the switch would be significantly larger than the capacitor itself. Deactivating the trim capacitor would place the trim capacitor in series with the switch capacitance. As the switch capacitance is significantly larger, the series capacitance is similar to the trim capacitor itself, which means that the active value does not really change.

A novel implementation, which is based on the idea that parasitic capacitances will reduce the signal amplitude of V_{SD} in the LSB CDAC as discussed above, is shown in Fig. 2.31. Now as the charge distribution of the LSB array into the MSB array is reduced by the scale-down capacitor (e.g. after the bit 6 capacitor), an LSB in the LSB CDAC is represented by 39 fF for the particular implementation instead of 0.61 fF in the MSB CDAC. Fortunately, 39 fF is significantly larger than the capacitance of a trim switch. So adding parasitic capacitance in the LSB CDAC as shown in Fig. 2.31 will effectively reduce the signal amplitude of LSB DAC and therefore the charge, which is coupled through C_{sd} into the MSB array. In this way, the LSB array can be adjusted to match the MSB array. This method is called scale-down trim.

Without the scale-down trim, some additional trim range needs to be foreseen for the trims of the bit capacitor. In a particular 16 bit implementation, four binary weighted trim capacitors were chosen for bit 6 (see Table 2.1), starting from 0.5 LSB, 1 LSB, 2 LSB, 4 LSB plus a sign bit. Five binary weighted capacitors were chosen for bit 5, six capacitors for bit 4, seven capacitors for bit 3, eight for bit 2 and finally nine for bit 1.

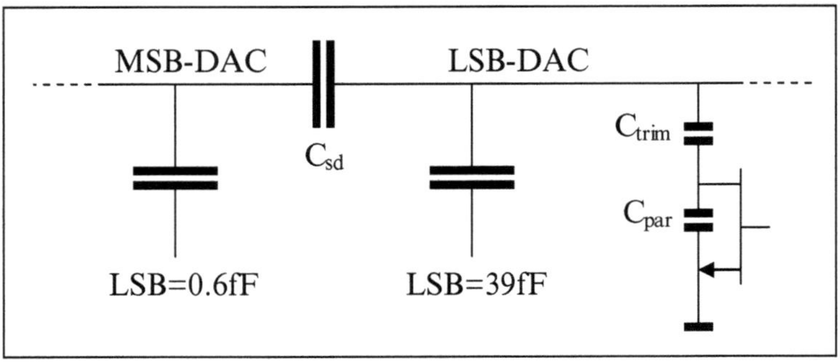

Fig. 2.31 Trim capacitor added to adjust the signal amplitude of the LSB CDAC

Table 2.1 Overview of typically required trim capacitors for a 16 bit SAR ADC

Trim	$C_{Tx\text{-}0}$	$C_{Tx\text{-}1}$	$C_{Tx\text{-}2}$	$C_{Tx\text{-}3}$	$C_{Tx\text{-}4}$	$C_{Tx\text{-}5}$	$C_{Tx\text{-}6}$	$C_{Tx\text{-}7}$	$C_{Tx\text{-}8}$	Sum
Unit	LSB	LSB	LSB	LSB	LSB	LSB	LSB	LSB	LSB	LSB
bit 6	0.5	1	2	4						7.5
bit 5	0.5	1	2	4	8					15.5
bit 4	0.5	1	2	4	8	16				31.5
bit 3	0.5	1	2	4	8	16	32			63.5
bit 2	0.5	1	2	4	8	16	32	64		127.5
bit 1	0.5	1	2	4	8	16	32	64	128	255.5
Sum										501

In the following, some important considerations for the trim capacitor array are discussed. If the weight of the C_{Tx-0} capacitor is identical to 0.5 LSB, then the DNL can be adjusted to ±0.25 LSB. Note that the trim of a less significant bit capacitor will affect the matching of more significant capacitors. If for example C_6 is increased by 1 LSB, then C_1–C_5 need to be increased by 1 LSB as well as they need to match to the sum of all lower bit capacitors as discussed above. More significant bits therefore require an increased trim range.

In the example of Table 2.1 the total trim capacitors sum up to 501 LSB, while the LSB capacitors in the scale-down array have the total size of 1023 LSB. The capacitors act as a voltage divider. If a capacitor with the size of an LSB in the trim-scale-down array switches from ground to reference, then the generated voltage change is divided by less than a change generated by a similar capacitor in the LSB-DAC. The signal is not divided strong enough.

To receive an exact change of one LSB, parasitic capacitors, which are connected to ground constantly, have to be added to the trim-scale-down array with the size of 1023 LSB − 501 LSB = 522 LSB. Alternatively, the trim-scale-down capacitor can be lower in size than the scale-down capacitor. This however would differ from the size of a unity capacitor in the layout and would lower the general CDAC matching.

Unused trim capacitors, which remain at ground all the time as they are not selected through the trim register, will lower the input voltage range of the comparator, but will not influence the DNL or the gain. In the design of the new generation of ADCs, the sum of trim capacitors have the same weight like the bit 6 capacitor and will therefore lower the comparators input by $2^{-6} \cdot 100\,\% = 1.56\,\%$. Generally, a reduction of the comparator input voltage lowers the SNR, but 1.56 % causes an insignificant drop of the SNR by 0.135 dB.

The DNL trim requires 45 signals, which control the switches. Thirty nine signals are required to select the 39 trim capacitors, six additional signals are required to program the sign of the trim. A small and reliable memory cell is required that is easy to program during the final test in production. Often electrically blown fuses are chosen as discussed below. OTPs (One-Time-Programmable ROM) are an alternative.

The trims of interest are set through a trim register, which can be programmed through the standard interface of the ADC, which here was chosen to be serial. The serial input word is separated into an address and data. With the 3 bit address, the trim bit of interest is chosen, and the data will enable or disable the trim capacitors. By default, all trim capacitors are deactivated, which means that they are permanently connected to ground. Table 2.2 shows the according register map:

If a trim capacitor is chosen through the trim register, then it is activated, but the fuse is not blown. This way, the trim program has the chance to double check, whether all trim capacitors are set correctly. If not, they can get corrected.

After setting the trim capacitors correctly, the fuses are actually blown. Burning one fuse requires a current in the range of 10 mA per fuse. The maximum current has to be limited, so that fuses are typically blown sequentially. The sequential blowing of selected fuses starts automatically by setting the blow register to ‘1010101010’. With address 111, a master bit can be set. The master fuse is blown and disables the trim register for ever.

The block diagram of a fuse cell is shown in Fig. 2.32. The fuse resistor and a power transistor T_P are located in the middle. The power transistor connects the fuse between the supplies during the blowing process. The respective current flow will melt the fuse. The power switch is controlled with active *SELECT* and *BLOW* signals. *SELECT* is directly connected to a particular bit in the trim register of Table 2.2. *BLOW* is generated in a separate cell, which sequentially activates the fuses for blowing, as only one fuse should be blown at a time. The value of a blown fuse is higher than 500 kΩ, an un-blown fuse is in the order of 50 Ω. Electrically blown fuses might re-grow to some extent. Having the recognition level two orders of magnitude above the un-blown value and two orders of magnitude below the blown value should provide the best reliability. A read-out circuitry is required to measure a blown fuse and to store the result in an output latch. The signal *TRIM_CNTL* is then connected to a trim switch and controls the same.

The layout of an electrical fuse is shown in Fig. 2.33. In difference to the laser link, it does not require latch-up protection, but a low resistance connection. The middle of the fuse has to be thin, so that the heat can build up here for burning. A significant space is consumed by the power transistor and the associated circuitry.

Table 2.2 Possible implementation of a trim register with address and data definition

Trim bit	add2	add1	add0	data9	data8	data7	data6	data5	data4	data3	data2	data1	data0
6	**0**	**0**	**0**	*x*	*x*	*x*	*x*	*x*	*sign*	*trim 3*	*trim 2*	*trim 1*	*trim 0*
5	**0**	**0**	**1**	*x*	*x*	*x*	*x*	*sign*	*trim 4*	*trim 3*	*trim 2*	*trim 1*	*trim 0*
4	**0**	**1**	**0**	*x*	*x*	*x*	*sign*	*trim 5*	*trim 4*	*trim 3*	*trim 2*	*trim 1*	*trim 0*
3	**0**	**1**	**1**	*x*	*x*	*sign*	*trim 6*	*trim 5*	*trim 4*	*trim 3*	*trim 2*	*trim 1*	*trim 0*
2	**1**	**0**	**0**	*x*	*sign*	*trim 7*	*trim 6*	*trim 5*	*trim 4*	*trim 3*	*trim 2*	*trim 1*	*trim 0*
1	**1**	**0**	**1**	*sign*	*trim 8*	*trim 7*	*trim 6*	*trim 5*	*trim 4*	*trim 3*	*trim 2*	*trim 1*	*trim 0*
Blow	**1**	**1**	**0**	*1*	*0*	*1*	*0*	*1*	*0*	*1*	*0*	*1*	*0*
Master	**1**	**1**	**1**	*0*	*1*	*0*	*1*	*0*	*1*	*0*	*1*	*0*	*1*

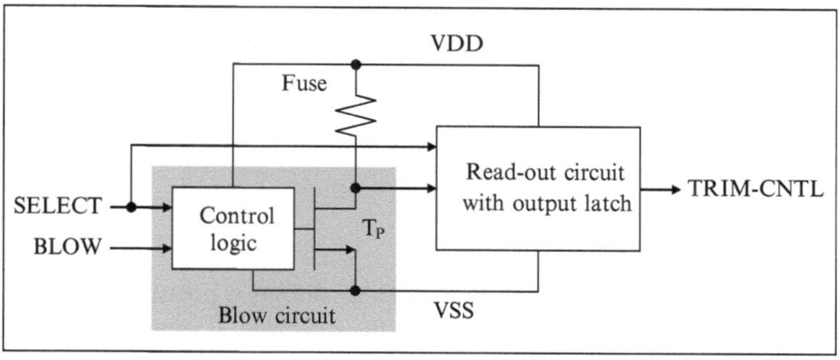

Fig. 2.32 Block diagram of a fuse cell with blow and read-out circuit

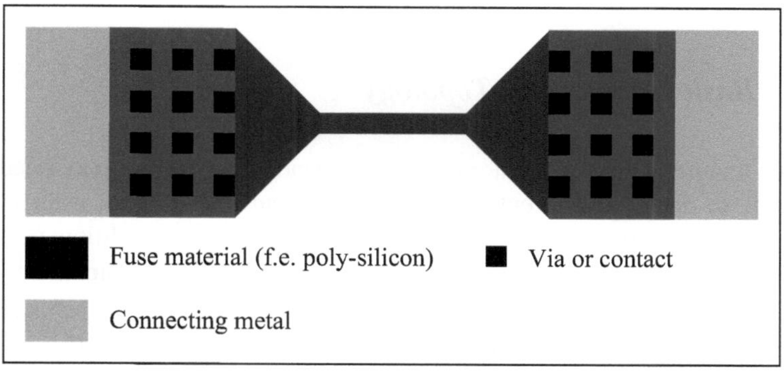

Fig. 2.33 Exemplary Layout of an electrical fuse

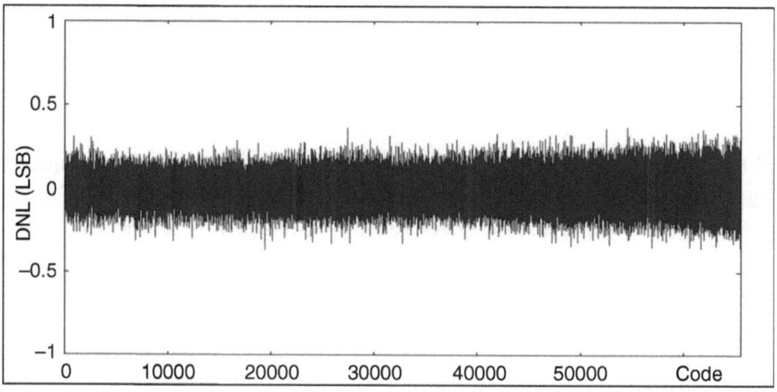

Fig. 2.34 Typical DNL versus code for a SAR ADC using in-package trim

The in-package trim solution helped to clean up the DNL. A typical DNL of all codes can be seen in Fig. 2.34. The minimum DNL is above −0.4 LSB, the maximum DNL is below +0.4 LSB. Unfortunately, the in-package trim did not solve all the DNL problems. Dynamic effects at the codes 49,151/52 and 57,343/44 remained initially and are discussed in the next chapters. Furthermore a novel digital trim technique is presented in Sect. 3.3.

2.3 Comparator

The comparison, whether the differential CDAC output is positive or negative, will be done by the comparator. It is gaining up the differential input to a digital signal level and latches the value with every clock cycle. In addition, low noise and low power are expected.

2.3.1 *Basic Comparator Topology*

The architecture of the comparator is mostly dependent on the required speed. It has to be able to detect a differential voltage of half an LSB or less, even if it was overdriven very hard during the prior bit decision. This condition occurs for example, if the input voltage is $\frac{REF}{4} + \frac{LSB}{2}$. During the MSB decision, the DAC is set to $\frac{REF}{2}$. The differential input of the comparator is therefore $\frac{REF}{4} - \frac{LSB}{2}$, so that the comparator totally overdrives. During the next bit decision, the CDAC is switching to $\frac{REF}{4}$, so that the differential input of the comparator is only half an LSB. The comparator now needs to be fast enough to switch its state within half a clock cycle to allow the reference source some settling time as well. Figure 2.35 is illustrating the comparator input during the sampling and the conversion process.

If the input stage of the comparator is estimated to a first order system, then the comparator settling will be exponential. The comparator has to settle within half a clock cycle to an n-bit resolution. The other half of the clock cycle needs to be reserved for the digital switching and the following reference settling. Such settling occurs quite frequently in switched-capacitor circuits. For SAR ADCs, the reference settling after a switching CDAC is such an example. Therefore, the settling should be discussed more general. If a voltage jumps by the size of the reference voltage REF, then the settling error ε after the settling time T_s has to be less than half an LSB. This is illustrated in Fig. 2.36.

The error voltage $\varepsilon(t)$ is calculated in the equation below under the assumption that the time $t_0 = 0$.

$$\varepsilon(t) = REF - V(t) = REF - \left(REF - REF \cdot e^{-\frac{t}{\tau}}\right) = REF \cdot e^{-\frac{t}{\tau}} \quad \text{with} \quad \varepsilon(T_S) < \frac{REF}{2^{n+1}}$$

This equation can be solved for τ, which is the time constant of the exponential function. The time constant equals the product of resistance R and capacitance C for

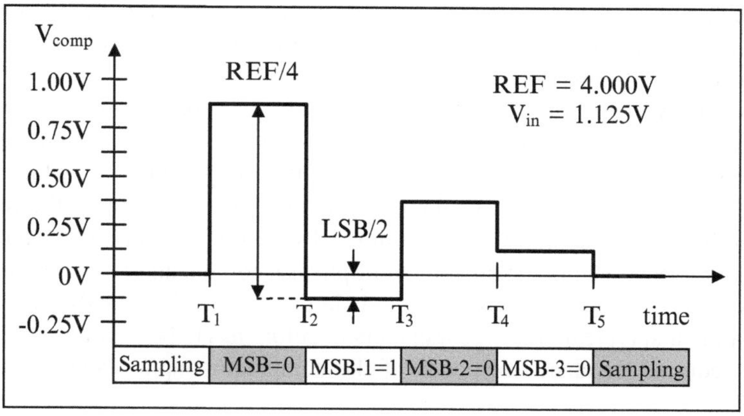

Fig. 2.35 Comparator input voltage V_{comp} during the conversion of a 4 bit ADC

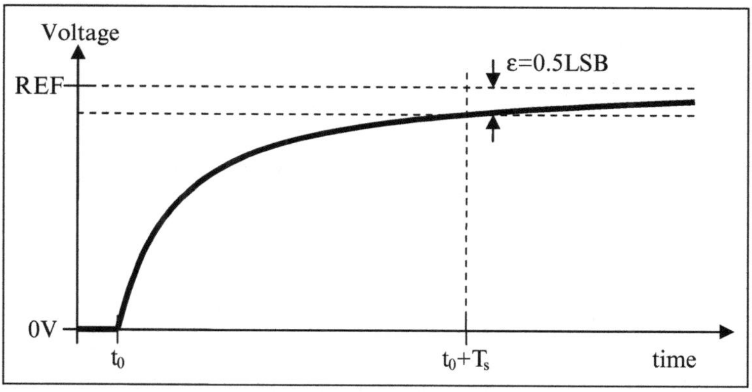

Fig. 2.36 Simplified exponential settling in switched capacitor designs

an RC low pass filter. Equation (2.12) is therefore important to calculate resistors and capacitors.

$$\tau = R \cdot C < \frac{T_s}{\ln(2) \cdot (n+1)} \tag{2.12}$$

The time constant τ is also directly related to the -3 dB frequency of the transfer function of the active circuitry, which here equals the bandwidth of the comparator.

$$f_{-3\text{dB}} = \frac{1}{2\pi \cdot \tau} > \frac{ln(2) \cdot (n+1)}{2\pi \cdot T} \tag{2.13}$$

The maximum available settling time for the comparator is in the range of half a clock cycle, regarding that some time is required for the switching process and for the CDAC settling time.

A 16 bit SAR converter uses typically 20 clock cycles per complete conversion. These are 16 clock cycles for the conversion process and 4 for the sampling period. A 500 kSPS converter therefore requires a 10 MHz clock frequency, which leaves 50 ns for the settling process of the comparator. For this example, the required bandwidth of the comparator can be estimated with Eq. (2.13) to 37.5 MHz. This bandwidth needs to be maintained over all process corners and temperatures, over varying supply voltages and common-mode input voltages. The bandwidth at room temperature and typical process parameters therefore has to be significantly higher.

The required frequency responds of the comparator input stage is drawn in Fig. 2.37.

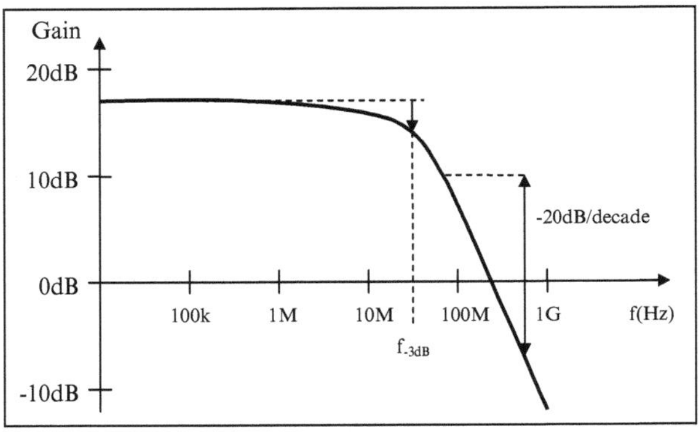

Fig. 2.37 Typical AC-responds of the comparator input stage

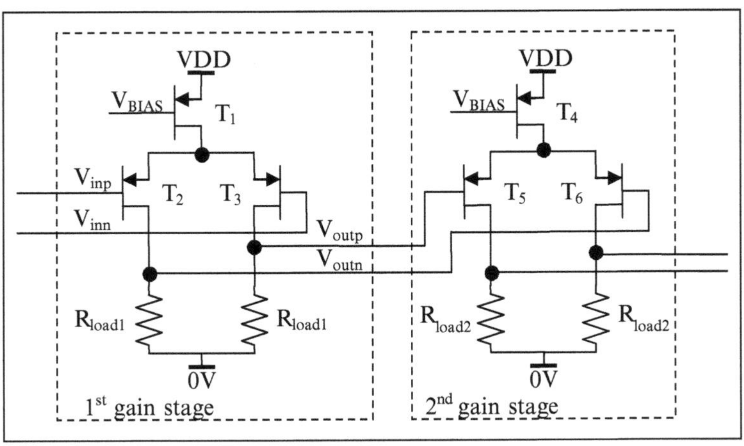

Fig. 2.38 Typical differential input pair used inside a comparator

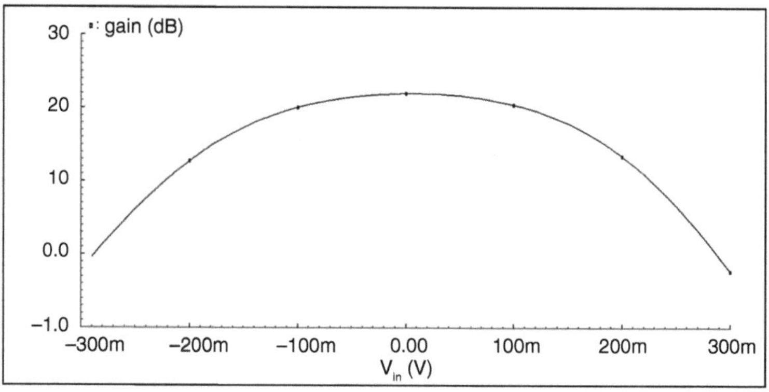

Fig. 2.39 Typical gain of a differential pair versus its differential input voltage V_{in}

The only structure, which can support this bandwidth at a reasonable power on a standard CMOS process with a gate length of 0.6 μm is a differential pair shown in Fig. 2.38 [13].

The width of the input transistors T_2 and T_3 of the first differential pair and their quiescent current are chosen to satisfy the noise requirements of the comparator. Such a differential pair is typically operating in weak inversion. If so, then the thermal noise is process independent and is only a function of the bias current. Due to wide bandwidths, the low frequency flicker noise can be neglected, particular as it is suppressed by an offset cancellation function, which is described below. The noise of the second stage is insignificant, because it is divided by the gain of the first stage, if the noise is related to the comparator input.

The impedance of the load R_{load1} is then used to set the bandwidth of the first gain stage. Resistors might be used as load. They often offer the same impedance at a lower voltage drop compared to load transistors. This allows more headroom for the common-mode operating point of the differential inputs V_{inp} and V_{inn}.

The gate capacitances of T_5 and T_6 are typically the most significant parasitic capacitors at V_{outp} and V_{outn}. As the noise of the second gain stage is insignificant, their width is typically chosen much smaller.

If several differential pairs are placed in series, then the offset is getting important. Not only that the offset will be reflected in the ADCs transfer function, but also the consecutive gain stages will fall into a non-ideal operating point and loose a lot of gain (see Fig. 2.39). If the first gain stage for example has an input related offset of 10 mV associated and the gain of each stage is estimated with 20 dB, then the input of the third stage is already seeing 1 V at the input and is outside a reasonable operating point.

Therefore, an offset cancellation (auto-zero stage) is required [14, 15]. Typically, the differential inputs of a gain stage are shorted during the sampling phase and the output offset is stored on auto-zero capacitors, which are called C_{az1p} and C_{az1n} for the first gain stage in Fig. 2.40.

They are placed in series to the signal path. One side is connected to the output of a particular gain stage and will see the input offset voltage multiplied by the gain of

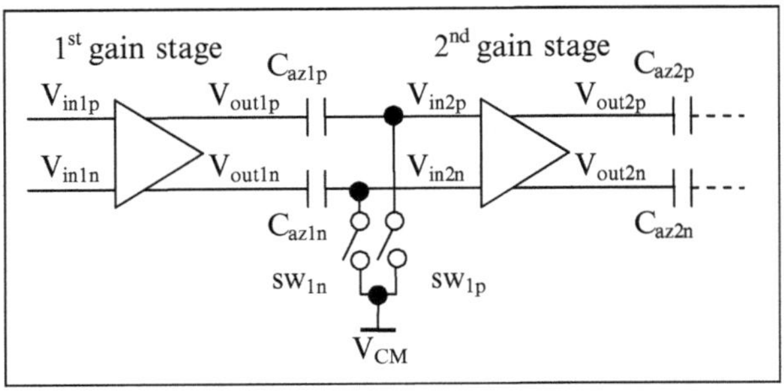

Fig. 2.40 Typical auto-zero stage inside a comparator

the stage during the sampling phase. The other side of the capacitors is connected to the input of the next gain stage and to the switches (sw_{1n} and sw_{1p}), which will short the capacitors to a common-mode voltage V_{CM}. In the end of the sampling process, the switches open, so that the offset voltage is stored on the capacitors.

The auto-zero capacitors are in series with the signal and will not reduce the comparators bandwidth directly. Unfortunately, the parasitic capacitance to ground will do so. The capacitance should therefore be limited. On the other side, the parasitic input capacitor of the following gain stage will form a capacitive divider with the auto-zero capacitor. From this perspective, the capacitance should be chosen large. Values, which are typically used, are in the range of 1 pF. Sampling noise is a minor issue. The effective noise needs to be divided by the gain of previous gain stages to be related to the comparator input.

The total required gain of the comparator is dependent on the output latch. A fully differential latch might only require a 200–500 mV input voltage. If a standard CMOS flipflop (FF) is used, then correct digital voltage levels are required. Therefore, the comparator needs to gain up half an LSB to the appropriate threshold voltage of the latch. For a 16 bit converter with a 5 V input range and a standard FF at the comparator output, 38 μV has to be gained up to above 3.8 V, which is a gain of factor 100,000 or 100 dB.

If the comparators bandwidth requirements are low enough, then an architecture, which differs from differential pairs, can be chosen as gain stage such as an open loop operational transconductance amplifier (OTA), which is shown in Fig. 2.41.

The OTA provides sufficient gain, so that the signal will reach digital signal levels and a standard flipflop can be used to strobe the signal. In addition, it converts the fully differential input into a single ended output.

A good alternative to the OTA is a differential cross-coupled latch, which is shown exemplarily in Fig. 2.42. First, the signal *LATCH* is low, so that the cross-coupled latch is disconnected from the supplies *VDD* and *GND*. *OUTP* and *OUTN* are charged to the differential voltage of the inputs *INP* and *INN*. The input is connected to the output of the last differential pair of the comparator.

The analog voltage is sampled with the rising edge of the *LATCH* signal, which disconnects the inputs of the capacitors and connects the cross coupled structure to *VDD* and *GND*. The NCH device (*MN*1 or *MN*2) with the higher gate voltage will pull down its drain voltage. Accordingly, the PCH device (*MP*1 or *MP*2) with the

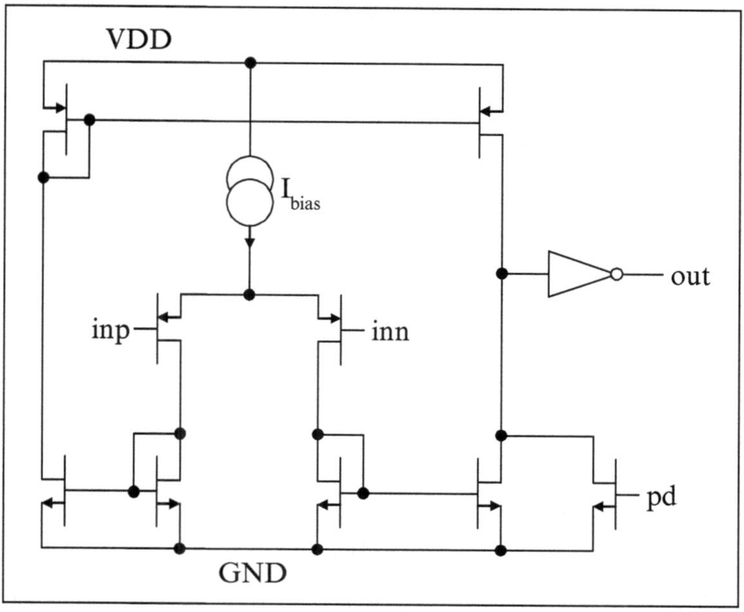

Fig. 2.41 Open-loop OTA used as last comparator gain stage

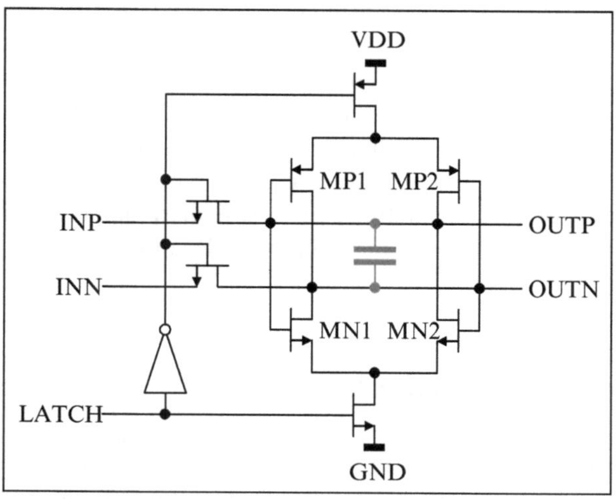

Fig. 2.42 Cross coupled latch used as last comparator stage

lower gate voltage will pull its drain voltage up. The outputs will be flipped to *GND* and *VDD* very quickly, which is shown in Fig. 2.43 at the time points 50 and 150 ns.

The first curve in Fig. 2.43 shows the differential input voltage to the cross coupled latch of Fig. 2.42. In the beginning, the latch is in the sampling phase, it triggers at 50 ns, starts sampling at 100 ns and triggers at 150 ns again. During the

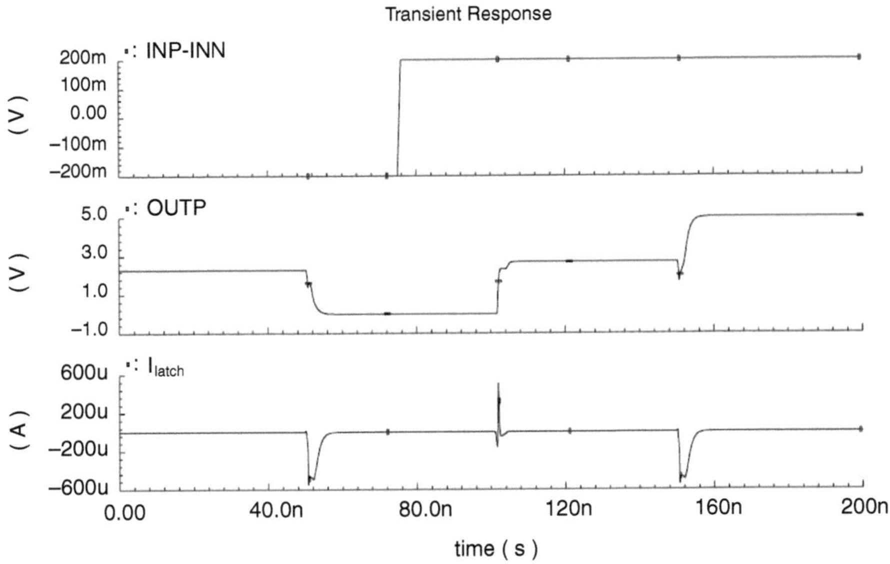

Fig. 2.43 Response of the cross coupled latch in the time domain and its current dissipation I_{latch}

sampling process, the capacitors (*OUTP* and *OUTN*) are charged to the input voltages *INP* and *INN*. When the differential input ($IN = INP - INN$) is negative, then *OUTP* will be pulled to *GND* in less than 3 ns (see at time point 50 ns). At 150 ns, where the input is positive, *OUTP* is pulled to *VDD* again in less than 3 ns.

Both structures, the OTA and the cross coupled latch, have their advantage. The cross coupled latch is working without any DC current and saves power (see third curve of Fig. 2.43). The OTA is adding a low corner frequency and reduces the noise contribution of the comparator. This is particularly important, if the comparator is the most significant noise source in the SAR converter.

2.3.2 Layout Considerations

The layout should be driven by the following considerations. Offsets should be avoided to minimize gain loss and the required amount of auto-zero stages. The sensitive analog input voltage should be separated from digital signals to avoid distortion. Finally, parasitic capacitance by wiring should be avoided to remain the high bandwidths.

As a first step, the differential input transistor pairs should be drawn interleaved. The transistor connections can cause a significant parasitic source-drain capacitance. Vias and metal routing on top of the differential pairs should be minimized. An example is drawn in Fig. 2.44. Points of particular interest are listed below.

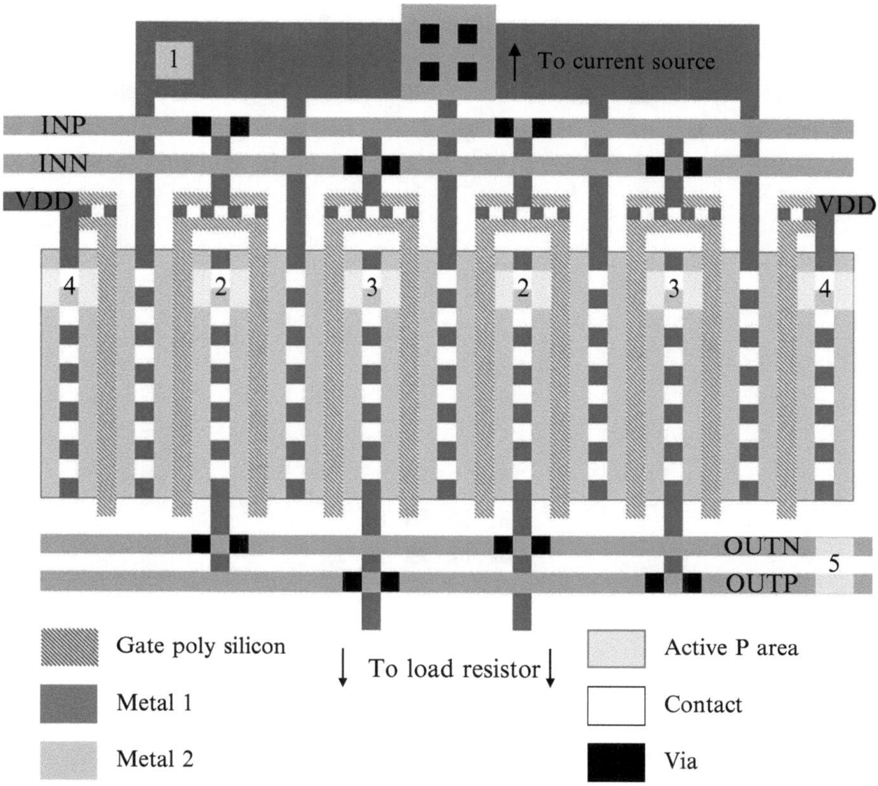

Fig. 2.44 Proposed layout of the differential input pair

1. The common node of the differential pair, which here is utilizes PCH devices, should be of low resistance, so that the voltage gradient in the wiring remains low. It might be beneficial to connect the common node symmetrical (for example in the middle). Note that the NWELL connection is not drawn for an better overview. Best matching was achieved, if the NWELL connection occurs on the top and the bottom of the differential pair.
2. Positive input transistor. The input transistors 2 and 3 are interleaved for best matching.
3. Negative input transistor. Parasitic extraction should be used to minimize parasitic capacitance. For example it might be beneficial to reduce the number of contacts, which on the other side should not be minimal to provide a low and solid contact resistance.
4. Dummy transistors are added on the sides to avoid mismatch of the fringe transistors, which can be caused by the etching process.
5. The outputs should be routed differentially on the shortest route to the next comparator stage.

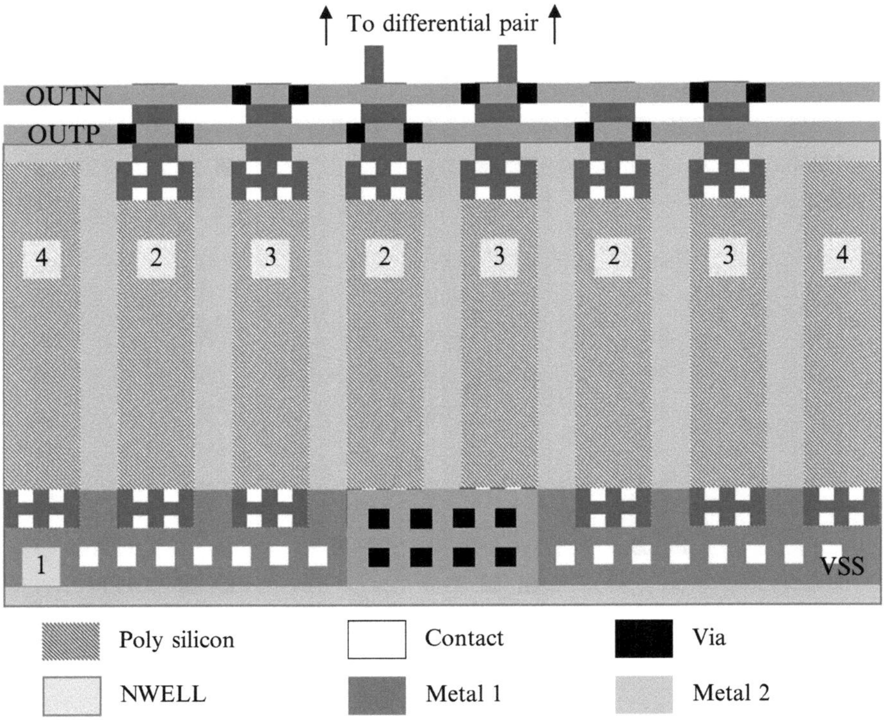

Fig. 2.45 Proposed layout for the load resistors of a differential input pair

A proposed layout of the load resistors is drawn in Fig. 2.45. The drawing is based on the following considerations:

1. Again, the ground connection should be solid to avoid voltage gradients in the wiring. The metal 1 connection to the next metal layer should be done symmetrically for symmetrical voltage drops. A NWELL connecting to VSS can be placed underneath the resistors to shield the resistors from substrate noise. The VSS connection from the first stage might be routed back to the pad separately from the VSS connections of the following stages to avoid kick-back through parasitic feedback loops. The settling of the following stages might cause distortion on VSS, which feeds back to the inputs via the load resistors of the input differential pairs. This might increase the settling time.
2. Positive load resistance. In this example, poly resistors were chosen. Generally, the resistors should provide a low temperature coefficient. More than the minimum amount of contacts should be used to minimize mismatch from contact resistance. For better matching, the poly should extend over the inner contacts. Here, the extension was used to add another row of contacts. The resistors should not have minimum width to reduce resistor variance and increase resistor matching due to etching effects on the fringe.

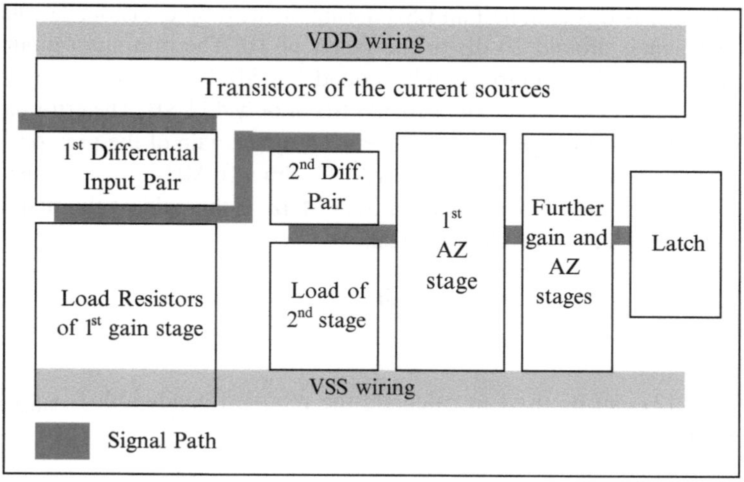

Fig. 2.46 Proposed floor plan for the comparator

3. Negative load resistance. It can be seen that the positive and negative load resistors are interleaved for better matching but also to reduce a heating effect, which is explained in Sect. 2.3.4.
4. Again, dummy resistors are added on both sides to reduce mismatches caused by etching.

Figure 2.46 finally shows a proposed floor plan for the comparator. The focus should be on a short signal path. Also the digital latch and the sensitive analog input are place on separate sides. Shielding should be used, if digital signals have to cross the analog signal path. Crossings should generally be avoided, but might be required at the auto-zero stages.

2.3.3 *Noise Considerations*

The comparator is one of the dominant noise sources beside the reference buffer, the sampling and the quantization. Therefore, an investigation was started to lower the comparator bandwidth without losing settling accuracy. A bandwidth calculation was developed, which is based on the idea that consecutive stages need less accuracy due to the gain of previous stages and therefore less bandwidth based on Eq. (2.13).

$$f_{-3\mathrm{dB}} = \frac{ln(2) \cdot (n+1)}{2\pi \cdot T}$$

The comparator input stage (first gain stage) needs to detect half an LSB. The output needs to settle fast, even if the comparator is fully overdriven at the previous bit decision and is less than half an LSB during the next bit decision. The gain of a differential pair is around 20 dB or the factor of 10. The minimum input of the second stage is 5 LSB; its output will be around 50 LSB. It is not really important, if the output of the second stage settles to 50 LSB or only 45 LSB. The error band ε of the settling function in Fig. 2.36 can be relaxed by the gain of the previous stage.

In the example of the 16 bit, 500 kSPS ADC with a 10 MHz clock frequency, the first gain stage required a bandwidth of 37.5 MHz. If the gain of the first stage is 20 dB, then the effective resolution equals

$$n_1 = \frac{20\ \text{dB}}{6\ \frac{\text{dB}}{\text{bit}}} = 3.3\ \text{bit}.$$

Now Eq. (2.13) can be used to calculate the required bandwidth $f_{-3dB,2}$ for the second gain stage

$$f_{-3\text{dB},2} = \frac{1}{2\pi \cdot \tau} = \frac{ln(2) \cdot (n - n_1 + 1)}{2\pi \cdot T} = \frac{ln(2) \cdot (16 - 3.3 + 1)}{2\pi \cdot 50\ \text{ns}} = 30.2\ \text{MHz}$$

Generally, if a comparator has k gain stages, then the required bandwidth of the gain stage j with $j\epsilon[2, k]$ can be calculated to

$$f_{-3\text{dB},j} = \frac{ln(2) \cdot \left(n + 1 - \sum_{i=1}^{j-1} n_i\right)}{2\pi \cdot T} \tag{2.14}$$

The reduction of the bandwidth reduces the power consumption and the total output noise of the comparator. If several gain stages are used, then the bandwidth of the last stage can be in the order of 4 lower than the bandwidth of the first stage. This will reduce the noise, which is added by the comparator by a factor of 2. Note that the analog signal is latched at the comparator output and that the effective noise bandwidth is therefore determined by the lowest bandwidth in the comparator.

The comparator noise is calculated in the following. Flicker noise is suppressed by the auto-zero functionality and the thermal noise is effective twice during the conversion process. It can be seen in Fig. 2.35 that there are two decisions in the conversion process, where the comparator input voltage is less than an LSB. Other decisions are not affected as the noise is significantly lower than the comparator input voltage. Of interest is the noise, which is related to the comparators input. This noise however is only effective to the lowest corner frequency of the gain stages, which can be much less than the bandwidth of the first gain stage. Also, the factor $\frac{\pi}{2} \cdot f_{-3\text{dB}}$ is only true, if the second corner frequency is far away of the first one, which for the comparator is typically not the case due to the higher number of required gain stages. A 12 bit converter might utilize around 4 gain stages, a 16 bit ADC around 6–8 gain stages.

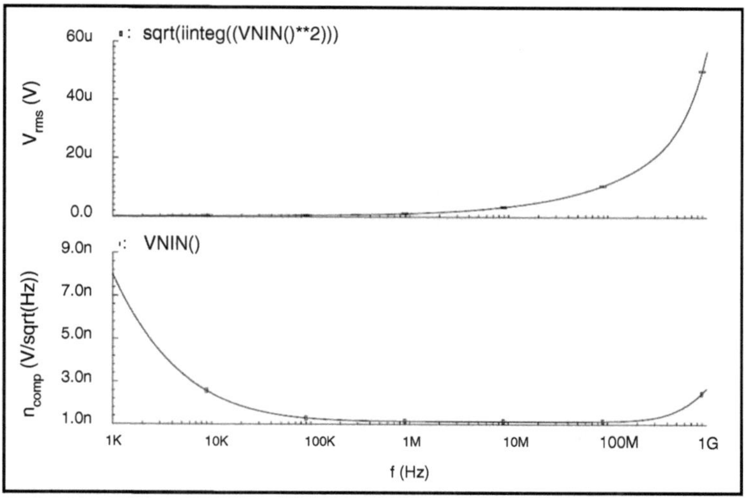

Fig. 2.47 Input related noise density and the associated rms-voltage of a comparator

The lower curve of Fig. 2.47 is showing the input related noise density n_{comp} of a 16 bit comparator. It is clearly visible that the noise increases at low frequencies (flicker or 1/f-noise). Then it remains constant at around $n_{comp} = 1.1\ \frac{\mathrm{nV}}{\sqrt{\mathrm{Hz}}}$ up to 100 MHz. 100 MHz is the bandwidth of the first comparator stage. From there, it is losing gain and the noise of the other gain stages is added. The upper curve shows the integral of the noise density. The value at the effective frequency equals the comparator noise.

The comparator noise is not only sampled twice during the conversion process, but the noise is also stored on the auto-zero capacitors. In difference to the noise of the gain stages, the sample noise of the first auto-zero capacitors cannot be divided by the gain of previous stages, because the dominating noise is generated by the first gain stage and is gained up by the same factor, before being stored on the capacitor.

The auto-zero noise might be very essential, because the bandwidth is not limited by the corner frequency of the following gain stages either. Though, the bandwidth is reduced, as the AZ capacitors are connecting to the common-mode voltage V_{CM}, which might be ground, during sampling instead of being in series between the gain stages, where only the parasitic capacitors (roughly 10 %) are reducing the bandwidth of the prior gain stage.

Fortunately, the sampling is typically performed over several clock cycles, so that the reduction in bandwidth is possible when choosing the auto-zero capacitor correctly. Further auto-zero stages such as C_{az2p} and C_{az2n} in Fig. 2.40 are less critical, as the inputs of the gain stages are shorted to V_{CM} by sw_{1p} and sw_{1n}. In this way, the gain of further auto-zero stages (e.g. gain stage 2) can be divided by the gain of the previous gain stages (gain stage 1), so that the noise is getting insignificant again. The total comparator noise can therefore be estimated to

$$V_{rms,comp} \approx \sqrt{\pi \cdot n_{comp}^2 \cdot f_{last} + \frac{\pi}{2} \cdot n_{comp}^2 \cdot f_{AZ}} \tag{2.15}$$

f_{last} is the −3 dB frequency of the lowest corner frequency and f_{AZ} is the −3 dB frequency of the lowest corner frequency between the comparator input and the first AZ stage during the AZ phase. Equation (2.15) takes the factor $\frac{\pi}{2}$ for the effective bandwidth into account, but is neglecting the kT/C noise of the AZ capacitors.

2.3.4 Dynamic Errors in the Comparator

The DNL performance was limited by the shift of capacitor values during the assembly process. Figure 2.48 shows a typical DNL curve over the output code. The capacitive matching is independent of the conversion rate, so that the resulting DNL errors are independent of speed. However, a small amount of spikes at particular codes change with varying conversion rate, so that these errors were called 'dynamic errors'. These errors always appeared in pairs: a long code followed by a short code [16].

An essential amount of research was required to allocate the error source. At these particular codes (40959/40960, 49151/49152 and 57343/57344) the critical bit decisions occur at bit 2 or bit 3, or in other words always when big capacitors are switching. This first pointed to a settling problem of the reference, but modifications in the reference path did not show any improvement. The root cause is the comparator input stage.

Code 49152 for example has the binary format '1100 0000 0000 0000'. The CDAC output and therefore comparator input is shown for the four MSBs in

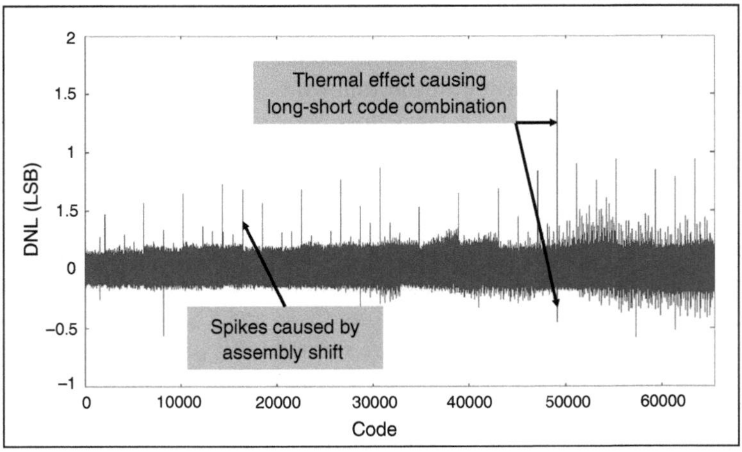

Fig. 2.48 Typical DNL of a SAR ADCs of the generation of the year 2000

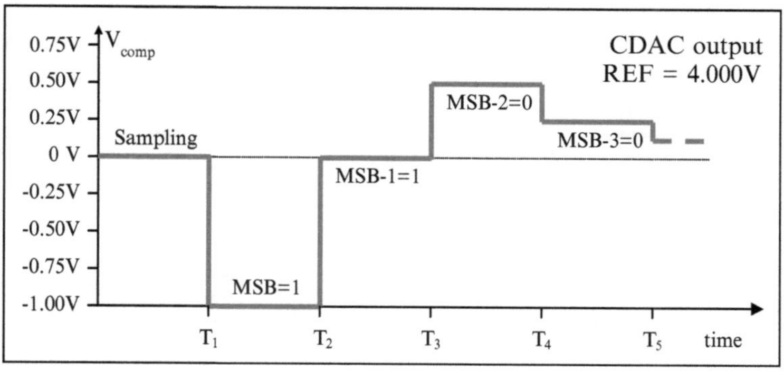

Fig. 2.49 Comparator input voltage V_{comp} during the conversion of code '1100...'

Fig. 2.49. The comparator input stage is overdriven during the MSB decision and settles to 0 V during the MSB-1 decision.

The effect for the comparator input stage is illustrated in Fig. 2.50. A current source is driving the current I_{bias} through the differential input pair. The transistor with the lower gate potential will carry more current and will have a higher voltage drop across the resistor.

During the MSB conversion in Fig. 2.49, the comparator input voltage is strongly negative. Therefore, I_{bias} is fully routed through the resistor R_p, which as a consequence dissipates a higher power and heats up while the resistor R_n, which carries no current, cools down. Due to the temperature coefficient, the resistors change their value. One resistance increases the other decreases.

The next bit decision in Fig. 2.49 features a 0 V input. The current I_{bias} is equally distributed to R_p and R_n. However, the resistors still have a different temperature and therefore a different resistance. This will cause the output voltage of the comparator input stage V_{out} to be different from 0 V. The heat of an integrated resistor caused by an increased current flow can build up quickly, but unfortunately it disappears slowly as it is well isolated by the silicon oxide. This effect is quantized in the following calculation.

The power, which is dissipated inside a resistor can be calculated to

$$P_I = I \cdot V = I^2 \cdot R.$$

The energy, which is dissipated within the resistor during the clock cycle of the MSB decision, is

$$E_I = P_I \cdot T_{clk} = I^2 \cdot R \cdot T_{clk}.$$

The electric energy is converted into thermal energy E_{th}

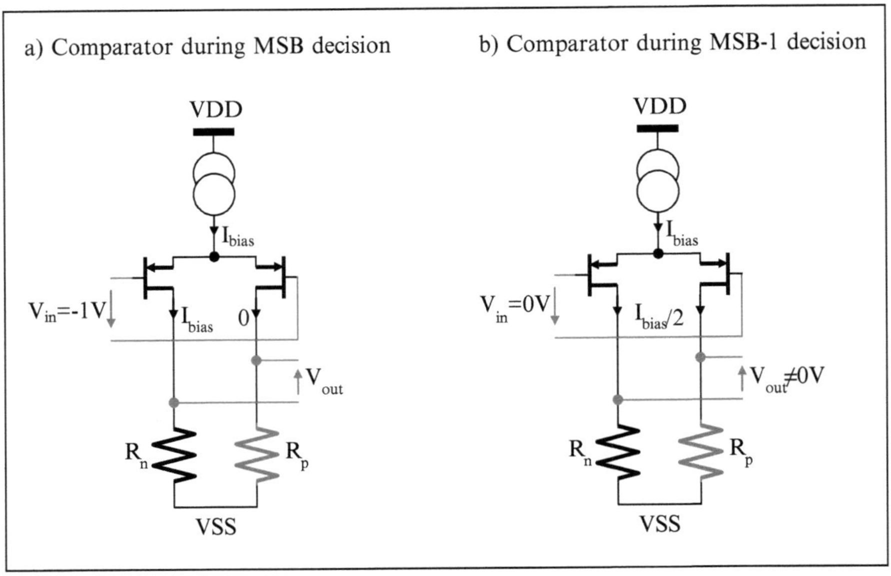

Fig. 2.50 Comparator input stage during (**a**) the MSB decision and (**b**) the MSB-1 decision

$$E_{th} = C_p \cdot m \cdot \Delta T,$$

where C_p is the specific heat of the material of the resistor, m the mass of the resistor and ΔT is the change in temperature. The mass can further be extracted by the width W, the length L and the thickness T together with the specific density ρ. For poly resistors, the specific heat is $C_p = 0.7\ \frac{\mathrm{J}}{\mathrm{gK}}$ and the specific density is given by $\rho = 2.328\ \frac{\mathrm{g}}{\mathrm{cm}^3}$. The above equations can now be solved for the change in temperature

$$\Delta T = \frac{I^2 \cdot R \cdot \Delta t}{C_p \cdot W \cdot L \cdot T \cdot \rho}. \tag{2.16}$$

The difference in temperature will now cause a drift of the resistance R_p

$$\Delta R_p = R_p \cdot TC \cdot \Delta T \tag{2.17}$$

where TC is expressing the temperature coefficient. Finally, the erroneous output voltage of the comparator input stage ΔV_{out} is calculated below. This dynamic offset voltage ΔV_{out} can further be referred to an input offset error ΔV_{in} by dividing ΔV_{out} by the gain of the differential stage g.

$$\Delta V_{out} = \frac{I_{bias}}{2} \cdot \Delta R_p \tag{2.18}$$

$$\Delta V_{\text{in}} = \frac{\Delta V_{out}}{\text{g}} \tag{2.19}$$

Now, an example should prove the relevance of the thermal effect. For low noise, the current could be up to $I_{bias} = 4$ mA. The typical load resistance ranges around 500 Ω. Length and width of the resistor might be in the range of 4 μm over 50 μm. The poly thickness of poly resistors is often in the area of $T = 0.2$ μm. Inserting these values into Eq. (2.16) will result in a change in temperature of

$$\Delta T = \frac{I^2 \cdot R \cdot \Delta t}{C_p \cdot W \cdot L \cdot T \cdot \rho} = 6.1 \text{ K}.$$

Typical poly resistors have a temperature coefficient around $TC = 800$ ppm/K. Together with Eq. (2.17), the drift of resistance calculates to

$$\Delta R = R \cdot TC \cdot \Delta T = 2.4 \ \Omega$$

Finally, the output and input related dynamic offset ΔV_{out} and ΔV_{in} assuming a gain of $g = 10$ result to

$$\Delta V_{out} = \frac{I}{2} \cdot \Delta R = 4.8 \text{ mV}$$
$$\Delta V_{in} = \frac{\Delta V_{out}}{\text{g}} = 0.480 \text{ mV}$$

If this comparator is used inside a 16 bit SAR ADC with a 4 V input range, then the LSB size is 61 μV, so that the thermal effect causes an error of 7.8 LSB. In a particular case, the above calculation matched the measured DNL error very well. The dynamic effect can be minimized with an adequate design and layout, which are presented in Sect. 2.3.2.

Unfortunately, reducing the current through the comparator input stage will also increase the noise density of the comparator. A certain amount of current and therefore some error have to be accepted.

The utilized process provides a variety of resistors. The poly resistor was therefore exchanged with a thin film resistor, which has a temperature coefficient close to zero. Unfortunately, this resistor is very thin. The mass is much lower and the thin film resistor can heat up much more than the poly resistor, so that the resistor drift can get significant again.

Therefore the layout needs adjustments. Both 490 Ω load resistors were replaced with 20 load resistors in parallel each with the size of $20 \cdot 490$ Ω. This adds three advantages. First, the current density is lower $\left(\frac{1}{20}\right)$ and the thermal effect is reduced. Second, the area is higher by 20^2, so that the temperature dissipation is improved.

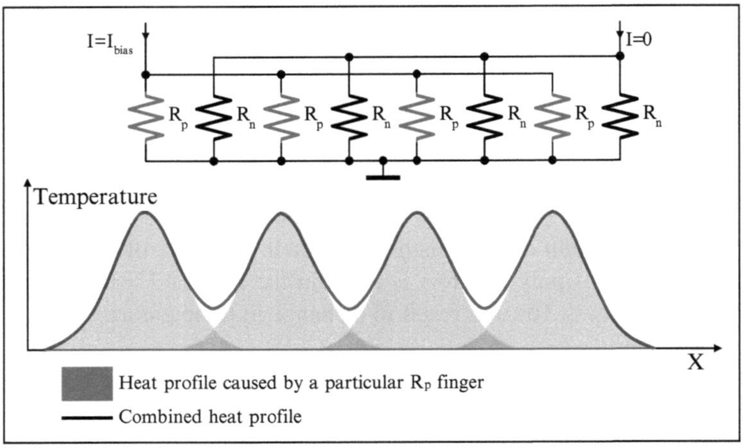

Fig. 2.51 Fingered resistors in a comparator input stage with Gaussian temperature distribution

Finally, the two load resistors can now be fingered, so that a 'cold' finger is in the middle of two 'hot' fingers. The thermal distribution after one clock cycle will also warm up the current free resistor, which is illustrated in Fig. 2.51, where the temperature distribution T is drawn over the X-coordinate at the resistor array.

The large resistor area could generate a high capacitance to the substrate. Fortunately, all the oxides between the thin film and substrate are about four times thicker than poly to substrate. Also, the influence of the parasitic capacitor to the bandwidth of the input stage can still be neglected if compared to the parasitic gate capacitance of the second differential input pair. Disadvantageously, additional masks are required for the thin film resistor.

Layout optimizations will only help to some extent. Additionally, a dynamic error correction was introduced, which is not only correcting the dynamic error of the comparator, but also settling effects.

2.4 Dynamic Error Correction

The general idea of the error correction [17–19] is to allow a certain amount of error during the first bit decisions, where the comparator has the highest overdrive and the reference the toughest settling requirements. The comparator input will converge successively to 0 V with each bit decision, so that the thermal effect does not occur during the LSB decisions. Furthermore, the reference only has to recharge very small capacitors during the LSB decisions. Additional errors during the LSB decisions are therefore not expected. The dynamic error correction shall correct errors, which are produced during the MSB decisions, when the LSBs are converted. The example of an ideal and an erroneous conversion is given in Fig. 2.52.

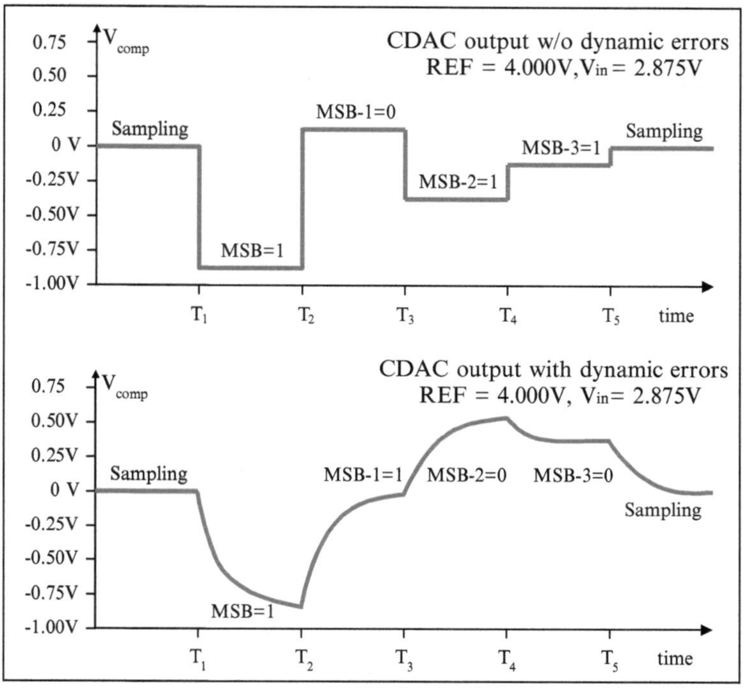

Fig. 2.52 Ideal CDAC settling and CDAC settling with dynamic error

The comparator input is not settling within the clock cycle of the MSB-1 conversion. The comparator outputs a one instead of a zero. Therefore, the MSB-1 capacitor will faulty reset from the reference voltage to ground and the CDAC output will not converge to 0 V $\pm$ 1 LSB at the end of the conversion.

The dynamic error correction allows such errors and is correcting for them later during the conversion process. If an error is allowed in the first phase of the conversion, then Eq. (2.13)

$$f_{-3\text{dB}} = \frac{ln(2) \cdot (n+1)}{2\pi \cdot T}$$

can be relaxed and the required bandwidths can be reduced. If for example a first error correction is performed after the first 4 bits, then an error is allowed with the size of the input voltage range divided by the factor of $2^4 = 16$, which is 312 mV for a 5 V input range. By using such error correction, either the bandwidth of the comparator and the reference path can dramatically be reduced to save power consumption or the conversion rate can be increased.

The bandwidth of a 16 bit SAR ADC can be reduced by a factor of (4 + 1)/(16 + 1), if an error correction after bit 4 is added. This way, the comparator and reference noise might be reduced by a factor of $\sqrt{5/17}$.

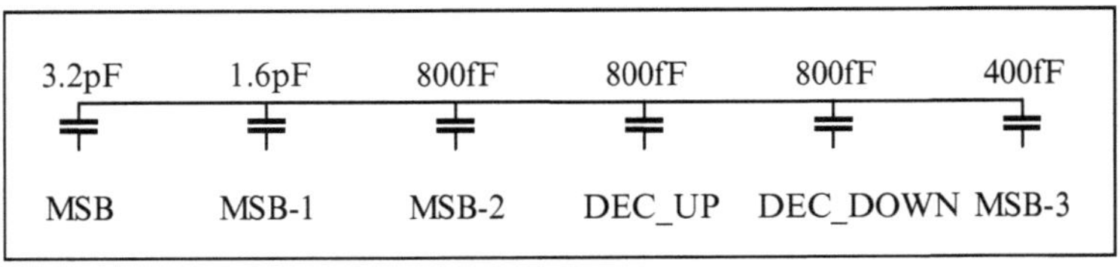

Fig. 2.53 Simplified capacitor array inside a CDAC with dynamic error correction

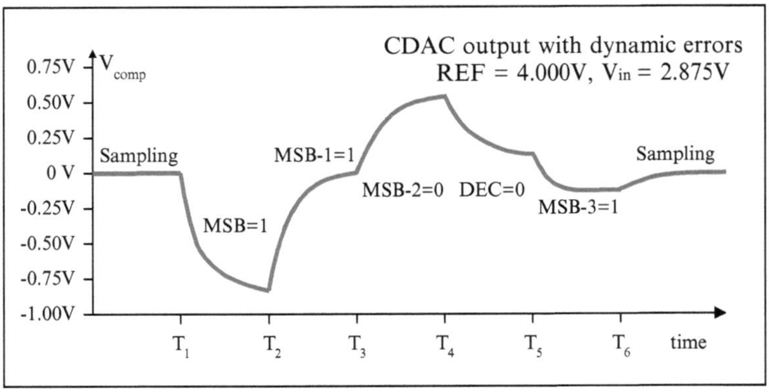

Fig. 2.54 Erroneous CDAC settling with dynamic error correction

The exemplary implementation of such an error correction with ±2 LSB is shown in Fig. 2.53. The 800 fF capacitor is placed 3-times in the schematic. One is used for the MSB-2 decision and the other two for the dynamic error correction, which can increase and decrease the voltage at the comparator.

The MSB down to the MSB-2 decisions are converted as before, but an error caused by inaccurate settling is allowed. In this way, a capacitor might be set into the wrong direction, but this can be corrected with the capacitors DEC_UP and DEC_DOWN.

Figure 2.54 shows this functionality based on the example of Fig. 2.52. Again, the ADC makes the wrong MSB-1 decision. Due to the MSB-2 being zero, the dynamic error correction will now check, if the CDAC output can be decreased by another 2 LSB. In this example, the MSB-3 decision is the LSB decision. A normal conversion without error correction would estimate that the CDAC output after the MSB-2 decision is within ±1 LSB. The dynamic error correction therefore checks, if this estimate is true. Due to the normal CDAC algorithm, the MSB-2 capacitor will reset from the reference voltage to ground for the next decision due to MSB-2 = 0. However, the MSB-3 capacitor will remain at ground. In this example, the CDAC output remains above 0 V again, which indicates that a dynamic error exists. For the next decision, the CDAC output needs to be lowered by another LSB. This is achieved by switching the DEC_DOWN capacitor from reference to ground and setting the MSB-3 capacitor. The ADCs output in this example is 1101, but

2 LSBs need to be subtracted due to the recognized dynamic error, so that the corrected output is 1011 again.

Depending on the MSB-2 decision in this example, the dynamic error correction will increase or decrease the CDAC output by the weight of the MSB-2 capacitor, which above was 2 LSB. If the MSB-2 is 1, then the error correction will increase the CDAC output. If the comparator output remains 1, then it indicates that a dynamic error exists and leaves the *DEC_UP* capacitor set to the reference. The digital output needs correction. If the comparator output switches to zero during the decision of the dynamic error correction, then no error exists. The *DEC_UP* capacitor is reconnected to ground and the digital output is correct. This is illustrated in Fig. 2.55.

The evaluation for dynamic errors, if the MSB-2 would have been zero, works as following. The MSB-2 capacitor normally resets to ground. This would already decrease the CDAC output by 2 LSB, so that all other capacitors remain in their state for the dynamic error correction step. If the comparator output still shows zero, then a dynamic error exists and the CDAC output is reduced by another LSB for the MSB-3 decision. This is done by switching the *DEC_DOWN* capacitor, which was pre-charged to the reference voltage, to ground. In addition, the MSB-3 capacitor is set to the reference voltage. Two LSBs need to be subtracted from the digital output word. If the comparator would have switched to one during the decision of the dynamic error correction, then no error exists, the DEC_DOWN remains coupled to the reference, the MSB-3 capacitor is connected to the reference as well and the digital output is correct. This is shown in Fig. 2.56.

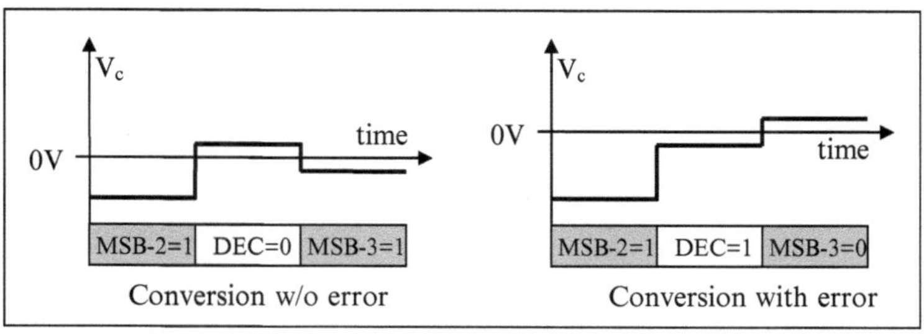

Fig. 2.55 Functionality of the dynamic error correction increasing the CDAC voltage

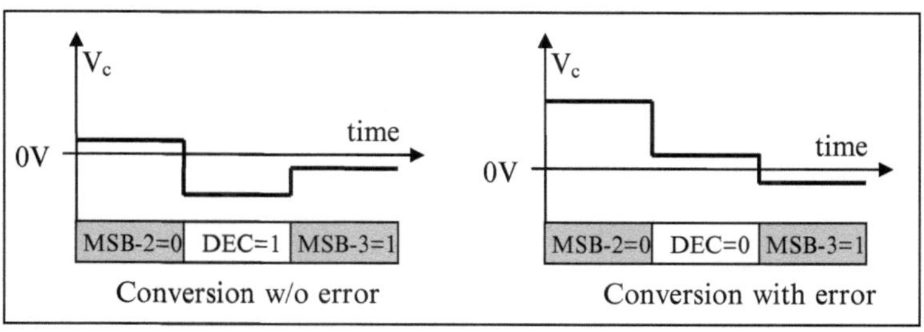

Fig. 2.56 Functionality of the dynamic error correction decreasing the CDAC voltage

By using a dynamic error correction, the linearity, the power and the speed of a converter can be improved. Indirectly, it will also reduce noise, as it allows a reduction of the bandwidth of the comparator and eventually of a reference buffer. Section 2.6.1 and Chap. 3 will show that the dynamic error correction is a very powerful tool and can be utilized by other algorithms as well.

2.5 Reference Voltage

The design of a low-drift reference should not be part of this book [20–25]. This chapter will concentrate on the influence of the reference on an ADC. There are two general reference architectures in use for ADCs. Some products provide an internal buffer (Fig. 2.57a), so that the reference input is high impedance. This buffer needs to have a wide bandwidth at low noise and will therefore consume a larger amount of power. The second solution (Fig. 2.57b) requires a large external capacitor C_{REF}, which provides the charge for the internal capacitors. This solution relaxes the amplifier's settling requirements with respect to bandwidth and noise.

However, if the application needs to adjust the reference externally using a resistive divider, then the impedance is getting too high, so that an additional external amplifier is required and the solution is more expensive. This is especially true, as most amplifiers and references become instable when driving large capacitors. The following chapters will discuss the requirements to the reference voltage and introduce a novel scheme, which is utilized by some new products on the market [26].

2.5.1 Internal Reference Buffer

It would be best from an application point of view, if the reference problem is handled completely inside the ADC. Therefore, several ADCs provide an internal amplifier called reference buffer (see Fig. 2.58).

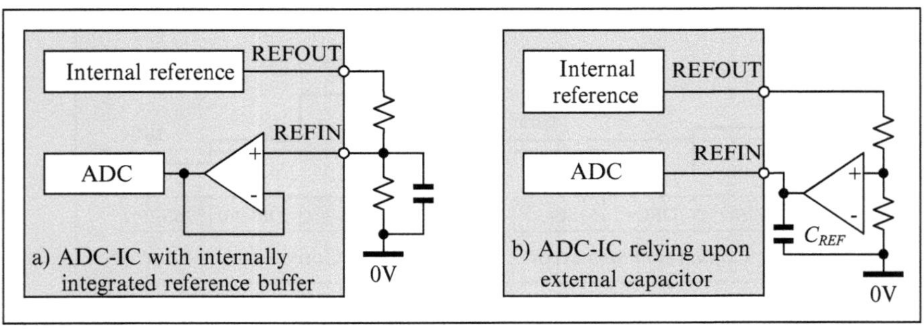

Fig. 2.57 Most commonly used reference topologies with IC external reference adjustment

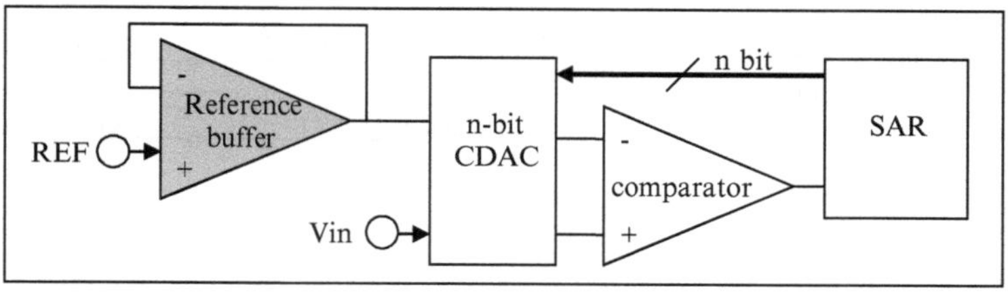

Fig. 2.58 SAR ADC with internal reference buffer

This scheme unfortunately has two disadvantages that both result from the fact that this scheme requires an amplifier with high bandwidth. The capacitor array of the CDAC has to be charged from 0 V to the reference within half a clock cycle to an n-bit resolution. This means the error at the CDAC capacitor array must be less than half an LSB. Equation (2.13) expresses the resulting bandwidth requirements.

A 16 bit SAR converter uses typically 20 clock cycles per complete conversion. These are 16 clock cycles for the conversion process and 4 for the sampling period. A 500 kSPS converter therefore requires a 10 MHz clock frequency, which leaves 50 ns for the settling process of the reference voltage. For this example, the required bandwidth of the amplifier can be estimated with Eq. (2.13) to 37.5 MHz.

If this converter has furthermore a 5 V input range and its peak-to-peak noise (6-times rms-noise) should be less than 3 LSB, then the rms-noise needs to be less than 0.5 LSB or 38 μV. The allowed noise density of the amplifier n_{ref} can then be calculated to

$$n_{ref} = \frac{38\ \mu\text{V}}{\sqrt{37.5\ \text{MHz}}} = 6.2\ \frac{\text{nV}}{\sqrt{\text{Hz}}}$$

Both low noise and high frequency can only be realized with high power consumption. This type of amplifier requires around 8 mA [27].

2.5.2 *Reference with External Capacitor*

These requirements can be relaxed, if a large capacitor C_{REF} is placed between the CDAC and the reference buffer, which is illustrated in Fig. 2.59. The size of this capacitor is chosen such that the voltage drop at the reference input *REFIN* is less than half an LSB during a complete conversion cycle. With such a capacitor, the time constant τ of the amplifier or reference can be in the range of the total conversion time.

The total capacitance C_{tot} that is switching between ground and the reference during the conversion process is dependent on the final CDAC architecture, but is in the order of the sampling capacitor $(C_{tot} \approx 3 \cdot C_s)$. The size of the additional capacitor C_{REF} can then be calculated to

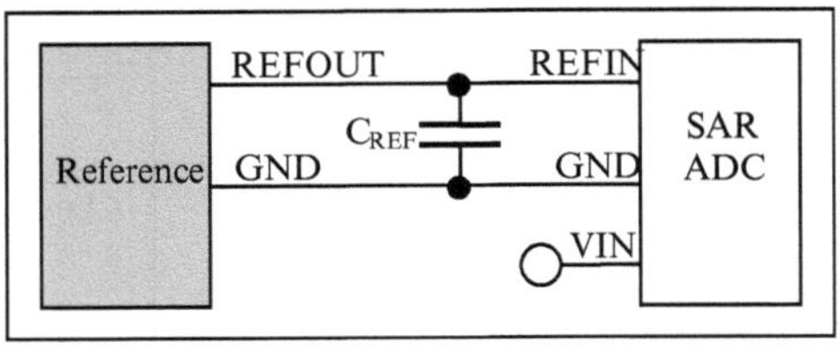

Fig. 2.59 Reference circuitry using an IC-external compensation capacitor C_{REF}

$$C_{REF} = \left(2^{n+1} + 1\right) \cdot C_{tot} \approx 2^{n+1} \cdot C_{tot} \tag{2.20}$$

If the CDAC capacitance is in the order of $C_{tot} = 50$ pF and the resolution of the ADC is 16 bit again, then the additional capacitor in the reference path can be calculated to 6.5 μF. A capacitor of this size cannot be realized on chip, so that an external capacitor is required.

Ideal are ceramic capacitors, where 22 μF are already available in a 0805 package size and a X5R quality, which means that the capacitance degrades less than 10 % temperature. The voltage coefficient is not influencing the ADC performance, as the reference is a DC voltage. However, please note that the effective capacitance could drop significantly if applying a DC bias voltage.

In first order, the big external capacitor will act as voltage source for the CDAC of the ADC. The resistance of the ADCs reference input (including ESD protection, wiring and switch resistance) will form a low pass filter together with the CDAC capacitors and cause a delay. This delay needs to be taken into consideration during the design phase.

The external capacitor will provide the charge for the current spikes coming from the CDAC, but it will discharge by less than to half an LSB. This charge needs to be provided from the reference buffer. The maximum current I_{max}, which the reference buffer has to provide over time, can be calculated to

$$C_{tot} = \frac{\Delta Q}{\Delta V} = \frac{I_{\max}}{f_{conv} \cdot REF} \Leftrightarrow I_{\max} = C_{tot} \cdot f_{conv} \cdot REF \tag{2.21}$$

If the conversion rate f_{conv} is 500 kSPS and the total capacitance C_{tot} of the CDAC 50 pF, which is charged from ground to a reference voltage REF of 5 V, then the average current is 125 μA. This current is typically input voltage dependent, so that the output of the reference source must not drop more than half an LSB at this load current. In other words, the load rejection $\frac{dREF}{dI}$ from the source, which is driving the capacitor, needs to be better than

$$\frac{dREF}{dI} < \frac{0.5 \cdot \mathrm{LSB}}{I_{\max}} = 300 \ \frac{\mu\mathrm{V}}{\mathrm{mA}} = 0.3 \ \Omega$$

Another consideration is that big capacitors tend to destabilize operational amplifiers or references. The external circuitry will generate costs and will consume space on the application in addition to the technical risk.

Ideally, this reference would be on-chip with the SAR converter. However, in many applications the reference input of the converter is used to adjust the input voltage range of the converter to the needs of the application. The reference voltage is often divided with a resistive divider and is then applied to the reference input (see Fig. 2.57).

2.5.3 An Improved Reference Scheme

The ideal reference source therefore needs to be stable over temperature and supply voltage, adjustable, low impedance and stable if driving large capacitors. This is provided by the architecture shown in Fig. 2.60.

A bandgap reference is used to generate a reference voltage, which is stable over temperature and supply. This reference voltage can be adjusted by an internal DAC. The adjusted voltage is then buffered by an amplifier, which can be optimized for a good load rejection and which is stable with large load capacitors. The buffer only requires a low bandwidth, which supports low power and low noise. The only required external component is the compensation capacitor [28].

This leaves the question open, which type of DAC is used in the reference path. The gain and the offset of an application are typically calibrated, so that the absolute value at the reference pin (REFIO) is not too critical or can even be regulated. Therefore, the integral linearity of the DAC in the reference path is uncritical, while the differential linearity is more important. The solution should also be cost efficient, so that trim or calibration should be avoided.

These are features of a string DAC (also see Chap. 7). A resistor string is dividing the reference voltage into equidistant voltages. A switch array is then selecting the voltage of interest. This structure is monotonic by default and offers good DNL, so that trimming is not required. The INL might suffer caused by voltage coefficients of the resistors or by the IC layout. The temperature coefficients

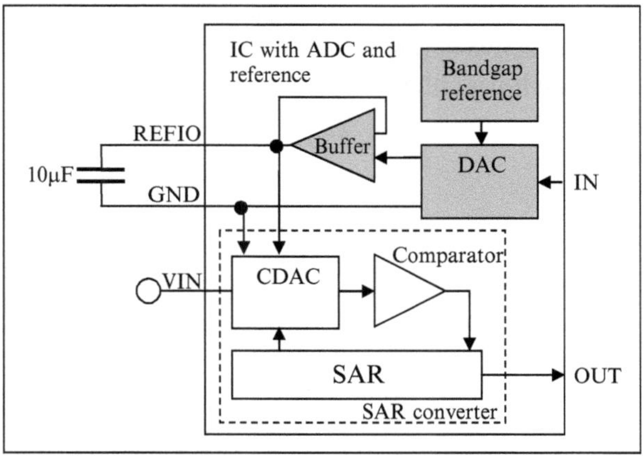

Fig. 2.60 IC with SAR ADC and adjustable reference circuit

of on-chip resistors typically match very well, so that the temperature stability is not an issue. The resistor string is also supply voltage independent, as active circuitry is not used beside the switches.

2.5.4 Noise of the Reference

Figure 2.57 shows two different structures for the reference path. In the first approach, a high speed reference buffer is directly connected to the CDAC. The second architecture works with an external capacitor, so that the bandwidth can be reduced by magnitudes.

In the first case, the bandwidth of the reference buffer is very similar to the bandwidth of the comparator and therefore both have to be designed with similar noise performance. The bandwidth for 16 bit accuracy requires a settling time higher than 12 τ (Eq. 2.12). If the settling time is half a clock cycle, then the time constant τ_1 can be calculated to $\tau_1 = \frac{1}{24 f_{clk}}$. The second solution, where the external capacitor is recharged from a maximum error of LSB/2 to maybe LSB/4, requires roughly a settling of one time constant τ_2 within a complete conversion cycle, which is typically 20 clock cycles for a 16 bit converter $\tau_2 = \frac{20}{f_{clk}}$.

Therefore, the bandwidth of the second solution is reduced by a factor of 480, which will reduce the rms-voltage of the noise by a factor of 21.9. This makes the reference noise insignificant to the comparators noise, especially as the rms-voltage is added geometrically. The noise of the reference path can be neglected for the version with an external capacitor. However, the solution with the pure amplifier needs further discussion.

Transistors include several noise mechanisms. Beside the thermal noise from the drain-source resistance, transistors also have flicker noise, which is also called 1/*f* noise, as it increases with lower frequencies. At applications with large bandwidths, the thermal noise is typically dominant.

This is especially true, if the buffer has an auto-zero function to cancel the offset. The auto-zeroing is typically done during every sampling period. Flicker noise can only get active during the conversion time T_{conv}. In this way, the auto-zeroing acts like a high-pass filter with a corner frequency of

$$f_{-3\mathrm{dB}} = \frac{1}{2\pi \cdot T_{conv}}$$

For the 500 kSPS converter, which uses 16 clocks with 10 MHz for the conversion, the corner frequency is around 100 kHz. The flicker noise will sufficiently be suppressed. The dominant noise is the thermal noise. The rms-voltage $V_{rms,buf}$ of the buffer noise can be calculated to

$$V_{rms,buf} = \sqrt{\int n_{ref}^2 df} = n_{ref} \cdot \sqrt{\Delta f}$$

Here, n_{ref} is the noise density at the buffer output. There are always two critical decisions during a conversion process, where the comparator input is less than an LSB. The noise has no effect during the conversion of the other bits. The noise is therefore sampled twice, so that the effective reference noise is $\sqrt{2} \cdot V_{rms,buf}$

$$V_{rms,ref} = \sqrt{2} \cdot n_{ref} \cdot \sqrt{\Delta f} \approx n_{ref} \cdot \sqrt{\pi \cdot f_{-3\,\mathrm{dB}}} \tag{2.22}$$

Equation (2.22) takes into consideration that the buffer will behave similar to a 1st order system, where the effective bandwidth is identical to $\frac{\pi}{2} \cdot f_{-3\mathrm{dB}}$.

Also note that reference noise is common to the negative and the positive CDAC side if the CDAC switches are in the same positions. The reference noise is suppressed by the CDAC's and comparator's common-mode rejection. For a unipolar ADC, this is typically the case at the negative FS and it is typically at mid-scale for the bipolar ADC. The total ADC noise is less at the zero input than at the full-scale, where the settings of the reference switches in the CDAC differ the most.

2.6 Noise Calculation

This chapter is summarizing the various noise sources. These noise sources are the sample and hold capacitor, the reference and the comparator. Also a bad DNL has an effect on the SNR as wrong results are generated at certain input voltages.

For best performance it is important to

(a) Remain the maximum signal amplitude at the comparator input. Avoid a reduction of the signal amplitude in the CDAC.
(b) Keep the sampling capacitance so large that the noise is not dominant.
(c) Reduce the bandwidth of the comparator as far as possible.
(d) Use a reference scheme with external capacitor.

The noise contributions of the sampling, the comparator and the quantization are already calculated in the previous chapters. They are listed below together with typical values for a 1 MSPS, 16 bit differential SAR ADC with a sampling capacitance of 40 pF.

$$V_{rms,samp} = \sqrt{\frac{2 \cdot kT}{C}} = 14.3\ \mu\mathrm{V}$$

$$V_{rms,comp} = 1.1\frac{\mathrm{nV}}{\sqrt{\mathrm{Hz}}} \cdot \sqrt{2 \cdot 40\ \mathrm{MHz} + \frac{\pi}{2} \cdot 24\ \mathrm{MHz}} = 11.9\ \mu\mathrm{Vrms} \quad \text{and}$$

$$V_{Nqu} = \sqrt{N_{qu}} = \frac{\mathrm{LSB}}{\sqrt{12}} = 22.3\ \mu\mathrm{Vrms}$$

The comparator has several gain stages with similar bandwidth. The effective bandwidth during the conversion is therefore estimated to $f_{-3\mathrm{dB}}$. The noise

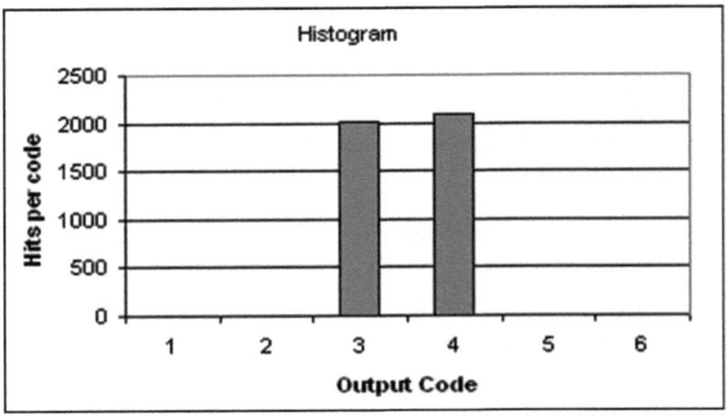

Fig. 2.61 Histogram of SAR ADC results of 4096 conversions at a DC input voltage

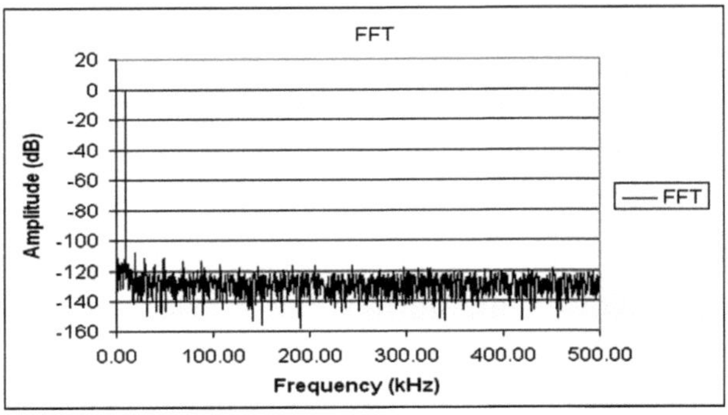

Fig. 2.62 FFT of a SAR ADC output with 4096 samples and a 10 kHz input signal

contribution of the reference is neglected. The total noise can be calculated with the geometrical sum.

$$V_{rms,tot} = \sqrt{V_{rms,samp}^2 + V_{rms,comp}^2 + V_{Nqu}^2} = 29\ \mu V_{rms} \tag{2.23}$$

The peak to peak noise is generally assumed to be 6.6 times the rms-noise and therefore $\frac{29\ \mu V \cdot 6.6}{76\ \frac{\mu V}{LSB}} = 2.5\ \text{LSB}$. A histogram of output codes at a DC input voltage is shown in Fig. 2.61.

The theoretical signal-to-noise ratio is calculated to 95 dB. The measured SNR with 10 kHz signal frequency only achieves 93 dB. Noise is added from the analog signal source and the analog input driver. Today's signal generators cannot achieve the performance of high performance SAR converter. Most high speed SAR converter are therefore only specified with a 2 kHz input signal, as it can be filtered well enough.

Figure 2.62 shows an FFT of the converter at a 10 kHz input signal and 4096 samples. The noise floor is at −130 dB.

2.6.1 A Novel Oversampling Method

If the noise distribution is white, which means equally distributed over the frequency, then a doubling of the oversampling ratio will lower the noise by the square-root of 2, which equals 3 dB. Unfortunately, oversampling will destroy the advantage of the SAR converter to deliver a snapshot of a signal at a particular point of time. To retain this advantage, the SAR converter must only sample once and convert the sampled signal several times [29, 30].

In this way, the sampling noise *kT*/*C* remains unchanged, but the conversion noise, which is generated by the quantization, the reference path and the comparator, will be reduced. This will modify Eq. (2.23) to

$$V_{rms,tot} = \sqrt{V_{rms,samp}^2 + \frac{\frac{LSB^2}{12} + V_{rms,ref}^2 + V_{rms,comp}^2}{OSR}} \tag{2.24}$$

This kind of oversampling with a factor *OSR* therefore only makes sense, if the conversion noise is dominating the sampling noise. The oversampling ratio is getting ineffective, if sampling noise is higher than the conversion noise. The factor *OSR* might therefore be chosen to

$$\frac{OSR = \frac{\mathrm{LSB}^2}{12} + V_{rms,ref}^2 + V_{rms,comp}^2}{V_{rms,samp}^2} \tag{2.25}$$

The oversampling consumes a lot of time, if the total conversion has to be repeated over and over again like shown in Fig. 2.63. If a 4-times oversampling would be performed for a 16 bit converter, then 4 clock cycles are typically used for the sampling process, 16 clocks for the initial conversion and 3 times 16 clocks for the oversampling, which leads to 68 total clock cycles instead of 20.

If the noise is less than eight codes peak-to-peak, it would be sufficient to repeat only the last 3 bit decisions. Unfortunately, this requires that the CDAC can increase and decrease its output voltage to adjust for the noise. If for example, the ideal conversion result is 01111111, but due to noise the output was 10000000, then all upper bits have to change during the next conversion, which normally cannot be performed by repeating the last 3 bit decisions.

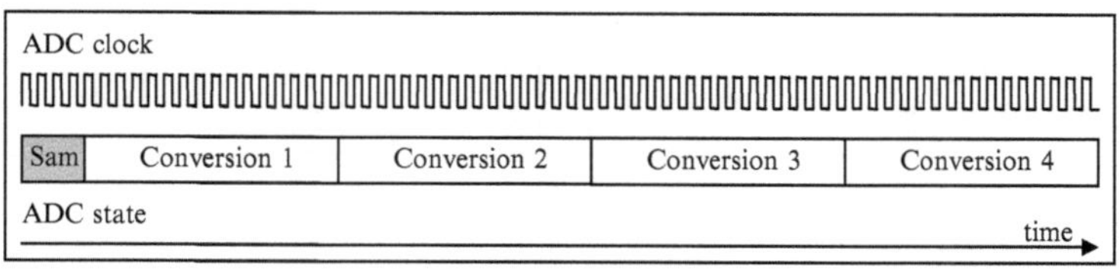

Fig. 2.63 Timing of a conventional 16 bit SAR ADC with integrated oversampling (OSR = 4)

The dynamic error correction helps, because it allows an adjustment of the CDAC and the digital output in both directions. The upper bit capacitors of the CDAC remain at their settings. The following conversions for the oversampling will start from the dynamic error correction bit and run through the last bit decisions.

Figure 2.64 shows that a total of 33 clock cycles are required for the same job. The sampling process takes four clock cycles again. The first conversion needs one additional clock cycle for the dynamic error correction and therefore a total of 17 clock cycles. The following three conversions only require four clocks each, one for the dynamic error correction and three for the remaining 3 bit decisions.

Figure 2.65 is showing an example. The noise is illustrated by different potentials of V_c during the decisions of the dynamic error correction.

In the example of Fig. 2.65, the dynamic error correction is placed at bit LSB + 3 and will adjust for 8 LSB. The first conversion is assumed to $8000_H - 0_H = 8000_H$, where the additional digit expresses the dynamic error correction, which is either 0_H

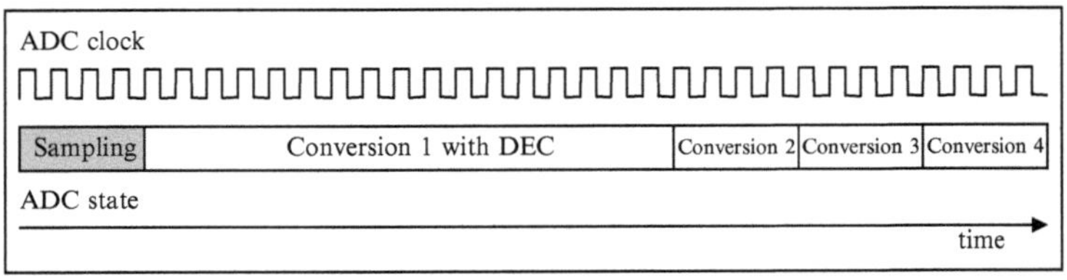

Fig. 2.64 Timing of a 16 bit SAR ADC with integrated oversampling (OSR = 4) using DEC

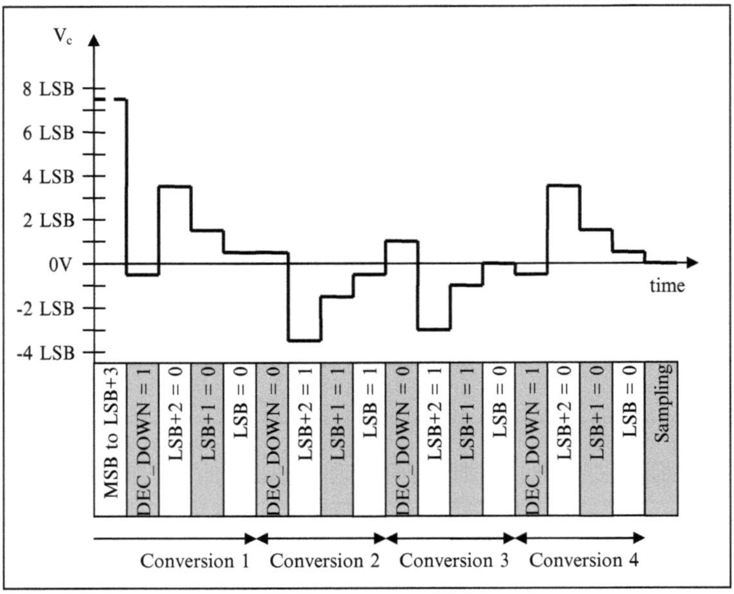

Fig. 2.65 Comparator input of a SAR ADC with DEC and OSR = 4 during a conversion

or 8_H. In Fig. 2.65, the output of the dynamic error correction has to be subtracted as the LSB + 3 decision was 0, which activates DEC_DOWN. As the DEC_DOWN decision is one during the first conversion, there is no dynamic error. The digital output is corrected by 0_H. During the second conversion, a dynamic error occurred and the result is $8007_H - 8_H = 7\ FFF_H$, the third conversion ends in $8006_H - 8_H = 7\ FFE_H$ and a fourth one in $8000_H - 0_H$ again. The sum is $1\ FFFD_H$ or the 16 bit average $7\ FFF_H$.

The dynamic error correction has to cover the peak-to-peak noise plus the dynamic error. It therefore requires more headroom.

This method of oversampling does not require additional analog circuitry, just additional digital circuitry for the additional state-machine and the calculation. It further requires 12 additional clock cycles if compared to a standard conversion without oversampling but dynamic error correction, which already has 21 clock cycles. The conversion rate only drops by 57 %.

The architectures, which are described in this chapter help to push the performance into a new generation.

2.6.2 Noise and Distortion Caused by the Power Supply

Noise and distortion of SAR ADCs are significantly influenced by the internal power supply. The issue is caused by parasitic elements. Isolated PWELLs generate a significant capacitance between the supply and ground. The chip-internal capacitance can reach around 100 pF. Both, the supply and the ground connection have a parasitic inductance to the external supply. This inductance is caused by the bondwires, the package leads, wiring on the PCB and also the external supply capacitor has parasitic inductance included. The inductance can quickly add up to 10 nH per pin. This parasitic LC components form a parasitic oscillation circuitry.

The digital gates in ADCs generate current and voltage spikes that stimulate the oscillation circuitry. Figure 2.66 shows an example, where voltage spikes on the ground with an amplitude of 600 mV were measured on channel 4 (Ch4) of an oscilloscope. Note that the scope had an analog bandwidth limitation of 100 MHz. Therefore, the real amplitudes can be even higher.

Frequencies on the internal supply around 100 MHz were measured on silicon. This oscillation typically takes several 10 ns to decay. The comparator will therefore make its decisions in the middle of the supply oscillation. Offset, offset drift and a wide DNL of ±1 LSB are the consequence. This is a relatively new issue in the design of SAR ADCs. Reasons are the following:

1. Older products typically did not use processes with isolated PWELL. Consequently the chip-internal supply capacitance increased over the last years. This reduced the oscillation frequency and therefore the decay time.
2. The conversion rates increased. Typical SAR ADCs in the 1990s operated at 100–200 kSPS. Today, 1 MSPS ADCs are the standard.

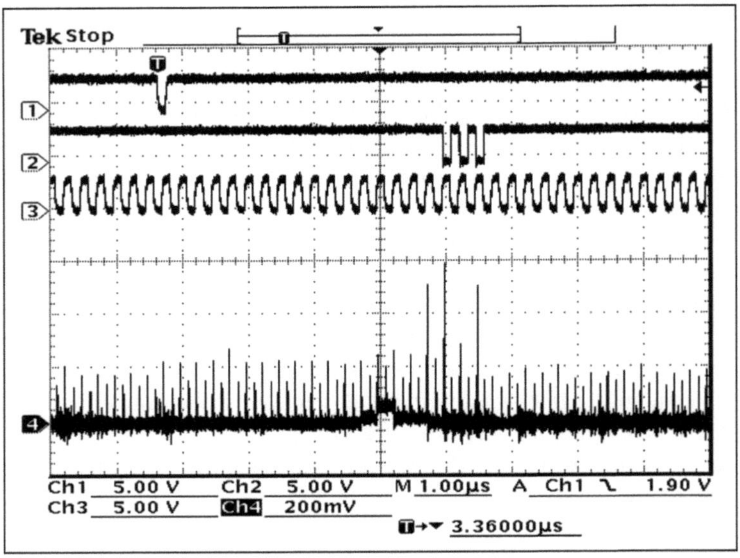

Fig. 2.66 Chip-internal voltage spikes on the ground caused by digital gate switching

3. The accuracy of SAR ADCs increased additionally. Previously, a wide DNL of ±1 LSB could be neglected compared to packaging shifts after laser trimming.

Designers should therefore add chip-internal decoupling circuits that decay the oscillation. They are typically a combination of resistors and capacitors in series between supply and ground. The actual values are dependent on the amount of capacitance that can internally be implemented and the parasitic inductance. Such a decay circuitry needs to be adjusted for each IC design in combination with the particular package in use.

Also note that the protection cells for electro static discharge (ESD) have some resistance in series. Even if the wiring only adds up to 0.2 Ω, it will generate a 4 mV drop if a DC current of 20 mA is used. If this ground is used for the negative reference at the CDAC, then the static voltage drop will show up as a gain error. Dynamic digital current spikes can be several 100 mA and would generate spikes of several 100 mV as can be seen in Fig. 2.66. Internal wire resistance of the supply and ground quickly add up to several Ohms. The supply of each component should therefore be wired separately back to the pad. High performing ADCs typically provide an extra current-free ground pin that is used as negative reference.

Distortion on the supply will also couple via parasitic capacitance into the substrate, which normally is the back-gate of transistors. Modern processes typically provide a buried N-layer that can be used to generate isolated PWELLs. Such isolated PWELLs should be used to reduce the coupling of digital noise into the substrate. It should also be used for sensitive analog circuitry to protect the transistors from such substrate noise.

2.7 Limitations and Next Steps

The previous chapters already discussed some newer approaches in SAR ADC design. Examples are the error correction, the trim of the scale-down capacitor or the oversampling with a minimized amount of clock cycles. The last one already provided a flavor of the possibilities that the error correction opens up. Further novel ideas and implementations are discussed in the next chapter.

5 V processes were standard for high performance SAR ADC designs over the last 15 years. If a process with smaller dimensions and lower voltage robustness is used, as for example a 2 V process, then the input voltage range is limited to that value as well. For this example, the LSB reduces by the factor of 2.5 and the sample capacitor C_s would have to increase by a factor of 6.25 to remain the ratio of the signal amplitude and the sampled noise constant, which reflects the SNR. Even if the power of active components like the comparator can be reduced with such a process, the larger capacitors need to be recharged between the input voltage, a positive reference level *REF* and a negative reference voltage level, which mostly equals to ground. Due to the larger capacitor size, the recharging process requires additional power, so that the total power consumption does not necessarily reduce. This relationship is shown in Fig. 2.67.

A novel circuit architecture called the 'floating CDAC architecture' is presented to overcome this issue. This architecture was optimized to the 'dual CDAC architecture'. These novel architectures can be beneficial in different ways. An ADC was implemented, which can convert large input voltage ranges with up to differential 40 V_{pp} with the ADC's core circuitry operating from a 5 V supply. This way, the signal-to-noise ratio can be improved by 15 dB beyond state-of-the-art converters enabling a 20 bit SAR ADC.

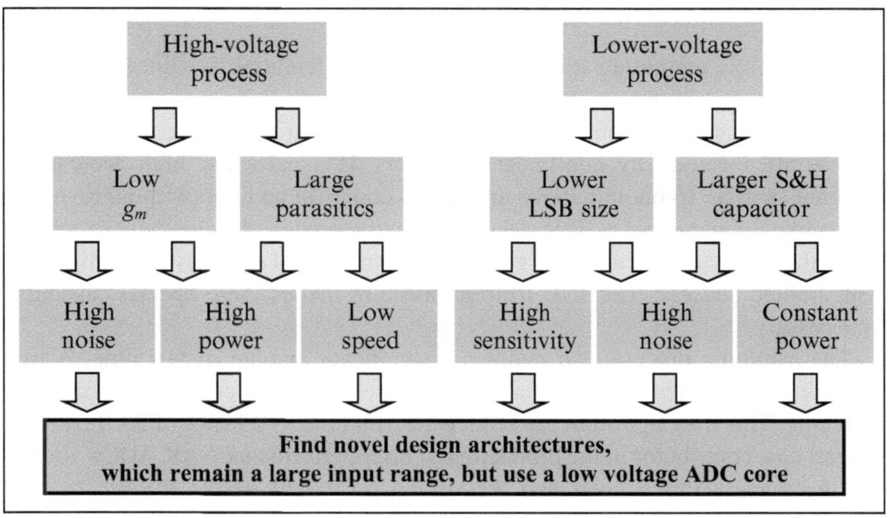

Fig. 2.67 Process selection disadvantageously influencing SAR ADC parameters

In another ADC implementation, the same architectures are used to achieve similar performance specifications as state-of-the-art ADCs with the difference in minimizing the power consumption into new dimensions and the additional benefits in reducing the die size and making the ADC easier to use in terms of robustness, which is particular important, if higher integration is anticipated. This is done by operating the core circuitry at a 1.2 V supply, but remaining the 5 V input range and therefore remaining the signal-to-noise performance.

As power is very significant, it would be a great advantage, if the circuit's power consumption could be turned off during sampling. Such a feature is called 'zero-power sampling'. The power consumption would then scale with the conversion rate with the full ability to continuously sample the input voltage. With long sampling times, the requirements for the input driver circuitry in the application would drop in terms of bandwidth or driver impedance (see Sect. 6.1.1). Such an architecture is achieved by connecting the sample capacitor to the input voltage on one side and to ground on the other side. Previous SAR ADCs had to sample to a common-mode voltage other than ground, which sets the operating point of the comparator input. The dual CDAC architecture fits naturally to implement this zero-power sampling feature.

A further reduction of the power consumption and the die size is achieved with a novel digital trim scheme, where the error is calculated and corrected in the digital domain. The state-of-the-art analog trim circuitry including trim capacitors with switches as described in Sect. 2.2.2 is not required anymore. Power is saved as the trim capacitors do not switch between the positive and the negative reference potential. This further reduces the driving requirements for the reference voltage. As a positive side effect, such a purely digital trim solution would also reduce the time required to calibrate the matching of the capacitors within the DAC. As the state-of-the-art trim solution utilizes capacitors that do not always match with high accuracy to the DAC capacitors, it is required until today to measure the matching twice or trice.

Chapter 2 is presenting trim methods for the differential non-linearity (DNL), offset, gain and common-mode rejection. Until now there was no method of trimming the integral non-linearity (INL). A novel trim scheme was invented. INL is often caused by non-linear capacitors. If a particular high accuracy is required such as 16 bit and above and if this comes in addition with high voltage inputs like the industrial ± 10 V standard, then the process development to achieve particularly linear capacitors can get time consuming and expensive and might end with limited success. The INL trim architecture makes these special capacitors unnecessary.

The trimming process also requires a significant amount of test time. A trim algorithm was implemented, so that the ADC itself is used to compare the capacitor matching. This trim algorithm can reduce the trim time by a factor of 50. Test is the biggest cost contributor in the production of high-performance SAR ADCs. Reducing the manufacturing costs and the selling price will enable the use of high performance ADCs in new applications and drive innovation.

References

1. J. McCreary, P. Gray, *All-MOS Charge Redistribution Analog-to-Digital Conversion Techniques - Part I*, in the *IEEE journal of Solid-State Circuits*, Volume 10, Issue 6, pp. 371–379, 1975.
2. Bernd M. Rundel, Timothy V. Kalthoff, *Capacitor array having reduced voltage coefficient induced non-linearities*, US Patent 6404376, Texas Instruments, 2002, URL: http://www.google.com/patents/US6404376, status of 3/11/2015.
3. Paul R. Gray, Robert G. Meyer, *Analysis and design of analog integrated circuits*, John Wiley & Sons Inc., New York/Chichester/Brisbane/Toronto/Singapore, 1993.
4. Matthew D. Felder, *Successive approximation analog-to-digital converter with current steered digital-to-analog converter*, US Patent 7209069, Sigmatel, 2007, URL: http://www.google.com.ar/patents/US7209069, status of 3/11/2015.
5. Frank Ohnhäuser, Martin Allinger, Mario Huemer, *Trim Techniques for DC specifications for A/D converters based on successive approximation*, AEU - International Journal of Electronics and Communications, Volume 64, Issue 8, pp. 790–793, 2010.
6. Wolfgang Knappe, *Fehlererkennung und Fehlerkorrektur bei Analog/Digital-Umsetzern*, PhD. Thesis, Technical University of Munich, 1992.
7. Khen-Sang Tan, Sami Kiriaki, Michiel de Wit, John W. Fattaruso, Ching-Yuh Tsay, W.E. Matthews, Richard K. Hester, *Error correction techniques for high performance differential A/D converters*, in the IEEE journal of Solid-State Circuits, Volume 25, Issue 6, pp. 1318–1327, 1990
8. Data Manual of the ADS7869, Texas Instruments, 2006, URL: http://focus.ti.com/lit/ds/symlink/ads7869.pdf, status of 3/11/2015.
9. Product datasheet of the ADS8556, Texas Instruments, 2011, URL: http://focus.ti.com/lit/ds/symlink/ads8556.pdf, status of 3/11/2015.
10. Product datasheet of the AD7656, Analog Devices, 2006, URL: http://www.analog.com/static/imported-files/data_sheets/AD7656_7657_7658.pdf, status of 3/11/2015.
11. Frank Ohnhaeuser, Miroslav Oljaca, *Offset error compensation of input signals in analog-to-digital converter*, US Patent 6433712, Texas Instruments, 2002, URL: http://www.google.com.ar/patents/US6433712, status of 3/11/2015.
12. Robert E. Seymour, *Method and circuit for gain and/or offset correction in a capacitor digital-to-analog converter*, US Patent 6922165, Texas Instruments, 2005, URL: http://www.google.com.ar/patents/US6922165, status of 3/11/2015.
13. Bruce Edward Amazeen, Michael C. W. Coln, Scott Wayne, Gerald A. Miller, Mick Mueck, *Multi-stage, low-offset, fast-recovery, comparator system and method*, US Patent 6429697, Analog Devices, 2002, URL: http://www.google.com.ar/patents/US6429697, status of 3/11/2015.
14. J.H. Atherto, H.T. Simmonds, *An offset reduction technique for use with CMOS integrated comparators and amplifiers*, in the IEEE journal of Solid-State Circuits, Volume 27, Issue 8, pp. 1168–1175, 1992.
15. Y.C. Huang, B.D. Liu, *A 1 V CMOS analog comparator using auto-zero and complementary differential-input technique*, in the proceedings of the IEEE Asia-Parcific Conference on ASICs, pp. 181–184, 2002.
16. Frank Ohnhaeuser, Mario Huemer, *Methods to Eliminate Dynamic Errors in High-Performance SAR A/D Converter*, in the proceedings of the International Symposium on Circuits and Systems (ISCAS2008), pp. 2398–2401, 2008.
17. Chakravarthy Srinivasan, Kiran M. Godbole, *Error correction architecture to increase speed and relax current drive requirements of SAR ADC*, US Patent 6747589, Texas Instruments, 2004, URL: http://www.google.com/patents/US6747589, status of 3/11/2015.
18. K. Bacrania, *A 12-bit ADC successive-approximation-type with digital error correction*, in the IEEE journal of Solid-State Circuits, Volume 21, Issue 6, pp. 1016–1025, 1986.

19. Seetharaman Janakiraman, Vikram Varma, Yujendra Mitikiri, *SAR ADC providing digital codes with high accuracy and high throughput performance*, US Patent 6958722, Texas Instruments, 2005, URL: http://www.google.com.ar/patents/US6958722, status of 3/11/2015.
20. David M. Jones, Jerry L. Doorenbos, *Bandgap reference curvature compensation circuit*, US Patent 6255807, Texas Instruments, 2001, URL: http://www.google.com.ar/patents/US6255807, status of 3/11/2015.
21. Jerry L. Doorenbos, *Method of curvature compensation, offset compensation, and capacitance trimming of a switched capacitor band gap reference*, US Patent 6060874, Burr-Brown, 2001, URL: http://www.google.com/patents/US6060874, status of 3/11/2015.
22. K.E. Kuijk, *A precision reference voltage source*, in the IEEE journal of Solid-State Circuits, Volume 8, Issue 3, pp. 222–226, 1973.
23. Paul Brokow, *A simple three terminal IC bandgap reference*, in the IEEE journal of Solid-State Circuits, Volume 9, Issue 6, pp. 670–674, 1974.
24. H. Banba, *A CMOS bandgap reference circuit with sub-1 V operation*, In the IEEE journal of Solid-State Circuits, Volume 34, Issue 5, pp. 388–393, 1999.
25. Keith Sanborn, Dongsheng Ma, Vadim Ivanov, *A sub-1 V low-noise bandgap voltage reference*, in the IEEE journal of Solid-State Circuits, Volume 42, Issue 11, pp. 2466–2481, 2007
26. Product datasheet of the ADS7863, Texas Instruments, 2011, URL: http://focus.ti.com/lit/ds/symlink/ads7863.pdf, status of 03/11/2015.
27. Product datasheet of the ADS8361, Texas Instruments, 2007, URL: http://focus.ti.com/lit/ds/symlink/ads8361.pdf, status of 03/11/2015.
28. Frank Ohnhaeuser, Mario Huemer, *Reference generation for A/D converters*, in the proceedings of the International Symposium on Signals, Systems and Electronics (ISSSE2007), pp. 355–358, 2007.
29. Seetharaman Janakiraman, Kiran M. Godbole, Surendranath Nagesh, *Increasing the SNR of successive approximation type ADCs without compromising throughput performance substantially*, US Patent 6894627, Texas Instruments, 2005, URL: http://www.google.com.ar/patents/US6894627, status of 3/11/2015.
30. Christopher Peter Hurrell, Gary Robert Carreau, *Analog-to-digital converter with signal-to-noise ratio enhancement*, US Patent 7218259, Analog Devices, 2007, URL: http://www.google.com.ar/patents/US7218259, status of 3/11/2015.

Chapter 3
Advanced SAR ADC Design

Chapter 2 discussed the basics of the SAR ADCs and their components such as the CDAC or the comparator in detail. Based on these fundamentals, Chap. 3 will present current research topics in the field of SAR ADCs. Generally, the core circuitry of the new SAR ADC should operate from a significantly lower supply voltage compared to the input voltage range. A novel architecture called the 'floating CDAC architecture' was developed, which is presented in Sect. 3.1. The 'floating CDAC architecture' had some technical disadvantages and was improved to the 'dual CDAC architecture'. This architecture is described in Sect. 3.2.

In a next step, trim solutions for these ADCs were evaluated. A purely digital trim solution is proposed in Sect. 3.3 followed by an INL-trim method in Sect. 3.4.

A novel method to measure capacitive matching, which is required for the trim algorithm, will further reduce the production costs. It is discussed in Sect. 3.5.

Now that the ADC core can run on a low voltage supply, adding the 'zero power sampling' aspect was added in Sect. 3.6.

Sections 3.7 and 3.8 are demonstrating the potential of the new architectures on two ADC examples for highest performance and lowest power.

3.1 Floating CDAC Architecture [1, 2]

The goal of this architecture is to convert a large input voltage range on an ADC core operating on a significantly lower supply voltage without dividing the signal amplitude internally. Two particular implementations are in focus. The first one should convert the ± 10 V signal range on a 5 V core supply to achieve a best in class noise performance. The second implementation will convert a 5 V signal on a 1.2 V core supply to minimize power and area consumption. Challenging is that the comparator input voltage can be in the range of $\pm \frac{REF}{2}$ during the MSB decision as shown in Fig. 2.35 and then converges to 0 V during the course of the conversion.

F. Ohnhäuser, *Analog-Digital Converters for Industrial Applications Including an Introduction to Digital-Analog Converters*, DOI 10.1007/978-3-662-47020-6_3

To prevent the hold switch from leaking and the comparator input from being destroyed, transistors have to be used with a voltage rating that is at least as large as the reference voltage REF. A work around is to divide the ADC internal signal level. It was shown that this would degrade the signal-to-noise ratio.

Summarized, the ideal architecture has to reduce the signal level during the MSB decisions. It has to be able to restore the full signal height for the conversion of the LSBs. An error correction scheme can be used in between to compensate for the different modes of operation of the MSBs versus the LSBs. A larger amount of error can be tolerated during the MSB decisions, if the error correction range is large enough.

The idea of this invention is that most of the charge in the CDAC remains unused during the MSB decisions. Only a small fraction is used, so that a small amount of additional capacitors can reduce the signal amplitude during the MSB decisions. Once these decisions are performed the full charge that was parked is activated. The signal amplitude increases to the usual level. Charge can be frozen on a capacitor, if one of the electrodes remains floating, so that the charge of this capacitor cannot redistribute.

For the MSB decisions, the charge can for example be frozen on the three major capacitors of bit 1, bit 2 and bit 3. As a nomenclature, bit 1 still describes the MSB, bit 2 the MSB-1 and so on. The bit 4 capacitor is not used for sampling and can be connected to ground during the MSB decisions. This way, it divides the signal amplitude. The bit 5 capacitor could act as a pure sampling capacitor. As such, it would connect to ground during the MSB decisions and will also reduce the signal amplitude. The bit 6, bit 7 and bit 8 capacitors are used to convert the MSBs by switching them between the positive and the negative reference level. These capacitors are not used for sampling.

The sum of the lower bit capacitors starting from the bit 6 capacitor are adding up to be identical to the bit 5 capacitor minus a capacitor representing the LSB. The bit 4 capacitor has twice the size of the bit 5 capacitor. Summarized, the charge of the bit 5 capacitor is distributed to four times this capacitance. The signal amplitude is therefore divided by a factor of 4. If a 5 V input range is converted, then the voltage remains in the range of $\pm\frac{REF}{8} = \pm 0.625$ V during the conversion.

The bit 4 capacitor and all lower bit capacitors starting from bit 6 were not used for the sample process. The ratio of the sample capacitance to the capacitance used for the conversion process, which is the sum of capacitors switching between the positive and negative reference, is setting the gain. To avoid gain errors, an additional sample capacitor with the size of the bit 4 capacitor plus the capacitors starting from bit 6 has to be added. This extra sample capacitor will also remain floating during MSB decisions.

The MSB decisions, which were achieved with the bit 6 through bit 8 capacitors, are now applied to the floating capacitors by switching them to the respective positive and negative ground levels. Also the charge of the additional sample capacitor is activated by switching it to ground for the rest of the conversion. This causes the signal amplitude to rise to the full level. The internal signal level gets only divided by the ratio of the sum of all sample capacitors and the total amount of capacitance. All capacitors were used for sampling beside the bit 4 capacitor, which equals 1/8th of the MSB capacitor, and the lower bit capacitors, which sum up to the bit 5 capacitance and therefore equal 1/16th of the MSB capacitance. If the MSB capacitor has the size C, then the internal signal amplitude $A_{sig,int}$ in ratio to the external signal amplitude $A_{sig,ext}$ can be calculated to

$$\frac{A_{sig,\mathrm{int}}}{A_{sig,ext}} = \frac{2C}{2C + \frac{3}{16}C} = 0.914 \tag{3.1}$$

Equation (3.1) shows that 91.4 % of the signal amplitude remains. The small loss in signal amplitude causes a minor loss in SNR of only 0.78 dB.

Figures 3.1, 3.2, 3.3, 3.4, 3.5 and 3.6 show the conversion process with the floating CDAC architecture starting with the sample phase in Fig. 3.1. In an example, an input voltage of 4 V should be sampled on C_{bit1}, C_{bit2}, C_{bit3}, C_{bit6} and C_S.

The sampling phase ends by opening the hold switch sw_{hold}. C_{bit1}, C_{bit2}, C_{bit3}, and C_S are switched into the floating mode. The sampled charge of C_{bit5} is activated by switching it to ground and C_{bit6} is switched to the positive reference voltage of 5 V for the MSB evaluation as shown in Fig. 3.2. The activated sampled charge for the MSB decisions $Q_{MSB,S}$ can be calculated to

$$Q_{MSB,S} = C_{bit5} \cdot V_{in} \tag{3.2}$$

The charge gets redistributed to C_{bit4} and all lower bit capacitors. All of these capacitors are connected to ground, while only the C_{bit6} connects to the reference for the bit 1 decision. The charge Q_{bit1} during the MSB decision can therefore be calculated by Eq. (3.3), taking into consideration that all LSB capacitors add up to C_{bit6}.

$$Q_{bit1} = C_{bit6} \cdot (REF - V_c) + (C_{bit4} + C_{bit5} + C_{bit6}) \cdot (0\ \mathrm{V} - V_c) \tag{3.3}$$

As the charge is frozen, the charges $Q_{MSB,S}$ and Q_{bit1} are identical and as the capacitors are binary weighted,

$$2 \cdot C_{bit6} \cdot V_{in} = C_{bit6} \cdot REF - 8 \cdot C_{bit6} \cdot V_c \qquad \text{(or)}$$

$$V_c = \frac{\frac{REF}{2} - V_{in}}{4} < 0 \tag{3.4}$$

Based on Eq. (3.4), the comparator is deciding, if the input voltage is larger than half the reference voltage, which is true here. Also note that the absolute value of the comparator input voltage remains below $\frac{REF}{8}$, which is less than 0.625 V. With $MSB = 1$, C_{bit6} remains at the reference voltage REF and C_{bit7} connects to REF for

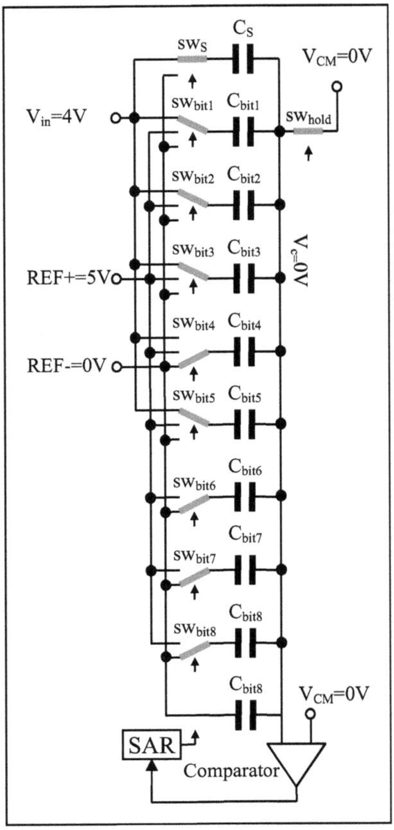

Fig. 3.1 SAR ADC with floating CDAC architecture in sampling mode

the bit 2 decision (see Fig. 3.3). Based on this the charge for the bit 2 decision Q_{bit2} can be calculated to

$$Q_{bit2} = (C_{bit6} + C_{bit7}) \cdot (REF - V_c) + (C_{bit4} + C_{bit5} + C_{bit7}) \cdot (0 \text{ V} - V_c) \quad (3.5)$$

Adding the binary weighting of the capacitor and the charge equity leads to

$$4 \cdot C_{bit7} \cdot V_{in} = 3 \cdot C_{bit7} \cdot REF - 16 \cdot C_{bit7} \cdot V_c \quad \text{(or)}$$

$$V_c = \frac{\frac{3 \cdot REF}{4} - V_{in}}{4} < 0 \quad (3.6)$$

The comparator evaluates, if the input voltage of 4 V is larger than three quarters of the reference, which is 3.75 V. This is true, so that the MSB-1 also equals 1. Therefore, C_{bit7} remains connected to REF and C_{bit8} connects also to REF for the bit 3 decision as shown in Fig. 3.4.

Following the calculations above, the comparator input voltage during the bit 3 decision can be calculated to

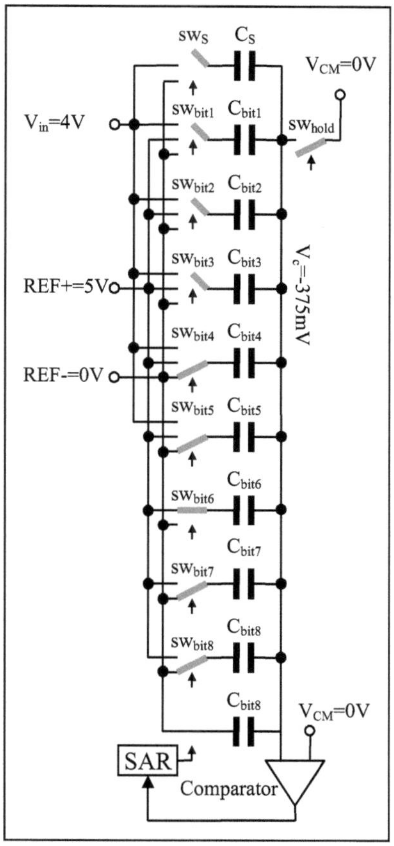

Fig. 3.2 SAR ADC with floating CDAC architecture during bit 1 decision

$$V_c = \frac{\frac{7 \cdot REF}{8} - V_{in}}{4} < 0 \tag{3.7}$$

Equation (3.7) is not true, so that the MSB-2 is set to 0. With knowing the first bit decisions, the frozen charge can be reactivated by switching the MSB capacitors accordingly to the previous decision. Therefore, C_{bit1} and C_{bit2} have to connect to REF and C_{bit3} to ground. Also the charge of C_S gets activated, by switching the capacitor to ground. C_{bit4} connects to REF as the conversion continuous with the bit 4 decision, while all the other LSB capacitors are reset to ground. This state is shown in Fig. 3.5.

Now, the charge of all sample capacitors is activated. The sample capacitors were actually chosen to sum up to twice the bit 1 capacitor C_{bit1}. The sampled charge for the LSB decisions $Q_{LSB,S}$ can therefore be calculated to

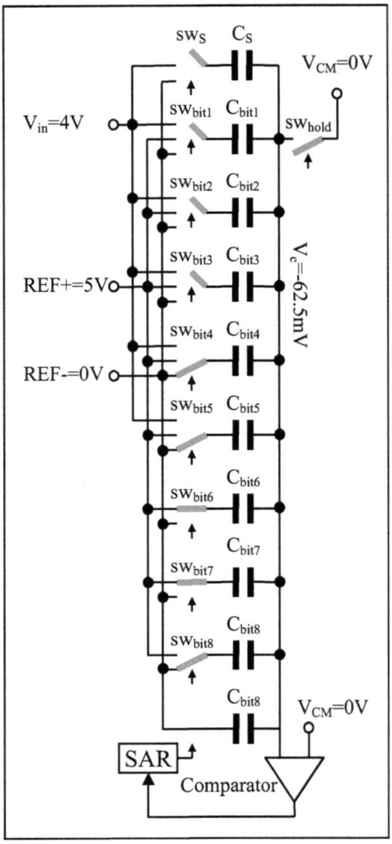

Fig. 3.3 SAR ADC with floating CDAC architecture during bit 2 decision

$$Q_{LSB,S} = 2 \cdot C_{bit1} \cdot V_{in} \tag{3.8}$$

The charge distribution of the bit 4 decision Q_{bit4} is now calculated in Eq. (3.9), which again takes the binary weighting into consideration.

$$\begin{aligned} Q_{bit4} &= (C_{bit1} + C_{bit2} + C_{bit4}) \cdot (REF - V_c) + (C_{bit3} + C_{bit4} + C_S) \cdot (0\ \text{V} - V_c) \\ Q_{bit4} &= C_{bit1} \cdot \left[\left(\frac{3}{2} + \frac{1}{8} \right) \cdot REF - \left(2 + \frac{C_S}{C_{bit1}} \right) \cdot V_c \right] \end{aligned} \tag{3.9}$$

As the charge is frozen, $Q_{LSB,S}$ equals Q_{bit4}, so that V_C results in

$$2 \cdot C_{bit1} \cdot V_{in} = C_{bit1} \cdot \left[\left(\frac{3}{2} + \frac{1}{8} \right) \cdot REF - \left(2 + \frac{C_S}{C_{bit1}} \right) \cdot V_c \right]$$

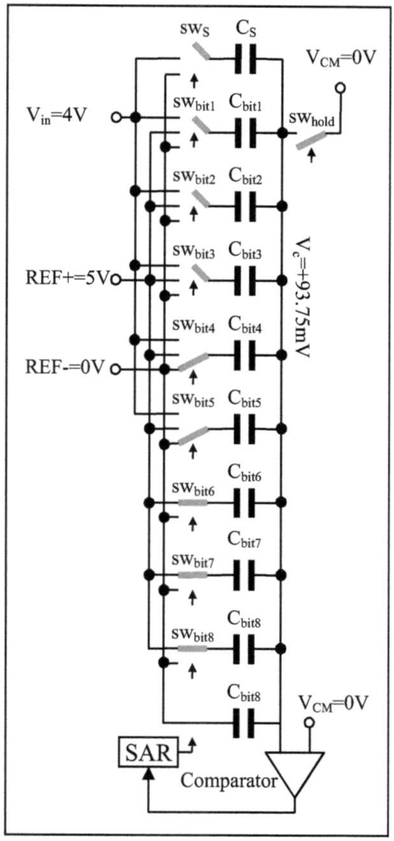

Fig. 3.4 SAR ADC with floating CDAC architecture during bit 3 decision

$$V_c = \frac{\left(\frac{3}{4}+\frac{1}{16}\right)\cdot REF - V_{in}}{\left(1+\frac{C_S}{2C_{bit1}}\right)} = \frac{\left(\frac{3}{4}+\frac{1}{16}\right)\cdot REF - V_{in}}{\left(1+\frac{3C_{bit5}}{32C_{bit5}}\right)} = \frac{\left(\frac{3}{4}+\frac{1}{16}\right)\cdot REF - V_{in}}{1.09375} < 0 \tag{3.10}$$

Equation (3.10) proves that the normal bit 4 decision $\left(\frac{3}{4}+\frac{1}{16}\right)\cdot REF < V_{in}$ is performed. With $V_{in}=4$ V and $REF=5$ V, the MSB-3 is 0 again, C_{bit4} is connected back to the negative reference potential and for the next conversion C_{bit5} is connected to REF as shown in Fig. 3.6.

Equation (3.10) also shows that the residual comparator input voltage is only divided by 9.4 % caused by the extra sample capacitor C_S. The normal residual comparator input voltage for the bit 4 decision can range within $\pm\frac{REF}{16} = 312.5$ mV for the complete input voltage range. Due to the attenuation, it remains below $\pm$286 mV.

Note that the maximum absolute value of the comparator input will remain below 625 mV throughout the complete conversion. If the ADC should run from a

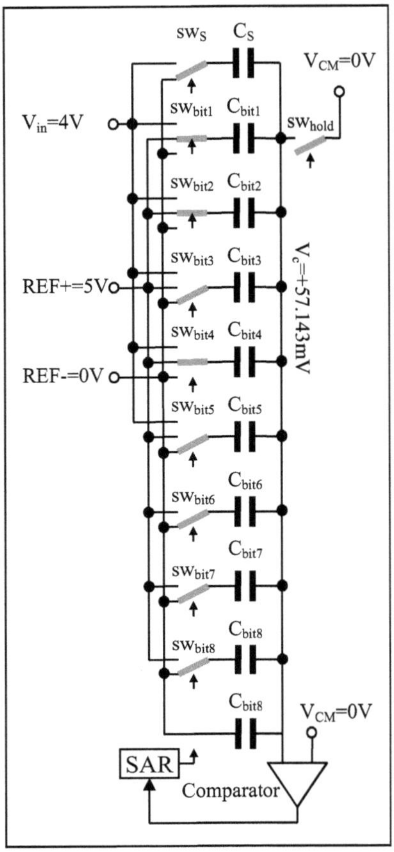

Fig. 3.5 Floating CDAC architecture during bit 4 decision

single supply voltage, then the comparator input voltage should remain above ground to prevent leakage of the hold switch sw_{hold}. Setting the common-mode voltage V_{CM} to 0.625 V instead of the exemplary 0 V would be a good choice. The maximum comparator input voltage would then be 1.25 V. Based on that, the supply voltage could be lowered all the way down to 1.25 V.

The comparator input voltage for the example above is illustrated in Fig. 3.7. Again, the signal amplitude is divided by a factor of 4 during the MSB decisions and restores to nearly full signal amplitude for the bit 4 and all following decisions.

As promising as the floating CDAC algorithm looks on paper, it includes some major problems, which turned up during the realization. The goal was an ADC that converts a fully differential signal, where a negative input signal V_{inn} is inverted to the positive input signal V_{inp}. This conversion runs from a 5 V supply without dividing the signal level internally. Each the positive and the negative input voltage have a swing of $\pm$10 V, so that the differential voltage range results in 40 V_{pp} (also see Fig. 3.8). Doing so, a novel highest performance SAR ADC should be created.

For this application, the sample switches (sw_S, sw_{bit1}, sw_{bit2}, sw_{bit3}, sw_{bit5}) had to be realized with 36 V CMOS transistors, which come with a significant amount of

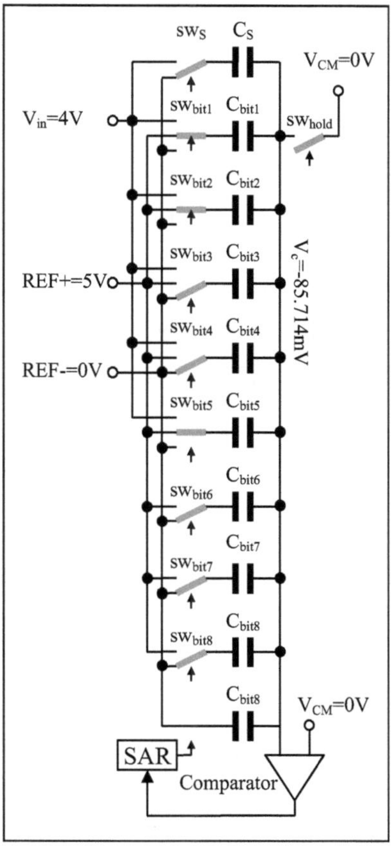

Fig. 3.6 Floating CDAC architecture during bit 5 decision

parasitic capacitance. When these transistors are turned off to set the capacitors in the floating state, then their charge injection will affect the comparator input voltage. This is including the high switching gate voltage that couples through the gate-source and the bit capacitor into the node V_c. This is particularly critical, as the on-resistance of these switches needs to be as low as possible to optimize the THD. The transistors therefore require large widths causing a large amount of charge injection.

This charge injection shifted the comparator input voltage by ±200–300 mV during the floating phase while the MSBs are converted with reduced signal level. This error is limiting the conversion with attenuated signal of the first 3 bits. The divider ratio was chosen to 8 to remain well in the 5 V supply range. Once the floating capacitors are activated, the errors, which are done during the floating phase, are gained up by the divider ratio causing errors in the range of up to $\pm 300\ \text{mV} \cdot 8 = \pm 2.4\ \text{V}$. If the comparator common-mode voltage is set to 2.5 V, which is half the supply, then the comparator input can already vary from 0.1 V to 4.9 V after activating the sampled charge of all capacitors. Even without the error caused by charge injection, the un-attenuated comparator input voltage during the

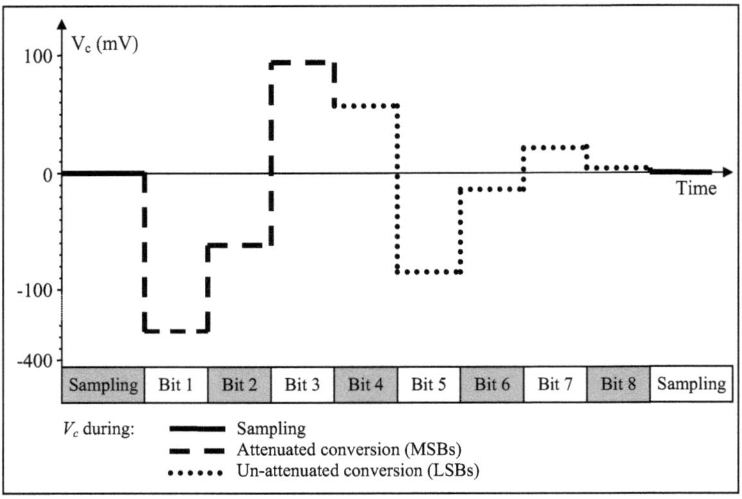

Fig. 3.7 Exemplary comparator input voltage of the floating CDAC architecture during a conversion

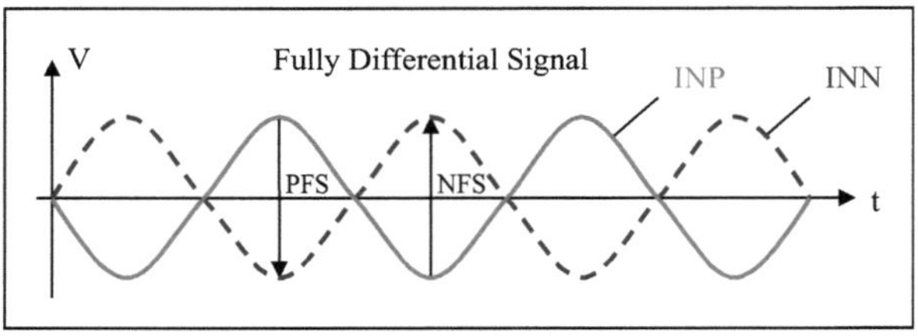

Fig. 3.8 Fully differential input signal

bit 4 decision for a ± 10 V input signal would still be in the range of ± 1.25 V by itself. Leakage of the hold switches would be the consequence of the additional error through the charge injection.

And finally on top of this, large voltage spikes are generated at the comparator input, when the floating capacitors are activated. Figure 3.9 shows an equivalent circuitry for the CDAC. In a first step, the voltage applied to each bit capacitor is replaced by ideal voltage sources V_1, V_2 and V_3. Path 1, which is formed by V_1, R_{sw1} and C_1 might represent any bit of interest. V_2, R_{sw2} and C_2, which below will be named path 2, combine all the switches and bit capacitors in parallel, which are connected to the negative reference voltage, and finally V_3, R_{sw3} and C_3, which below are referenced as path 3, combine all switches and bit capacitors in parallel, which are coupled to the positive reference voltage.

To evaluate the influence of V_1 to the comparator input voltage V_c, all other voltage sources can be connected to ground (V_2 and V_3). Again the parallel circuitry of path 2 and path 3 can be combined by an equivalent circuitry (R_{sw4} and C_4). The equivalent circuitry on the right side in Fig. 3.9 clearly shows that the CDAC is a

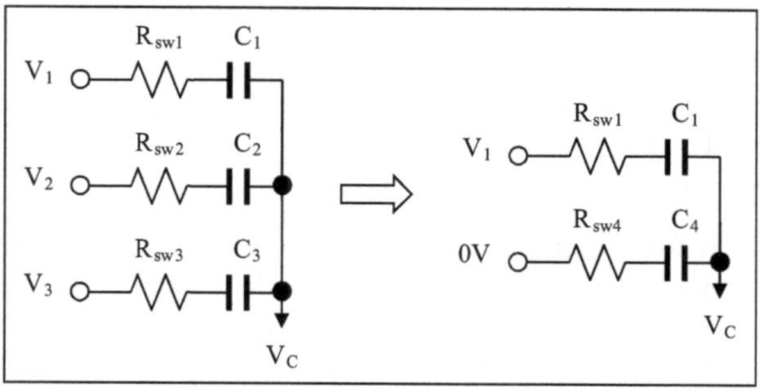

Fig. 3.9 Equivalent circuitry of the CDAC to simplify calculations

divider circuitry, when V_1 switches from the negative reference potential to the positive reference potential. This is described in Eq. (3.11)

$$\frac{V_C}{V_1} = \frac{\left(R_4 + \frac{1}{j\omega C_4}\right)}{\left(R_1 + \frac{1}{j\omega C_1}\right) + \left(R_4 + \frac{1}{j\omega C_4}\right)} \tag{3.11}$$

If the switch impedance to the negative reference potential matches the switch impedance to the positive reference potential and if the switch dimensions are binary weighted as the capacitors are, then $R_1 = \frac{R_4}{a}$ with $C_1 = a \cdot C_4$. If this is placed into Eq. (3.11), then

$$\frac{V_C}{V_1} = \frac{\left(R_4 + \frac{1}{j\omega C_4}\right)}{\left(\frac{R_4}{a} + \frac{1}{a \cdot j\omega C_4}\right) + \left(R_4 + \frac{1}{j\omega C_4}\right)} = \frac{\left(R_4 + \frac{1}{j\omega C_4}\right)}{\left(\frac{1}{a} + 1\right) \cdot \left(R_4 + \frac{1}{j\omega C_4}\right)} = \frac{a}{1+a} \tag{3.12}$$

Equation (3.12) shows that the CDAC is actually an all-pass if parasitic capacitors to ground are neglected and if the reference is assumed as an ideal voltage source. Ideally, the comparator input voltage would follow a jump of V_1 immediately. This fact would be good for the normal reference settling behavior. Unfortunately it causes an issue for the floating CDAC architecture at the point of time the capacitors are activated. Parasitic delays of the digital control logic might cause the floating capacitors to be activated at times that differ by a few nanoseconds. This difference in activation time will cause the comparator input voltage to follow the reference potential of the capacitor that is activated first. Due to the all-pass behavior of the CDAC, this will happen with high speed. In simulations, spikes were visible all the way up to 2 V. These spikes degrade again, once the other capacitors get activated also.

Above, the assumption was done that the on-resistance of the switch to the positive reference voltage is matched to the on-resistance to the negative reference potential. The design can try to accomplish this condition by choosing the best transistor dimensions. However, the on-resistance is changing with parameters, like

- Reference voltage
- Supply voltage
- Supply voltage matching for bipolar supplies
- Temperature or
- Process variation

If for example the impedance of all switches to the positive reference is lower than to the negative reference, then the comparator input voltage will first be pulled into the positive direction, before all capacitors are settled.

The effects above decrease with an increasing switch resistance as any parasitic capacitance at the comparator input node will generate a low pass behavior. Higher resistances will consequently also increase the settling time and reduces the conversion rate.

A complicated error correction scheme was found [2] to make this architecture work. In worst case the MSB capacitor has to be switched several times between the positive and the negative reference voltage potential, which generates a difficult load for the reference driver circuitry. In the particular test chip, which had bipolar supplies, the supply voltage variation had to be limited to ±5 % compared to the standard of ±10 %. The test chip still relied on a good delay matching of the digital control signals. For the case that the estimated matching is not achieved, spikes could occur, which might destroy the comparator input over life time and generate a reliability issue.

Therefore, a more robust algorithm is required, which increases the accuracy during the attenuated MSB decisions. This way, the error correction scheme could be relaxed on one side and V_c converges closer to 0 V after activating all capacitors, which would leave more room for spikes during the charge activation process. It would further be advantageous, if the switch resistance during the activation phase could be increased. This would certainly increase the settling time, but if the activation could occur in parallel to further bit decisions, than this would not matter. A solution is offered by the dual CDAC architecture described in the next chapter.

3.2 Dual CDAC Architecture [3]

The floating CDAC architecture from Sect. 3.1 allows us to freeze charge during the MSB conversions, but it causes so much error by charge injection that it is difficult to run the MSB conversions on the same CDAC. An alternative solution is to use an additional second CDAC for the MSB conversions. This solution is called the 'dual CDAC architecture'. A block diagram is shown in Fig. 3.10.

The switching scheme is explained in the following, where Fig. 3.11 shows the device in sample mode. *CDAC*1 contains the full CDAC also called the fine CDAC with floating capacitors and undivided comparator input voltage. It is sampling on the capacitors C_{bit1} through C_{bit5}. The lower bit capacitors starting from C_{bit6} remain at the negative reference potential, which here should be ground, for the

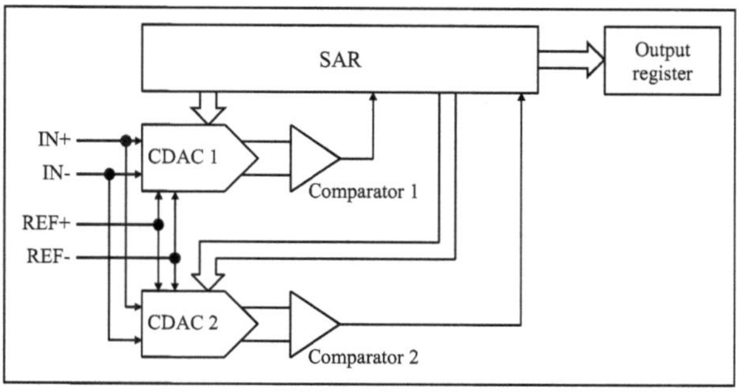

Fig. 3.10 Block diagram of a SAR ADC with dual CDAC architecture

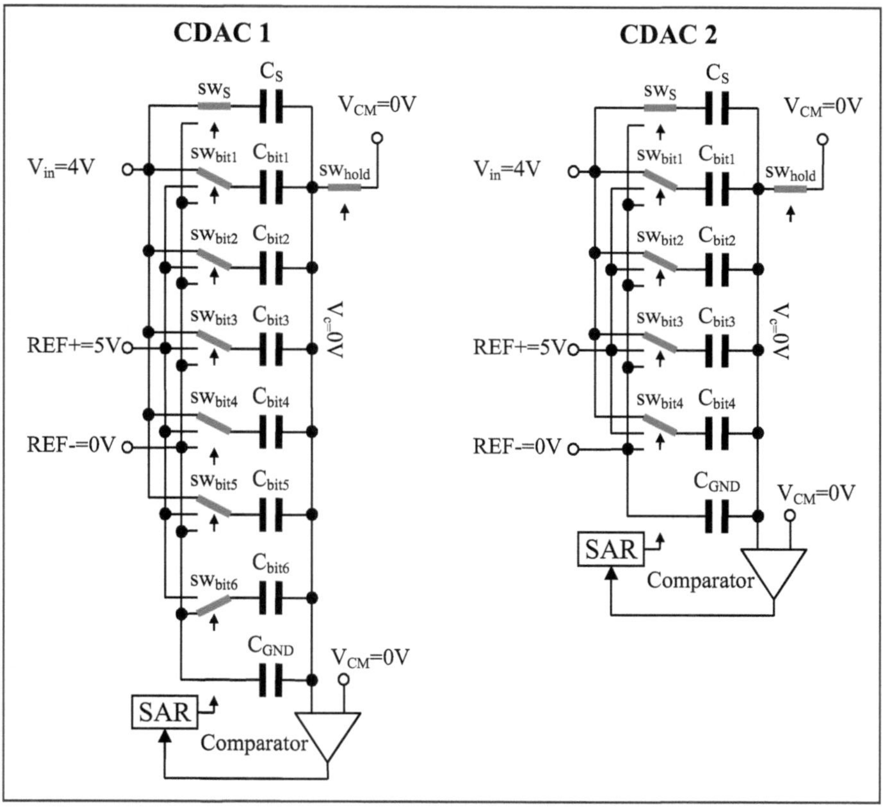

Fig. 3.11 SAR ADC with dual CDAC architecture in sample mode

two following reasons. First, the switching process of the sampling capacitors into the floating state is still coming with charge injection. Only capacitors that are continuously at a fixed potential (C_{bit6} and C_{GND}) as well as all parasitic capacitors to ground or reference will keep the comparator input voltage V_c close to $V_{CM} = 0$ V. Additionally, the sample switches are huge and come with significant overhead such as level shifters and gate drivers. It is therefore helpful to limit the sampling process to the largest capacitors.

To avoid gain errors, the sum of sample capacitance has to be identical to the amount of capacitance switching between the reference potentials in the course of the conversion. An additional sample capacitor C_S is added to make up for all lower bit capacitors starting from C_{bit6}, which are not used for sampling. C_S therefore has to be identical to C_{bit5}. The capacitors C_S and C_{GND} will reduce the internal signal amplitude, but C_S is only 3.125 % of the total capacitance. The size of the two capacitors together might be 5 % of the total capacitance and will therefore influence the SNR by less than 0.5 dB.

*CDAC*2, also called the coarse CDAC, is used for the MSB decisions and divides the signal level internally with the capacitor C_{GND}. If the internal signal amplitude should be divided by 4, then C_{GND} needs to be three times the size of the sampling capacitance or in this case six times the MSB capacitance. The code transitions generated by *CDAC*2 have to remain at the same input voltages as generated by *CDAC*1, so that the sampling has to be done on twice the MSB capacitance. Therefore, the additional sample capacitor C_s is added, which here needs to have the size of C_{bit4}.

For the MSB decision, the two CDACs switch into the state as shown in Fig. 3.12. The process starts with opening the hold switches sw_{hold} of *CDAC*1 and *CDAC*2 in parallel. *CDAC*1 will place the sampling capacitors into the floating mode, while *CDAC*2 connects the MSB capacitor C_{bit1} to the positive reference voltage and all other capacitors including C_S to the negative reference potential for the MSB conversion.

The total sampled voltage Q_{S2} on *CDAC*2 can be calculated to

$$Q_{S2} = 2 \cdot C_{bit1} \cdot V_{in} \tag{3.13}$$

The charge Q_{bit1} after switching into the MSB decision mode is

$$Q_{bit1} = C_{bit1} \cdot (REF - V_c) + (C_S + C_{bit2} + C_{bit3} + C_{bit4} + C_{GND}) \cdot (0\ \text{V} - V_c) \tag{3.14}$$

Equation (3.14) can be modified taking into consideration that the extra sample capacitor C_S together with all the LSB capacitors of *CDAC*2 starting with the bit 2 capacitor (C_{bit2} through C_{bit4}) adds up to the MSB capacitance C_{bit1}. Also C_{GND} is six times the MSB capacitance C_{bit1}.

$$Q_{bit1} = C_{bit1} \cdot (REF - V_c) - 7 \cdot C_{bit1} \cdot V_c = C_{bit1}(REF - 8 \cdot V_c) \tag{3.15}$$

Now the calculation can continue in the same fashion as done for the floating CDAC algorithm by considering that the charge is frozen and the sampled charge Q_{S2} is identical to charge during the MSB decision Q_{bit1}.

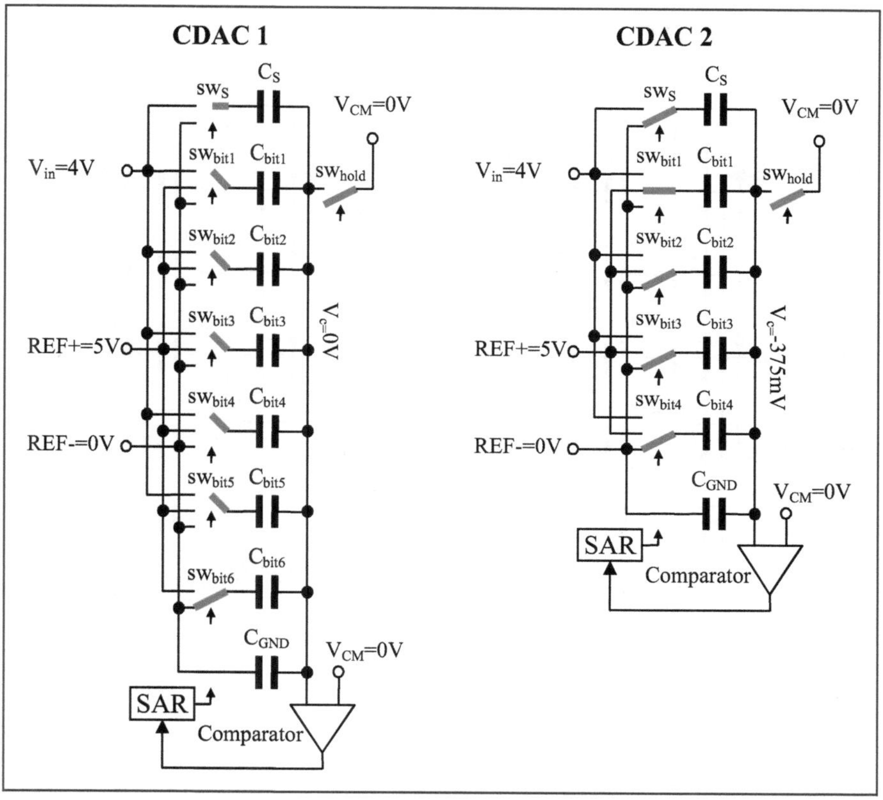

Fig. 3.12 SAR ADC with dual CDAC architecture during the MSB (bit 1) decision

$$2 \cdot C_{bit1} \cdot V_{in} = C_{bit1}(REF - 8 \cdot V_c)$$

$$V_c = \frac{\frac{REF}{2} - V_{in}}{4} < 0 \tag{3.16}$$

Equation (3.16) shows that input voltage is again compared to half the reference voltage. It further shows that the signal amplitude at the comparator input is divided by a factor of 4. The difference of half the reference voltage *REF* and the input voltage V_{in} is in the range of $\pm\frac{REF}{2}$. As this voltage range is divided by the factor of 4, the comparator input voltage remains for a 5 V reference within ±0.625 V.

For the particular example with $REF = 5$ V and $V_{in} = 4$ V, the MSB decision in respect to Eq. (3.16) is 1. The MSB capacitor of *CDAC*2 remains its connection to the reference voltage and the bit 2 capacitor is switched from ground to the reference voltage. This state is illustrated in Fig. 3.13. The new charge redistribution for the bit 2 decision Q_{bit2} is expressed in Eq. (3.17).

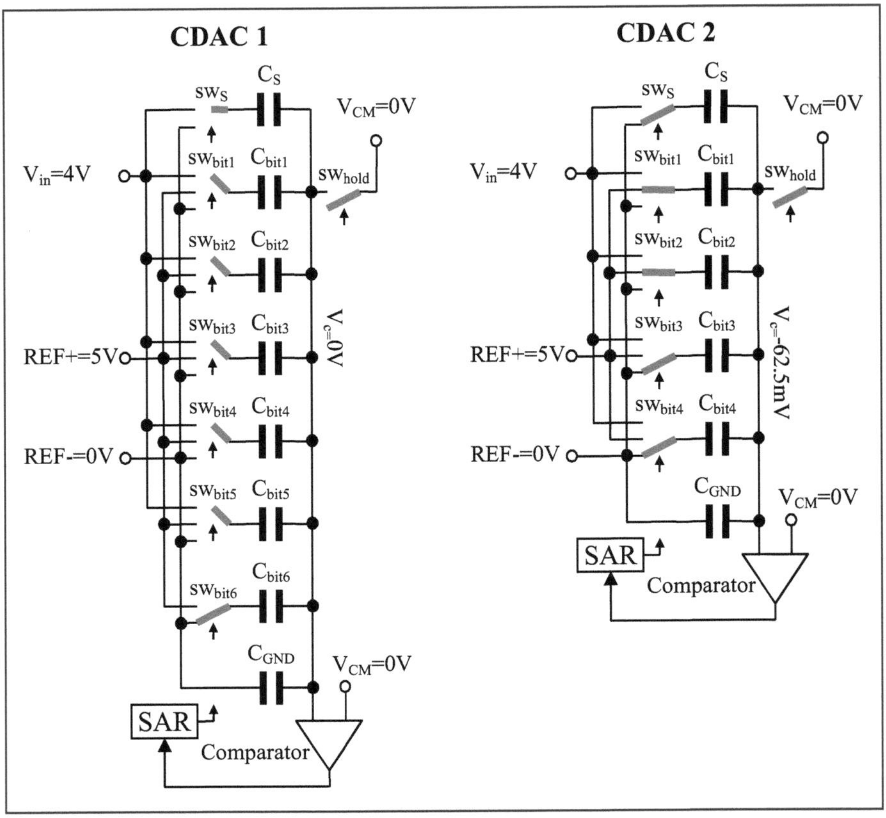

Fig. 3.13 SAR ADC with dual CDAC architecture during the MSB-1 (bit 2) decision

$$Q_{bit2} = (C_{bit1} + C_{bit2}) \cdot (REF - V_c) + (C_S + C_{bit3} + C_{bit4} + C_{GND}) \cdot (0\ \text{V} - V_c) \tag{3.17}$$

C_s, C_{bit3} and C_{bit4} sum up to C_{bit2} and C_{GND} equals 12 times C_{bit2}, so that

$$Q_{bit2} = 3 \cdot C_{bit2} \cdot (REF - V_c) - 13 \cdot C_{bit2} \cdot V_c = C_{bit2} \cdot (3 \cdot REF - 16 \cdot V_c)$$

With Q_{bit2} being equal to Q_{S2}, the comparator input voltage settles to

$$4 \cdot C_{bit2} \cdot V_{in} = C_{bit2} \cdot (3 \cdot REF - 16 \cdot V_c)$$
$$V_c = \frac{\frac{3}{4} \cdot REF - V_{in}}{4} < 0\ \text{V} \tag{3.18}$$

Equation (3.18) is true in the example with $REF = 5$ V and $V_{in} = 4$ V, so that the bit 2 is set to 1.

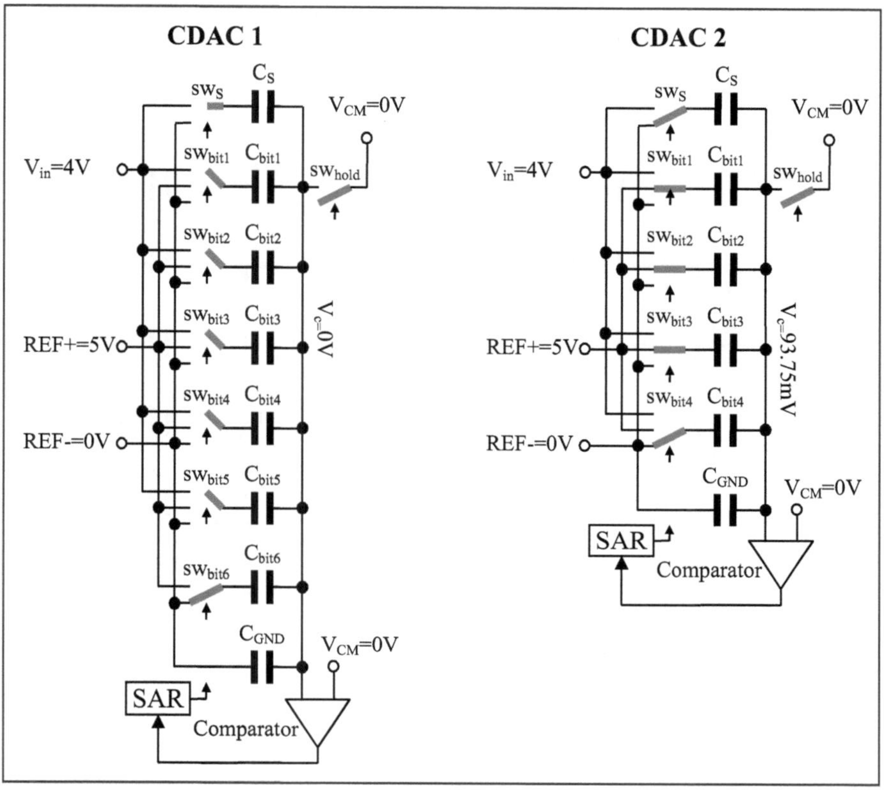

Fig. 3.14 SAR ADC with dual CDAC architecture during the MSB-2 (bit 3) decision

For the bit 3 decision, $CDAC2$ therefore remains C_{bit2} at the reference and additionally switches C_{bit3} to the reference. Even $CDAC1$ could now be activated as the comparator input voltage during the bit 3 decision would remain within the voltage range of $\pm\frac{REF}{8} = \pm 0.625$ V even with neglecting the above mentioned signal reduction of 5 %. However, to allow larger spikes during the activation process at the comparator input V_c, the activation is delayed to the bit 4 decision. The new CDAC states are shown in Fig. 3.14. The calculation process starts again with looking at the new charge redistribution Q_{bit3}.

$$Q_{bit3} = (C_{bit1} + C_{bit2} + C_{bit3}) \cdot (REF - V_c) + (C_S + C_{bit4} + C_{GND}) \cdot (0\ \text{V} - V_c)$$

$$\begin{aligned} Q_{bit3} &= 7 \cdot C_{bit3} \cdot (REF - V_c) - 25 \cdot C_{bit3} \cdot V_c \\ &= 7 \cdot C_{bit3} \cdot REF - 32 \cdot C_{bit3} \cdot V_c \end{aligned} \tag{3.19}$$

Comparing Q_{bit3} with Q_{S2} then results in

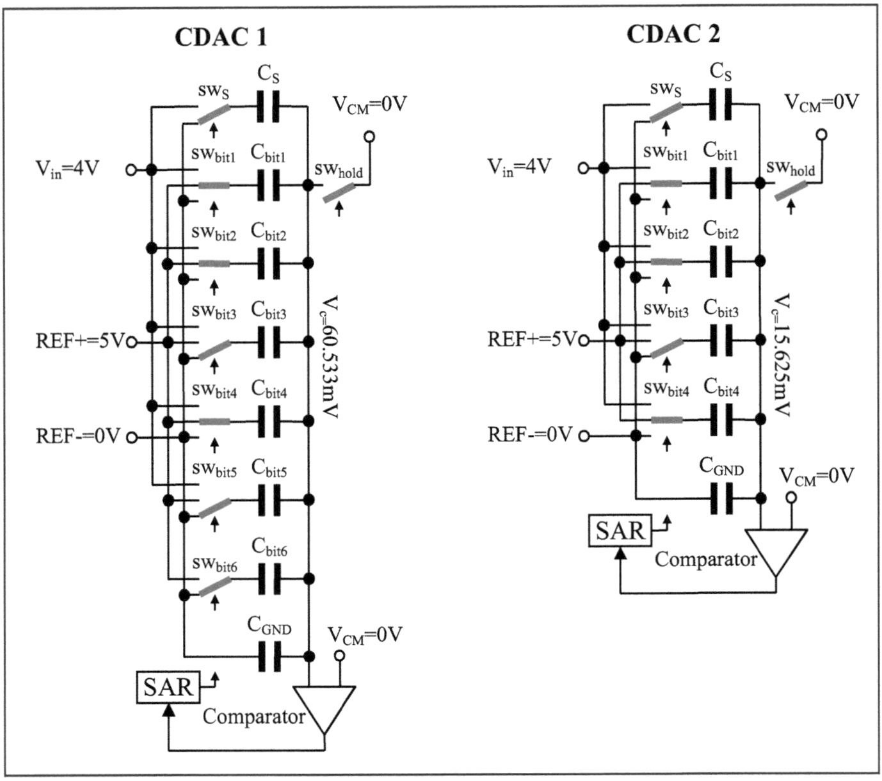

Fig. 3.15 SAR ADC with dual CDAC architecture during the MSB-3 (bit 4) decision

$$8 \cdot C_{bit3} \cdot V_{in} = 7 \cdot C_{bit3} \cdot REF - 32 \cdot C_{bit3} \cdot V_c$$
$$V_c = \frac{\frac{7}{8} \cdot REF - V_{in}}{4} < 0 \text{ V} \tag{3.20}$$

Equation (3.20) is not true, so that the MSB-2 (bit 3) is set to 0. In this respect, C_{bit3} in $CDAC2$ has to be reset to 0 V, while C_{bit4} is connected to the reference for the bit 4 decision (see Fig. 3.15).

Now it is time to activate $CDAC1$. However, the activation process is performed with high impedance switches. This attenuates the spikes at V_c of $CDAC1$ during the activation process, as C_{bit6}, C_{GND} and further parasitic capacitors are connected with significantly lower impedance to ground.

As described above, C_S in $CDAC1$ equals C_{bit5}, so that the total sampling capacitance also adds up to twice the MSB capacitance C_{bit1}. The sampled charge on $CDAC1$ Q_{S1} can therefore be calculated to

$$Q_{S1} = 2 \cdot C_{bit1} \cdot V_{in} = 16 \cdot C_{bit4} \cdot V_{in} \tag{3.21}$$

The charge distribution $Q_{bit4,1}$ after the activation period of $CDAC1$ is summarized in Eq. (3.22).

$$\begin{aligned} Q_{bit4,1} &= (C_{bit1} + C_{bit2} + C_{bit4}) \cdot (REF - V_c) \\ &\quad + (C_S + C_{bit3} + C_{bit5} + C_{bit6} + C_{GND}) \cdot (0\ \text{V} - V_c) \\ Q_{bit4,1} &= 13 \cdot C_{bit4} \cdot (REF - V_c) - (3 \cdot C_{bit4} + C_{bit6} + C_{GND}) \cdot V_c \\ Q_{bit4,1} &= 13 \cdot C_{bit4} \cdot REF - (16 \cdot C_{bit4} + C_{bit6} + C_{GND}) \cdot V_c \end{aligned} \tag{3.22}$$

Again, the sampled charge equals the redistributed charge, so that the comparator input voltage of $CDAC1$ after the activation process settles to

$$\begin{aligned} 16 \cdot C_{bit4} \cdot V_{in} &= 13 \cdot C_{bit4} \cdot REF - (16 \cdot C_{bit4} + C_{bit6} + C_{GND}) \cdot V_c \\ V_c &= \frac{\left(\frac{3}{4} + \frac{1}{16}\right) \cdot REF - V_{in}}{\left(1 + \frac{C_{bit6} + C_{GND}}{16 \cdot C_{bit4}}\right)} = \frac{\left(\frac{3}{4} + \frac{1}{16}\right) \cdot REF - V_{in}}{\frac{C_{CDAC1,tot}}{C_{CDAC1,samp}}} < 0\ \text{V} \end{aligned} \tag{3.23}$$

Equation (3.23) proves two statements from above. First, the internal signal level is divided by the ratio of the total $CDAC1$ capacitance by the $CDAC1$ capacitance that was used for sampling. This remains the internal signal amplitude nearly undivided, as nearly all the $CDAC1$ capacitors (beside C_{bit6} and C_{GND}) were utilized for the sampling. Second, the bit 4 decision of $CDAC1$ would be identical to the decision done by $CDAC2$, which is calculated accordingly to the previous bit decisions with the result shown in Eq. (3.24).

$$V_c = \frac{\left(\frac{3}{4} + \frac{1}{16}\right) \cdot REF - V_{in}}{4} < 0\ \text{V} \tag{3.24}$$

Equation (3.24) is not true for this example, so that bit 4 is set to 0. Now, $CDAC2$ is done with all decisions. It would generally be possible for $CDAC2$ to start the sampling process again. However, it might be advantageous to keep it in the current state until $CDAC1$ finished the conversion process. The switching process, of $CDAC2$ might add some additional distortion onto the reference potentials. Additionally, if the bandwidth of the sampling paths of $CDAC1$ and $CDAC2$ are matching and the ADC external input drivers are too week for correct settling, then the sample capacitors of $CDAC1$ and $CDAC2$ are still charged to identical voltages, which prevents additional errors. If $CDAC2$ has its own comparator, it could now be paced into a power saving mode.

$CDAC1$ will now be used for the further conversion process. Based on the bit decision according to Eq. (3.24), the bit 4 capacitor C_{bit4} is reset to ground and C_{bit5} switches to the reference voltage as shown in Fig. 3.16. The switches of the previously activated floating capacitors are now supported by additional low impedance switches. $CDAC1$ is now in the mode as a normal SAR ADC would

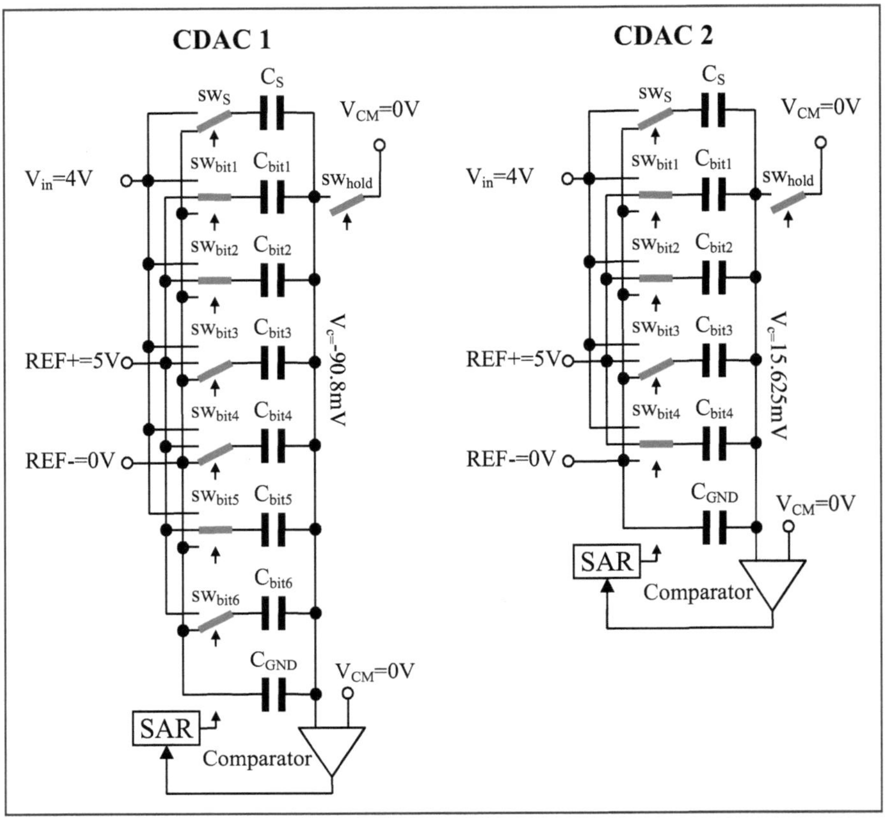

Fig. 3.16 SAR ADC with dual CDAC architecture during the MSB-4 (bit 5) decision

be and continues the conversion as such. Equation (3.25) shows the charge distribution for the bit 5 decision Q_{bit5} accordingly.

$$\begin{aligned} Q_{bit5,1} &= (C_{bit1} + C_{bit2} + C_{bit5}) \cdot (REF - V_c) \\ &\quad - (C_S + C_{bit3} + C_{bit4} + C_{bit6} + C_{GND}) \cdot V_c \end{aligned} \tag{3.25}$$

Equation (3.25) together with Eq. (3.21) will result in

$$V_c = \frac{\left(\frac{3}{4} + \frac{1}{32}\right) \cdot REF - V_{in}}{\frac{C_{CDAC1,tot}}{C_{CDAC1,samp}}} < 0 \text{ V} \tag{3.26}$$

The condition described in Eq. (3.26) is true for $REF = 5$ V and $V_{in} = 4$ V, C_{bit5} remains connected to the reference and C_{bit6} connects to the same.

The high voltage test chip described in Sect. 3.1 was adjusted to the dual CDAC architecture with the result that errors got significantly smaller. This actually allowed the usage of the dynamic error correction described in Sect. 2.4 to correct

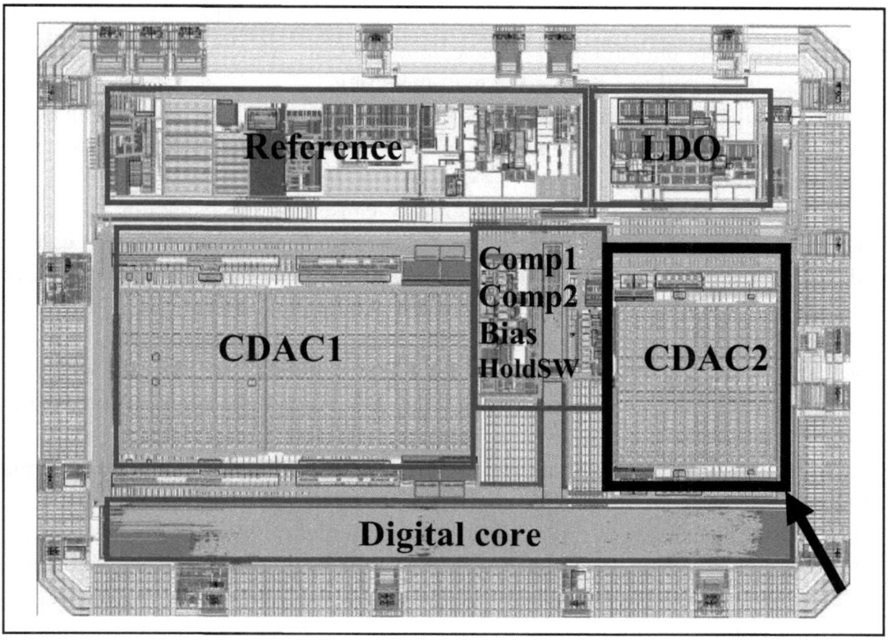

Fig. 3.17 Layout example of an SAR ADC with dual CDAC architecture

for the errors. More bits could be evaluated with *CDAC*2, which reduced the residual comparator input voltage to allow larger spikes. Additionally, the higher *CDAC*2 resolution enabled the soft turn on with high impedance switches, which again reduced the spikes. The dual CDAC architecture turned out to be a very robust solution, with the drawback of consuming more die area. An example is shown in Fig. 3.17. *CDAC*2 consumes 10 % of the total die size and roughly 25 % of the ADC core.

The size of the *CDAC*2 capacitor is dependent on the tolerable errors and the amount of available error correction. The smaller the capacitors are, the larger is the mismatch between *CDAC*1 and *CDAC*2 and the more error correction is required in *CDAC*1. Mismatch between the CDACs is caused by:

- Gain error
- Offset error and the
- Integral non-linearity

The gain error of a state-of-the-art SAR ADC as described in [4] is in the range of ±0.25–0.5 % of the full-scale range (FSR) and is mainly caused by the DNL trimming process. The offset error is typically in the range of ±1–2 mV. However, *CDAC*2 works with reduced signal amplitude, so that the FSR is reduce by the attenuation, which in the above example was 4. If the ADC's FSR is 5 V, then the *CDAC*2 internal FSR is only 1.25 V. The offset error can therefore also be expressed in ±0.08–0.16 % FSR. The integral non-linearity is typically less than

3 LSB for a 16 bit ADC and can therefore be neglected. The largest mismatch source is therefore the gain error. The scale-down trim, which is presented in Sect. 2.2.2, was developed to improve the gain matching.

For a dual CDAC ADC with state-of-the-art trimming, it is recommended to have an error correction range around ±1 % FSR. This would require a dynamic error correction step based on Sect. 2.4 after bit 6 or bit 7.

The size of *CDAC*2 in Fig. 3.17 is mostly caused by the capacitor C_{GND}. An alternative approach is explained in the following. The signal amplitude can be reduced by sampling only on a fraction of the sample capacitance [5, 6]. To achieve the factor of four, it would be sufficient to sample the input voltage only on C_{bit2} of *CDAC*2. This also requires the scaling of the reference voltage by the same factor. If for example the 5 V reference is used for *CDAC*1, then *CDAC*2 would run on a 1.25 V reference. As an additional advantage, low voltage switches can be used in *CDAC*2 without the need of level shifter.

Unfortunately, additional circuitry is required that divides and buffers the reference voltage with an amplifier before coupling the reference to the capacitors. If high accuracy resistors are available, then the divider ratio can be 0.1 % accurate. The amplifier however might have a ±5 mV offset. Summarized, the additional circuitry would add a gain error of up to ±0.5 %, which then would require additional error correction.

The dual CDAC architecture comes with some additional advantages. The capacitors of *CDAC*2 can be significantly smaller than the ones of *CDAC*1. Consequently, they lower the load for the reference voltage. The smaller capacitors also allow a fast settling of reference voltage during the bit decisions on *CDAC*2. As an increased noise level is not an issue during these decisions, a coarse comparator can be used in combination with *CDAC*2, which has smaller input transistors and less gain stages. The smaller transistors reduce the parasitic capacitors and increase the comparator bandwidth. The reduced amount of gain stages lower the throughput time of the comparator. Both help to accelerate the MSB decisions. It would surely be possible to share one comparator between the two CDACs, but the both test chips, which are introduced in Sects. 3.7 and 3.8 use fine and coarse comparators for *CDAC*1 and *CDAC*2 due to the above compelling reasons.

The gain error is the largest source of mismatch caused by the state-of-the-art trim scheme. It was minimized with the scale-down trim proposed in Sect. 2.2.2. Such a scale-down trim solution was successfully implemented in the high-voltage dual CDAC test chip. The result shows up in the matching between *CDAC*1 and *CDAC*2, which can be measured by running an all-code linearity test with turning the error correction off. This way, the mismatch errors between the CDACs cannot be corrected and show up in the ADCs transfer function.

Figure 3.18 visualizes that the offset error can be neglected. The error of the major transitions is increasing towards the full scale of the ADC, which is caused by the gain mismatch between the two CDACs. It is incredible that this error remains around 40 LSB on an 18 bit level, or in other words less than 0.01 % of the full scale range. Based on this information, an error correction range of 128 LSB range would be sufficient. The error correction could be moved from bit 6 or bit7 down to bit

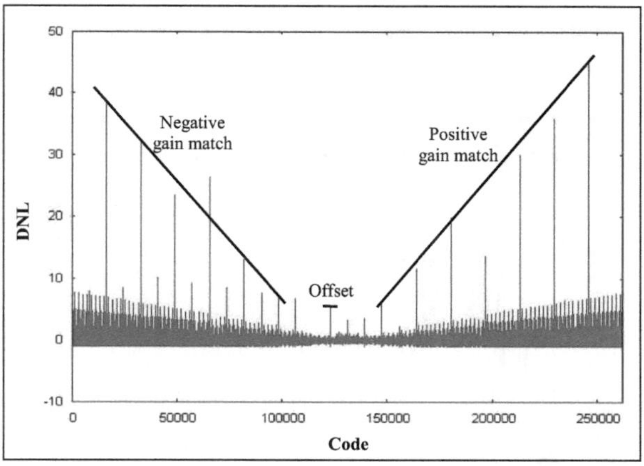

Fig. 3.18 Typical DNL of a SAR ADC with dual CDAC architecture with DEC being turned off

12 or bit 13. This means that the error correction would shift in the LSB DAC and therefore does not require trim.

As a side effect the scale-down trim will reduce the required amount of trim capacitors for the MSB CDAC and reduces the overhead for the trim circuitry. Note that the scale-down trim minimizes the bit trim and therefore the gain error. The matching of the two CDACs would further be improved, if the MSB capacitors will not get trimmed in the CDAC at all.

3.3 Digital Trim [7, 8]

Is it possible to get rid of the adjustment of capacitors at all? This could not only further improve the gain matching of the CDACs, but would also reduce the die size, the circuit complexity as well as the load on the reference source. Figure 3.19 illustrates the size of the analog trim circuitry for a 16 bit high voltage SAR ADC. The analog trim portion covers 10 % of the total die size.

The idea here is that all errors could be corrected digitally, if the result of the ADC is deterministic, which is the case if the comparator input voltage converges to less than an LSB at the end of the conversion and if the errors that are done during the conversion are known. The convergence can be forced with the dynamic error correction even if the capacitors do not match. Similar to the state-of-the-art ADCs, the mismatch of the bit capacitors are measured during production and stored in memory. The values in the memory are not used as previously to control the switching of trim capacitors, but are used in the digital domain to correct the errors with a digital state machine.

The calculation is actually pretty simple. If a certain bit is decided to be 1 during the conversion process, then its bit capacitor gets switched from ground to reference and remains there. The capacitor mismatch causes an error at the comparator input voltage in the conversion process. If the bit would have been 0, then the capacitor

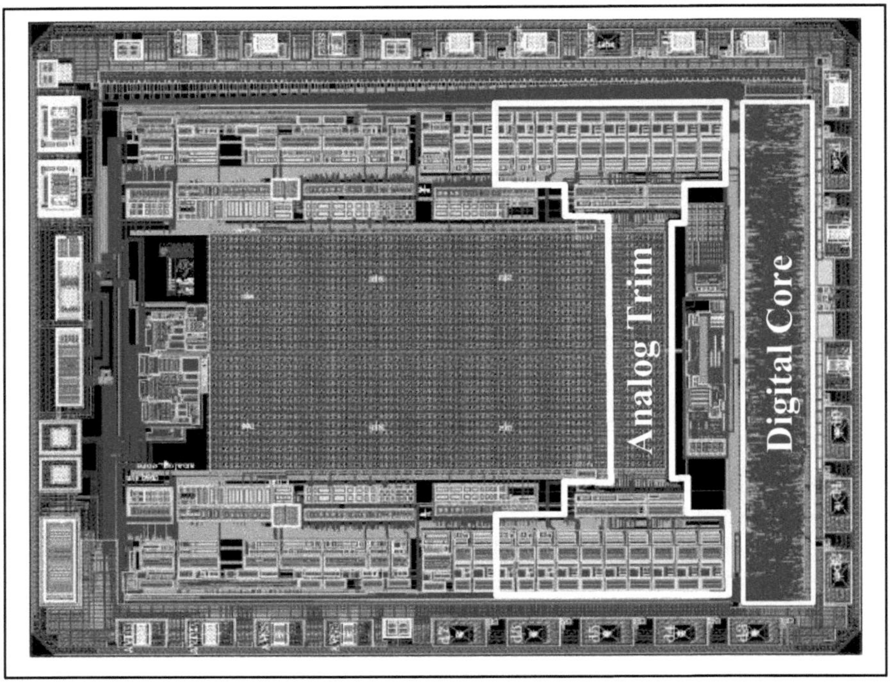

Fig. 3.19 High Voltage SAR ADC with analog trim consuming 10 % of the die area

would switch back to ground and has no further influence on the comparator input voltage. Consequently, it is not adding an error to the conversion. For every bit that is converted to 1 and that requires trimming (e.g. MSB capacitors), the previously measured and stored error is read from the memory and is added to a total calculated error. The algorithm is shown in Fig. 3.20.

The multiplexer *MUX*1 is selecting the memory with the trim information of the bit, which is currently evaluated, and couples the trim information to the multiplexer *MUX*2. If the comparator decision is 1, then the trim value or in other words error of the current bit decision is added by the adder *ADD*1 to the sum of errors done by the previous bit decisions, which is stored in the register *REG*. The digital trim algorithm can also emulate the trim of the scale-down capacitor, which here is implemented by *MUX*4.

If a scale-down capacitor does not match, then all MSBs share initially the same error (for example the respective code being 2 LSB too long). The state-of-the-art trim algorithm would now correct the least significant bit above the scale-down capacitor (for example bit 6) and decrease the bit capacitor by the 2 LSBs. This now lowers the sum of the bit 6 capacitor together with the LSB capacitors to be 4 LSBs too small. Therefore, the next bit capacitor to be trimmed (in this example bit 5) has to be adjusted by 4 LSBs, the next capacitor by 8 LSB and so on. This factor of 2 is implemented in the multiplexer *MUX*4 by shift operations. Note that the analog scale down trim is the more elegant solution, as it avoids the large gain error.

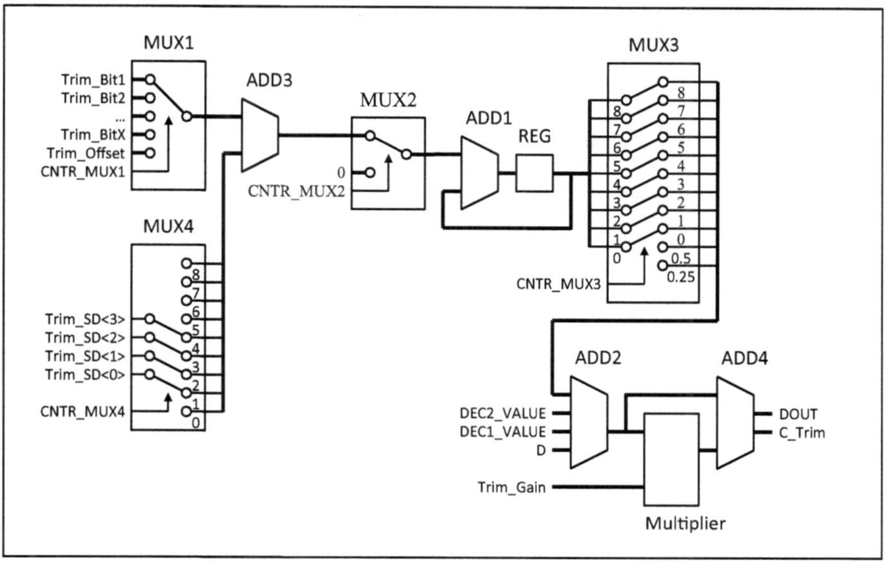

Fig. 3.20 Block diagram of an exemplary digital trim correction

The error register *REG* is now followed by another multiplexer, which again is shifting the register word. It is used to adjust the trim resolution. A good matching CDAC only requires a small amount of trim range. Therefore, the memory space could be used to achieve a higher resolution. For the case that the CDAC is highly unmatched and a large trim range is required, the register result can be shifted to larger ranges with less resolution. In other words, a 6 bit wide digital error correction value can express a trim range of 63 LSB with a trim resolution of 1 LSB or a trim range of 15.75 LSB with a resolution of 0.25 LSB.

The total error *TRIM_VALUE* after finishing all bit decisions that require trimming is added in the next step by *ADD*2 to an eventual correction value of a dynamic error correction. Here a dual dynamic error correction is implemented with the correction values *DEC*2*_VALUE* and *DEC*1*_VALUE*. After the conversion is finished, the total correction value needs to be added to the ADC result *D*.

If the gain trim is also implemented, then the *MULTIPLIER* multiplies the gain error with the result from *ADD*2. If for example the gain error is 1.2 %, then the multiplier would have to multiply the *ADD*2 result with the factor 1.012. The required digital gate count might be lower, if the *ADD*2 result is only multiplied with 0.012 and then added to its original value *D*1 as implemented with *ADD*4 in Fig. 3.20. Care must be taken that no overflow or underflow occurs. Also an offset trim can easily be implemented by adding a fixed value (*TRIM_OFFSET*) to the error word.

Unfortunately, calculations in the digital domain with a dedicated word width can only correct in 1 LSB steps, while the previous analog trim could add or subtract capacitors with the size of half or quarter LSBs. This can be achieved

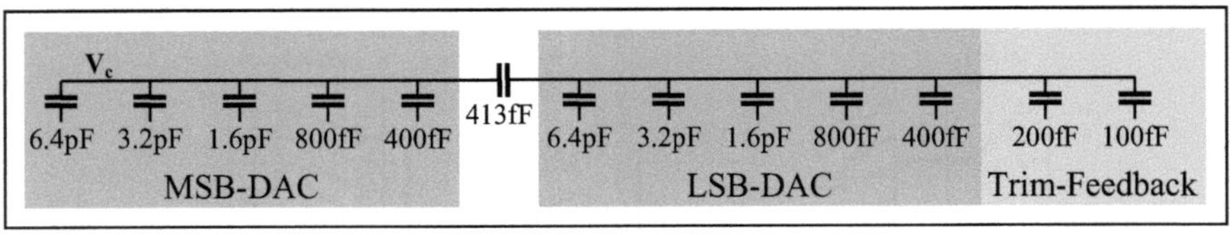

Fig. 3.21 CDAC extended with half and quarter LSB capacitors for digital trimming

with a feedback from the digital trim circuitry into the CDAC (*C_TRIM*). For this reason, two capacitors with the size of half and quarter LSB are added into the CDAC as shown in Fig. 3.21.

The digital trim algorithm will internally calculate with two additional digits. If the arithmetic unit for example evaluates a correction value of 17.75 LSB, then the comparator input voltage is increased with the two additional capacitors by 0.75 LSB and 17 LSB will later be added digitally. The switching of the fractional bits should occur after all calculations are finished to avoid multiple switching of the capacitors. The switching however has to occur prior to the last dynamic error correction step as the ADC would not converge otherwise.

If for an additional example a negative error of −12.75 LSBs has to be corrected, then either the quarter LSB capacitor is switched to the reference to round the error up to −13 LSB, which then gets later corrected digitally. If the CDAC is implemented fully differentially, then the same additional fractional capacitors can be added on the negative CDAC side. In this case, switching both fractional capacitors on the negative side would shift the comparator input voltage by −0.75 LSB and −12 LSB are corrected later in the digital domain. The second approach would slightly simplify the algorithm.

Note that the digital DNL correction will still generate a gain error. It is therefore advisable to combine the analog scale-down trim together with digital trim algorithm. This will also reduce the required trim range and therefore reduces the amount of memory and digital logic. This gain error is induced by dynamic error correction step, which is done after the MSB conversions. In other words, this gain error does not affect the matching of the MSBs between *CDAC*1 and *CDAC*2 in a dual CDAC approach and therefore further improves their matching. The improved matching will further reduce the required amount of dynamic error correction.

The digital trim can correct for all parameters, which don't vary over input parameters such as the reference voltage. This is for example true for all errors (e.g. DNL and gain) caused by the capacitor matching. Offset errors are easy to correct for, but they are caused by charge injection, which is dependent on the supply voltage. If the charge injection remains constant, while the reference voltage and therefore the size of an LSB is reduced, then the offset is changing in terms of LSBs with the reference voltage. The digital trimming of the offset will therefore not cancel the drift of the offset over temperature, supply or reference voltage.

Also the integral non-linearity might be dependent on the reference voltage. A correction scheme is discussed in the next chapter.

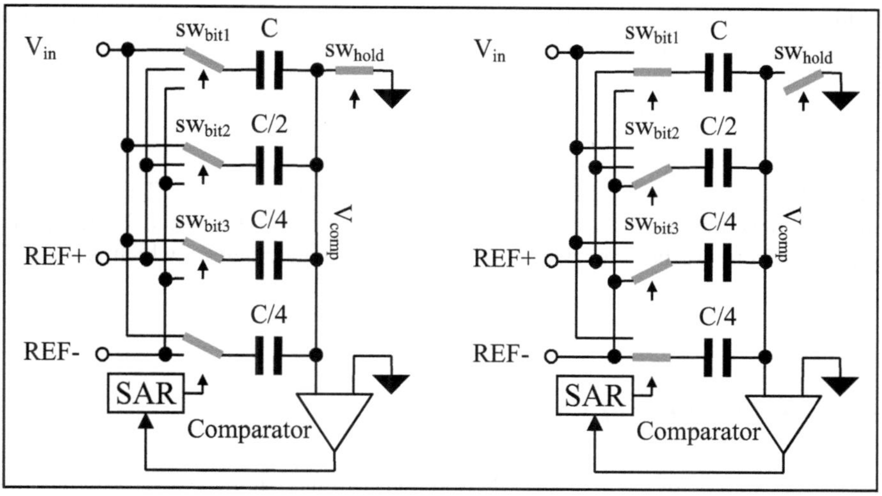

Fig. 3.22 SAR ADC during sampling and during the decision of the MSB used for INL calculation

3.4 INL Trim [9]

One of the main sources for INL of modern high-performance ADCs is the voltage coefficient of the sample capacitors, where the capacitance changes with the voltage across it. The capacitor value therefore differs during sampling and conversion. Most critical is actually the quadratic voltage coefficient, when the capacitors are used in high resolution (+16 bit) and high voltage (±10 V) SAR ADCs. Therefore, a novel INL trim circuitry was invented, which is implemented in the 20 bit test chip presented in Sect. 3.7.

The switching scheme of the CDAC, which is used as the base for the following calculations, is shown in Fig. 3.22. The input voltage can range from –*REF* to +*REF* and can be expressed by $V_{in} = a \cdot REF$. The factor '*a*' represents the digital output code of the ADC. Most SAR ADCs use the ground potential as the negative reference voltage $-REF$, so that $a \in [0, 1]$. Other schemes are possible, where the input voltage for example is only sampled on a fraction of the total capacitance. One possible reason is the conversion of a ±10 V input range by using a 2.5 V reference voltage [5, 6].

There might be several error sources for the INL, but all of them are a function of the digital output code '*a*'. The previously discussed charge injection of the hold switch is one source, but can be minimized with a differential approach and with a proper layout. The remaining error source, which is most critical, is caused by the voltage coefficient of the capacitors. The capacitance *C* is changing with the voltage V_{cap} across it. The linear voltage coefficient b_l causes the capacitor to change linear with the voltage across it. The quadratic voltage coefficient b_q expresses the quadratic change of the capacitor in respect to the voltage. Higher order coefficients can typically be neglected, so that

$$C = C_{nom} \cdot \left(1 + b_l \cdot V_{cap} + b_q \cdot V_{cap}^2\right) \tag{3.27}$$

Due to the voltage coefficients, there is an error in sampled charge. The goal is the elimination of the error caused by the capacitors voltage coefficients. To design an adequate circuitry, the error needs to be quantified in calculation, which is done in the next section.

3.4.1 Calculation of the INL

An initial example of an existing ADC should illustrate that the voltage coefficients are indeed an issue. The particular ADC, which provides a ±10 V input voltage range, does normally sample on a quarter of the total capacitance and then convert the signal using a 2.5 V reference. The voltage across the capacitors at the end of the conversion is much lower than during sampling, so that the error in charge is not compensated. Especially the quadratic voltage coefficient b_q could generate a sever INL error, which can be calculated the following way. First the charge Q_{samp}, which equals the sampled input voltage V_{in} times the sampling capacitor C_{samp}, is expressed by Eq. (3.28). Here, the linear voltage coefficient is neglected.

$$\begin{aligned} Q_{samp} &= C\c{C}_{samp} \cdot V_{in} = C_{samp}\left(1 + b_q \cdot V_{in}^2\right) \cdot V_{in} \\ &= C_{samp} \cdot V_{in} + C_{samp} \cdot b_q \cdot V_{in}^3 \end{aligned} \tag{3.28}$$

The charge is then distributed to the total sum of CDAC capacitors C_{tot}, which here is expected to be four times C_{samp}, so that the voltage will be divided by a factor of four. High performance processes [10] provide a quadratic voltage coefficient of $b_q = -1\frac{ppm}{V^2}$, so that the error voltage at a 10 V full-scale input voltage FS is

$$V_{err} = \frac{C_{samp}}{C_{tot}} \cdot b_q \cdot FS^3 = -250 \ \mu\text{V} \tag{3.29}$$

This error at the full-scale represents more than 3 LSB, when the ADC resolution is 16 bit. The INL is defined as the deviation of the ADC transfer function from a straight line through the endpoints, which therefore has an amplitude of 1.26 LSB.

$$INL = \frac{C_{samp}}{C_{tot}} \cdot b_q \cdot \left(V_{in}^3 - V_{in} \cdot FS^2\right) \tag{3.30}$$

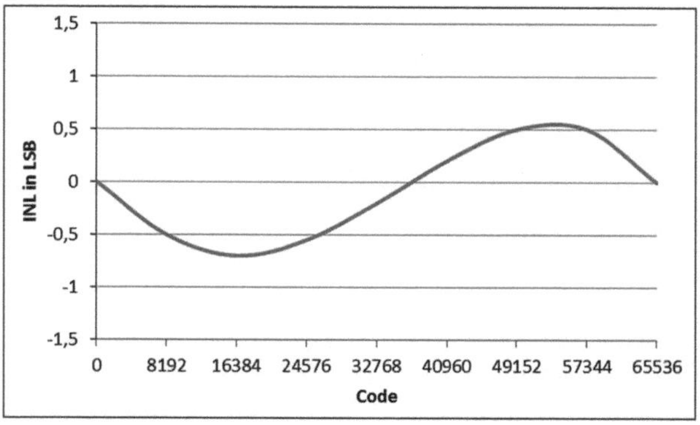

Fig. 3.23 Measured INL of a typical ADC with ±10 V input range [5]

The measured result shown in Fig. 3.23 confirms the expected shape. The amplitude remains around 0.6 LSB, which indicates that the actual voltage coefficient is slightly better than specified.

It can be seen that the error voltage is related to the cube of the input voltage, which can also be expressed as the digital output code '*a*' times the reference voltage *REF*, so that

$$V_{err} \sim REF^3 \cdot a^3 \tag{3.31}$$

If the reference voltage can vary as it is the case for most commercially available standalone ADCs, then a circuitry is required for the error compensation that generates the cube of the reference voltage.

The test chip of Sect. 3.7, which was designed to evaluate the INL compensation circuit, provides a different switching scheme than the example above. To achieve highest resolution (20 bit), the ±10 V input voltage range is sampled on 80 pF capacitors. Additional 10 pF are connected to *–REF*. If the input voltage is described with $V_{inp} = a \cdot REF$ with $a \in [-1, 1]$, then the sampled charge $Q_{samp,p}$ can be calculated to

$$\begin{aligned} Q_{samp,p} &= 80\ \text{pF}\left(V_{inp} - 0\ \text{V}\right) + 10\ \text{pF}(-REF - 0\ \text{V}) \\ &= a \cdot 80\ \text{pF} \cdot REF - 10\ \text{pF} \cdot REF \end{aligned} \tag{3.32}$$

During the conversion, the 10 pF capacitor is switched to the positive reference during the complete conversion. This adds a common-mode voltage if done symmetrically on a positive and a negative CDAC side. It allows the comparator to operate in a valid operating point even if the hold switch connects to ground. The 80 pF capacitors are switched between a positive reference of $+REF = REF = 10$ V and a negative reference potential of $-REF = -(+REF) = -10$ V. In the end of the conversion and at the critical code decision during the conversion, where the differential comparator input is zero, $\frac{1+a}{2}$ amount of the 80 pF are connected to the positive reference and $\frac{1-a}{2}$ to the negative reference, so that the charge distribution $Q_{crit,p}$ can be summarized to

$$Q_{crit,p} = \frac{1+a}{2} \cdot 80 \text{ pF}\left(REF - V_{cp}\right) + \frac{(1-a)}{2} \cdot 80 \text{ pF}\left(-REF - V_{cp}\right) + 10 \text{ pF}\left(REF - V_{cp}\right) \tag{3.33}$$

The charge is frozen, so that $Q_{samp,p} = Q_{crit,p}$. The Eqs. (3.32) and (3.33) are therefore leading to

$$V_{cp,crit} = REF \cdot \frac{20}{90} \tag{3.34}$$

Equation (3.34) describes the common-mode voltage, which is generated with the additional 10 pF capacitor. Note that such a capacitor exists in the positive and the negative CDAC side. They are in identical operating points during sampling as well as during a critical bit decision. They therefore do not influence the INL and can for simplicity be eliminated for the following calculations.

Additionally, the voltage coefficients from Eq. (3.27) should be considered. Equations (3.32) and (3.33) are changed into the Eqs. (3.35) and (3.36).

$$Q^*_{samp,p} = 80\,\text{pF}\left(1 + b_l \cdot a \cdot REF + b_q \cdot a^2 \cdot REF^2\right) \cdot a \cdot REF \tag{3.35}$$

$$\begin{aligned} Q^*_{crit,p} = &\frac{1+a}{2} \cdot 80\,\text{pF} \cdot \left(1 + b_l \cdot \left(REF - V_{cp}\right) + b_q \cdot \left(REF - V_{cp}\right)^2\right) \cdot \left(REF - V_{cp}\right) \\ &+ \frac{1-a}{2} \cdot 80\,\text{pF} \cdot \left(1 + b_l \cdot \left(-REF - V_{cp}\right) + b_q \cdot \left(-REF - V_{cp}\right)^2\right) \\ &\cdot \left(-REF - V_{cp}\right) \end{aligned} \tag{3.36}$$

Again, the charge is frozen and therefore equals each other $Q^*_{samp,p} = Q^*_{crit,p}$, so that

$$\begin{aligned} 0 = &-V_{cp} + b_l \cdot V_{cp}^2 - b_q \cdot V_{cp}^3 \\ &+REF \cdot \left(-2a \cdot b_l \cdot V_{cp} + 3a \cdot b_q \cdot V_{cp}^2\right) \\ &+REF^2 \cdot \left(b_l - b_l \cdot a^2 - 3 \cdot b_q \cdot V_{cp}\right) \\ &+REF^3 \cdot \left(a \cdot b_q - b_q \cdot a^3\right) \end{aligned} \tag{3.37}$$

The ADC is designed fully differentially, so that it has a negative side that needs consideration. The switch positions of the negative side during the conversion are inversed to the positive switches. Terms with '*a*'need to be replace by '–*a*'.

$$\begin{aligned} 0 = &-V_{cn} + b_l \cdot V_{cn}^2 - b_q \cdot V_{cn}^3 \\ &+REF \cdot \left(2a \cdot b_l \cdot V_{cn} - 3a \cdot b_q \cdot V_{cn}^2\right) \\ &+REF^2 \cdot \left(b_l - b_l \cdot a^2 - 3 \cdot b_q \cdot V_{cn}\right) \\ &+REF^3 \cdot \left(-a \cdot b_q + b_q \cdot a^3\right) \end{aligned} \tag{3.38}$$

Equations (3.37) and (3.38) have to be analyzed carefully. All terms without an '*a*' are input voltage independent and will result in an offset error. If this offset is applied to the positive and the negative side ($V_{cn} = V_{cp}$ at the critical decision), then it is cancelled by the fully differential architecture of the CDAC. Further with $V_{cn} = V_{cp} = V_c$, also the terms with a^2 are eliminated. The remaining terms express the following error:

$$Error = REF \cdot a \cdot \left(4 \cdot b_l \cdot V_c - 6 \cdot b_q \cdot V_c^2\right) + REF^3 \cdot \left(-2a \cdot b_q + b_q \cdot 2a^3\right) \tag{3.39}$$

The first term with $REF \cdot a$ is linear to the input voltage and is generating a fix gain error, which can be neglected. With the example of $VC_l = -4$ ppm/V and $VC_{qu} = -1$ ppm/V^2 the gain error would be $4 \cdot VC_l \cdot V_c - 6 \cdot VC_{qu} \cdot V_c^2 = 38.5$ ppm. In the term of Eq. (3.39) with REF^3, the first part is again linear to the input voltage and only causes an insignificant gain error, which is reference dependent. Only the second term with a^3 really causes an INL, which needs compensation. A compensation voltage $V_{INL,comp}$ (see Eq. (3.40) and Fig. 3.24) is required:

$$V_{INL,comp} = -(a \cdot REF)^3 \cdot 2b_q \tag{3.40}$$

During the running conversion, the MSBs can be used to generate the compensation voltage $V_{INL,comp}$, which then is applied to the capacitor array. Adding a voltage during a running conversion will cause that the ADC will not achieve convergence. An error correction step has to follow after applying the INL compensation voltage to the capacitor array. Consequently, the error correction range needs to be larger than 60 LSBs on a 20 bit level and must occur prior to the LSB-6 decision. It would be preferable to minimize the steepness and the maximum of $V_{INL,comp}$. Equation (3.41) provides an alternative compensation voltage. The additional linear term is reducing both amplitude and steepness, which can also be seen in Fig. 3.25.

$$V_{INL,comp} = -REF^3 \cdot 2 \cdot b_q\left(a^3 - a\right) \tag{3.41}$$

The next chapter will explain the INL compensation circuit in detail.

3.4.2 INL Compensation Circuit

Figure 3.26 provides a possible top level block diagram for an INL compensation circuit. A source generates a current, which is proportional to REF^3 and controls the amplitude of the compensation voltage $V_{INL,comp}$. This current is then applied to a digital-to-analog converter (DAC), which is controlled by the bit decisions from the ADC and more particular by its successive approximation register (SAR). The DAC has two stages, the string DAC, which provides supporting points reflecting Eq. (3.41), and a linear interpolating amplifier *AMP* providing another 3 bit resolution [11]. The CDAC provides an additional capacitor C_{INL} on both the positive

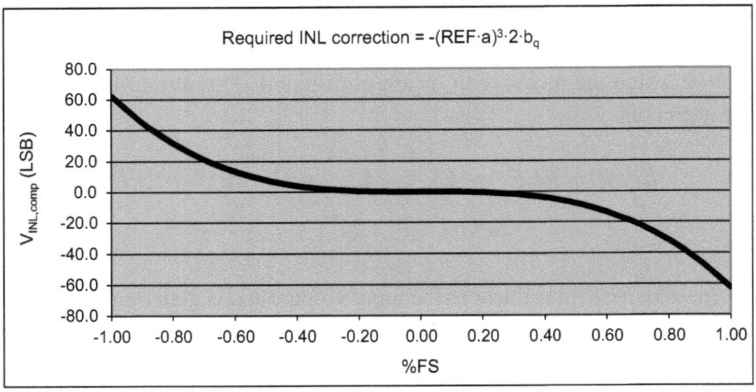

Fig. 3.24 Possible INL compensation voltage $V_{INL,comp}$ for a particular 20 bit SAR ADC

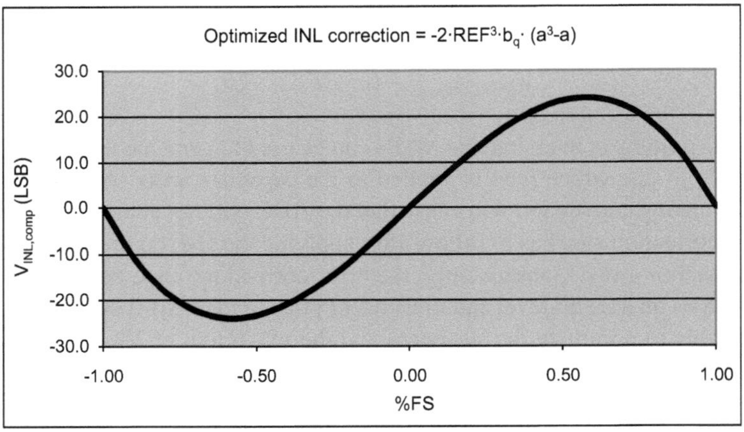

Fig. 3.25 Optimized INL compensation voltage $V_{INL,comp}$ in LSB for a particular 20 bit SAR ADC

and the negative side to compensate for the positive or negative portions of the error voltage $V_{INL,comp}$ dependent on the MSB decision.

3.4.2.1 Current Source

Potential functions can be realized by utilizing bipolar transistors that provide a current, which is exponentially dependent on the base-emitter voltage [12].

$$I_C = I_0 \cdot e^{\frac{V_{BE}}{V_{th}}} \tag{3.42}$$

V_{th} is the thermal voltage ($V_{th} = \frac{kT}{e}$), which is related to the Boltzmann constant k,

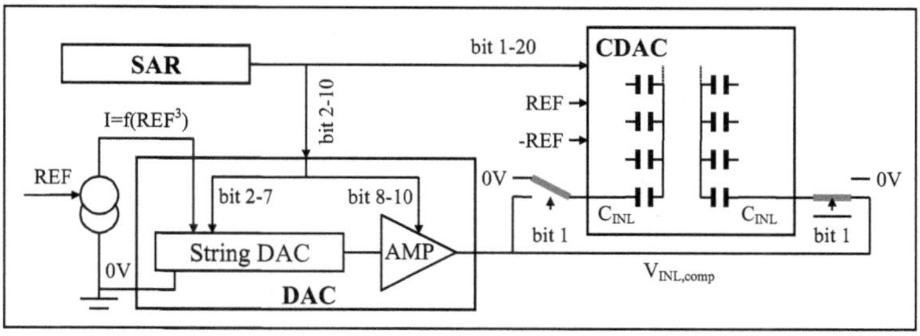

Fig. 3.26 Top-level block diagram of the suggested INL compensation circuit

the absolute temperature T and the unique charge e. The parameter I_0 is dependent on the process, the temperature and the collector-emitter voltage.

The idea is to generate a current I_{REF}, which is proportional to the external reference voltage ($I_{REF} \sim REF$). This current is then compared to an internal reference current $I_{1.25V}$, which is related to the constant internal reference voltage of 1.25 V. Voltage controlled current sources *VCCS* are used for this purpose. The ratio of I_{REF} to $I_{1.25V}$ is expressed by

$$\frac{I_{REF}}{I_{2.5V}} = e^{\left(\frac{V_{BE,REF}}{V_{th}} - \frac{V_{BE,1.25V}}{V_{th}}\right)} \tag{3.43}$$

The cube of this ratio will result in

$$\left(\frac{I_{REF}}{I_{1.25V}}\right)^3 = \left(e^{\left(\frac{V_{BE,REF}}{V_{th}} - \frac{V_{BE,1.25V}}{V_{th}}\right)}\right)^3 = e^{\frac{3 \cdot V_{BE,REF} - 3 \cdot V_{BE,1.25V}}{V_{th}}} \tag{3.44}$$

A similar function is realized with the simplified schematic in Fig. 3.27.

The circuit in Fig. 3.27 can be analyzed with the following Eq. (3.45):

$$V_{BE,Q5} = V_{BE,Q6} + V_{BE,Q4} + V_{BE,Q2} - V_{BE,Q1} - V_{BE,Q3} \tag{3.45}$$

The bipolar transistors Q_2, Q_4 and Q_6 are supplied with I_{REF}, Q_1 and Q_3 with $I_{1.25V}$. Equation (3.45) is therefore equal to

$$V_{BE,Q5} = 3 \cdot V_{BE,REF} - 2 \cdot V_{BE,1.25V}$$

The output current I_{Q5} is therefore

$$I_{Q5} = I_0 \cdot e^{\frac{3 \cdot V_{BE,REF} - 2 \cdot V_{BE,1.25V}}{V_{th}}} = \frac{I_0{}^3 \cdot e^{\frac{3 \cdot V_{BE,REF}}{V_{th}}}}{I_0{}^2 \cdot e^{\frac{2 \cdot V_{BE,1.25V}}{V_{th}}}} = \frac{I_{REF}^3}{I_{1.25V}^2} \sim I_{REF}^3 \tag{3.46}$$

The full schematic is shown in Fig. 3.28. The nominal 10 V reference *REF* is divided with a ratio of 8 resistively (R_1 and R_2). This way, identical voltage

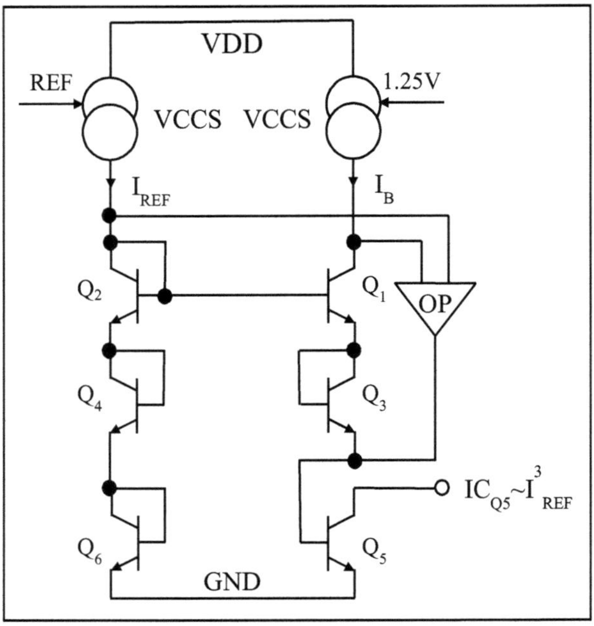

Fig. 3.27 Simplified current source with an output current I_{Q5} being proportional to REF^3

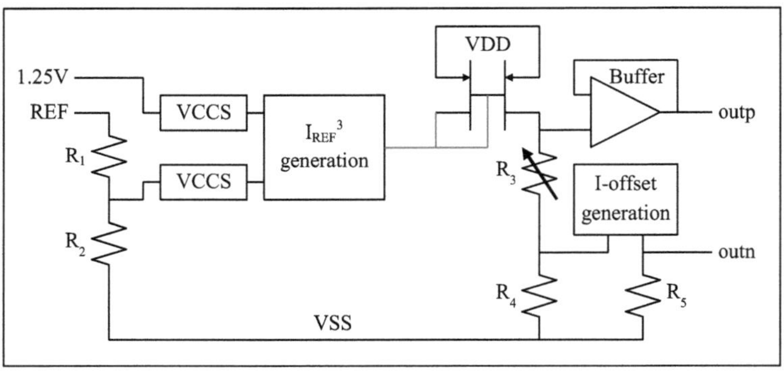

Fig. 3.28 Possible schematic generating an output voltage proportional to REF3

controlled current sources *VCCS*, can be utilized to generate I_{REF} from *REF* and $I_{1.25V}$ from an internal 1.25 V bandgap circuitry. These currents are mirrored into the circuitry proposed in Fig. 3.27. Its output current is proportional to the third power of the reference.

For faster settling, it was decided to buffer the output of this REF^3-stage before it connects to the string DAC. The output current is therefore converted back to a voltage with resistors R_4 and the adjustable R_3, which allows an amplitude trim. The voltage is then buffered with the amplifier *Buffer*. An offset voltage of 240 mV is added by driving DC currents (I-offset generation) through the resistors R_4 and R_5. With the added offset, simple amplifiers can be used, which do not have to drive

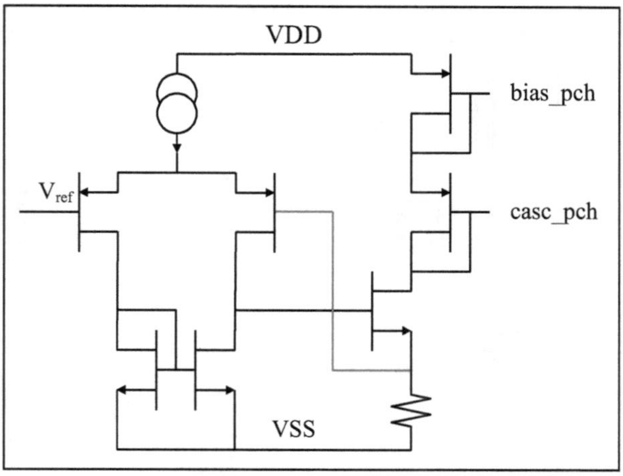

Fig. 3.29 Possible circuit to generate a current being proportional to an input voltage V_{ref}

down to 0 V. The INL compensation capacitor in the CDAC will be switched between *outn* and the shifted compensation voltage from the DAC.

The voltage controlled current source is basically a 2-stage amplifier, which regulates the voltage across a 1.2MΩ resistor to its input voltage, which nominally is around 1.2 V. The working current in this circuit is therefore around 1 μA. Figure 3.29 shows the schematic. The output buffer can basically be identical to the voltage controlled current source. The resistor would be replaced by the string DAC.

3.4.2.2 String DAC

The string DAC (see Chap. 7) has to provide the supporting points of Eq. (3.41). This could have been realized by using respective resistor values. However, this is limiting possible redesigns and reuses, where the shape might need adjustment. Therefore, a linear 7 bit string DAC was used with respective 128 equally sized resistors. Equation (3.41) is implemented with a digital encoder, which has bit 2 through bit 7 from the SAR as inputs. The particular ADC has an error correction step after bit 7, so that most of the voltage applied by the INL compensation should be corrected at this point.

However, the maximum voltage steps of the INL compensation needs to be significantly less than one LSB of the ADC. Unfortunately, this resolution cannot be provided after the first 7 bit decisions of the ADC caused by the steepness of Eq. (3.41). Further resolution is therefore achieved with the interpolating amplifier, which is controlled with bit 8 through bit 10. The influence of the interpolating amplifier can then be compensated by a second error correction stage after bit 13. More information regarding string DACs is provided Sect. 7.3.

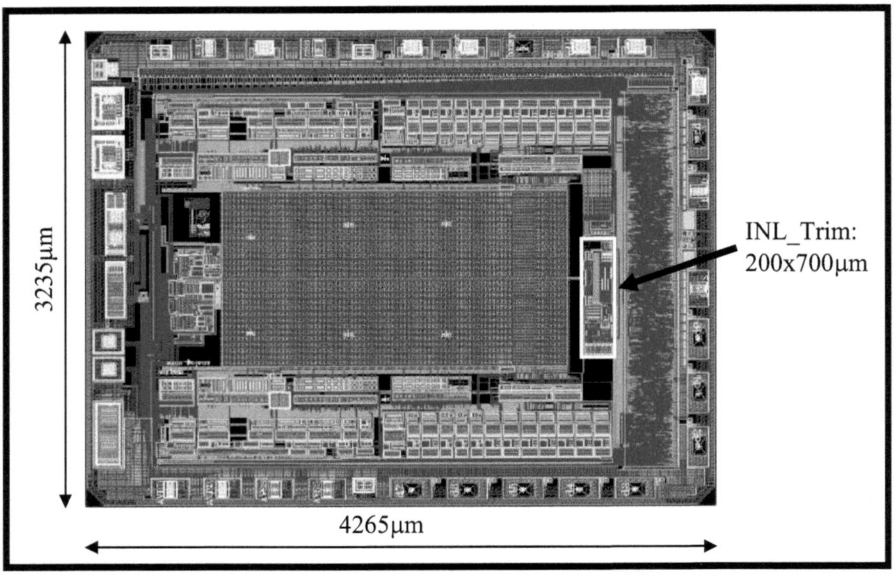

Fig. 3.30 Layout of a test chip including a SAR ADC with INL compensation circuit

3.4.3 Results and Measurements

The INL compensation scheme was realized on a 20 bit test chip. The layout of the complete ADC can be seen in Fig. 3.30. The INL compensation circuit consumes an area of 200 µm x700µm, which equals only around 1 % of the total die area.

Figure 3.25 above shows the theoretically required INL compensation voltage. The shape measured on silicon is plotted in Fig. 3.31. As a result, the graphs match very well.

In a next step, the amplitude of the error compensation curve was verified, which has to be proportional to the cube of the reference voltage *REF*. The result is displayed in Fig. 3.32.

Figure 3.23 already proves that the voltage coefficient can vary and the INL curve might differ in the amplitude from the expected value. The schematic in Fig. 3.28 therefore provides an amplitude trim with the resistor R_3. The ideal value of the amplitude over the trim value was simulated and measured on silicon. Figure 3.33 presents the results.

An essential time of the production is used to trim the high performance SAR ADCs. Particularly the DNL trim drives the production costs up essentially and avoids a wider usage of 16 bit SAR ADCs in applications. The innovation presented in the next chapter therefore concentrates on a 'design for test' solution that reduces the trim time dramatically.

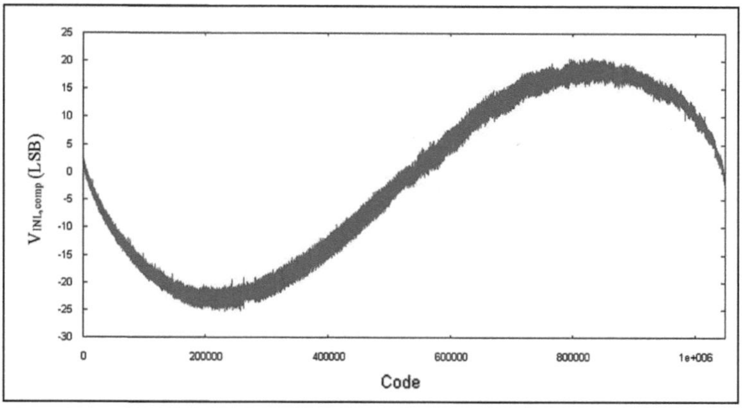

Fig. 3.31 Measured INL compensation voltage V_{INL_comp}

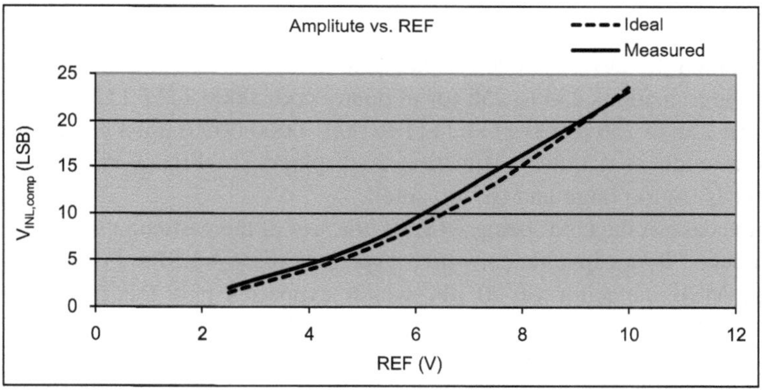

Fig. 3.32 Measured amplitude of the INL compensation voltage over *REF* compared to its ideal value

3.5 Novel Method to Measure Capacitive Matching [13–15]

Section 2.2.2 shows that the differential non-linearity is basically caused by mismatch of the bit capacitors. This mismatch can be fixed in an analog fashion by switching trim capacitors in parallel or anti-parallel with the particular bit capacitor or in a digital way, where the dynamic error correction is used to achieve the convergence of the converter and then corrects the errors in the digital domain (Sect. 3.3). For both cases, the mismatch of the capacitors has to be measured and stored in memory.

This is typically done by measuring the DNL of the particular code. Such a code has a lower and an upper code transition. The input voltage of both transitions needs to be measured. The DNL results from the difference of the input voltages. Details are explained in Sect. 6.2.1. If for example the bit 4 capacitor of a 12 bit ADC is

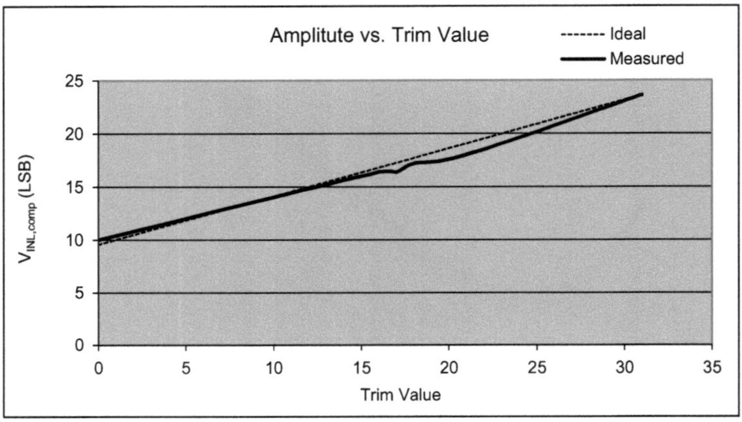

Fig. 3.33 Measured amplitude of the INL compensation over the trim code compared to its ideal value

trimmed, then the DNL of code 255 needs to be determined and the input voltages at the code transitions 254 to 255 (or in binary code 0000 1111 1110 to 0000 1111 1111) and 255 to 256 (0000 1111 1111 to 0001 0000 0000) must be measured. An erroneous code in a transfer function is displayed in Fig. 3.34, where the bit capacitor is (a) too large and (b) too small.

For the case of the DNL being +4 LSB, the size of the particular bit capacitor has to be reduced by a capacitance, which expresses these 4 LSBs. For a 16 bit ADC with an MSB capacitor of 20 pF, which expresses half the full scale range (32768 LSB), the capacitance for one LSB equals the 20 pF divided by 32768 or 0.6 fF. The capacitor, which is 4 LSB long, has to be reduced by 2.4 fF. Unfortunately, the measurement of the DNL is very time consuming. Section 6.2.1 is presenting the servo loop and the histogram measurement as typical test methods. For a production test solutions, where several ADCs are trimmed and tested in parallel, the histogram method is the more suitable solution.

This methodology is applying a linear ramp to the input of an ADC. If all codes in the transfer function have equal length, then each code should occur with the same amount of hits after collecting the ADC results. If a code occurs more often, then the respective DNL is positive. For the example above, the particular code would occur 5 times the average hits per code. The test time is now a factor of the amount of codes that are tested, the average hit per code and the conversion rate.

If for example, the amount of errors from the ADC add up to ± 2 mV and the errors of the test hardware to ± 10 mV then the linear input ramp has to start -12 mV below the expected transition voltage of the code of interest and it has to stop 12 mV above it. A range of ± 20 mV or in other words ± 250 LSB for a 16 bit ADC with 5 V input range is typical. The average number of hits per code is basically a function of the ratio of the system noise and the required accuracy considering that multiplying the hits per code by a factor of four will improve the accuracy by a factor of 2. For a 16 bit ADC with an rms-noise of 0.5 LSB and an

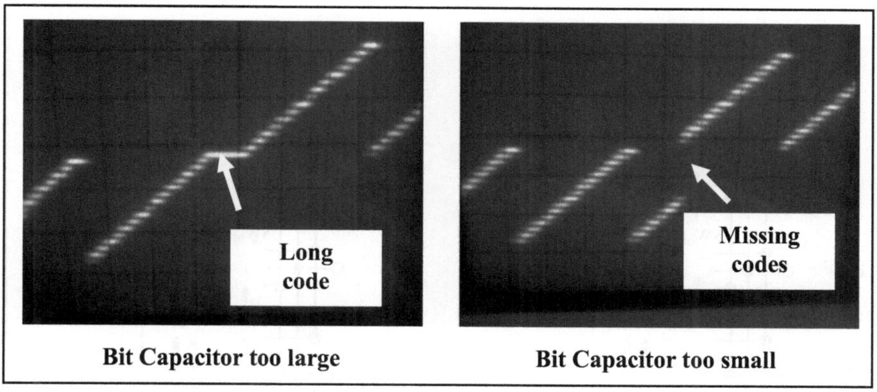

Fig. 3.34 Measured errors in the transfer function of SAR ADCs caused by capacitor mismatch

anticipated standard deviation for the DNL measurement of 0.05 LSB, 100 hits per code are required. Summarized, 50,000 conversions are required for one measurement. A state-of-the-art 16 bit SAR ADC, where the first 6 MSBs are trimmed, consumes around 8 s of trim time considering a reduced conversion rate (error correction is turned off), some retesting or setup and computing times of the test machine. For more information on test algorithms see Sect. 6.2.1.

The idea of the novel trim algorithm utilizes the fact that the SAR ADC normally converts an unknown input voltage with a known capacitive matching. In the same way, the converter could compare the capacitors, if the input voltage is known. Very suitable would be to sample the reference voltage on the capacitor of interest and run the conversion with the lower bit capacitors that are already matching well. Now this would not really work, as the sum of the lower bit capacitors is less than the bit of interest. This is another area, where the dynamic error correction comes in handy as it adds redundancy and enables the convergence.

In the example of Fig. 3.35, a CDAC is shown where the bit 2 capacitor should be trimmed and is connected to the positive reference voltage for sampling. All other capacitors are coupled to the negative reference. After opening of the hold switch, it is made sure that all upper bit capacitors including the bit capacitor of interest (here bit 1 and bit 2) are switched to the negative reference. The remaining capacitors are used for the conversion process. In this example the DEC capacitor equals the bit 4 capacitor. The sum of the bit 3, bit 4 and DEC capacitor ideally equals the bit 2 capacitor. Consequently they will redistribute as much charge when switched between the reference potentials as the bit 2 capacitor and the comparator input voltage V_C will converge after the conversion of the error correction.

The digital output voltage is then calculated as during a normal operation of the CDAC. In the example of Fig. 3.35, conversion result is 00110 and 00010 needs to be added as the error correction is turned on. This results in 01000, which equals the bit 2 decision and indicates that the capacitors match perfectly. The resolution is only one LSB in this example, but the noise of the ADC together with the oversampling technique will increase the resolution to the required value. Problems

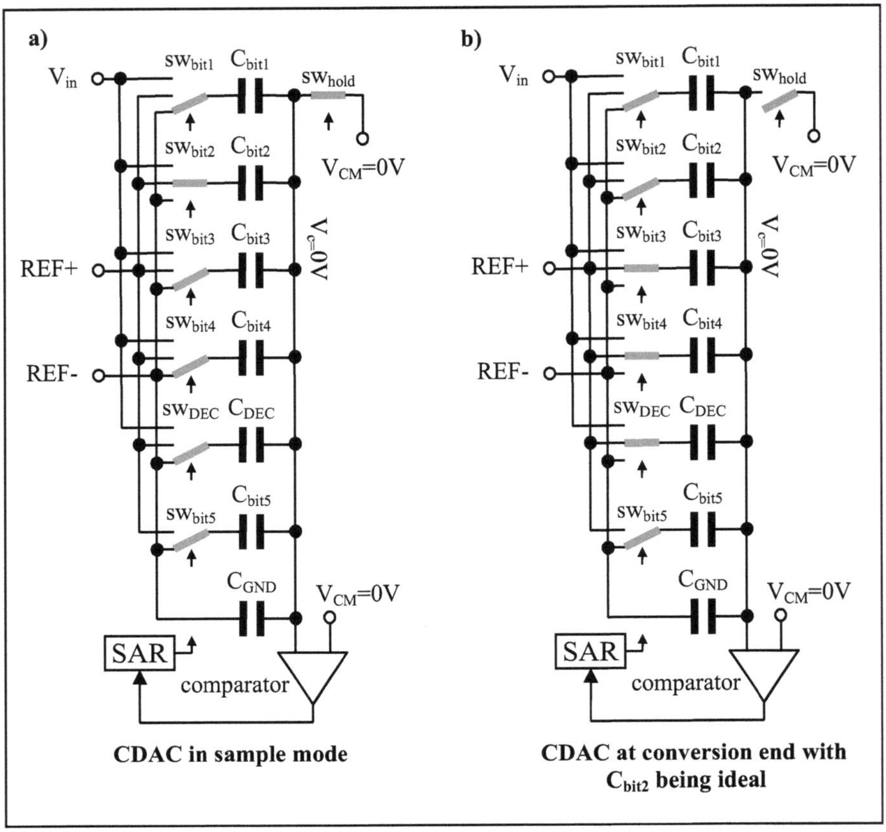

Fig. 3.35 Exemplary CDAC used for deriving a measurement algorithm for capacitive matching with ideal capacitors. (**a**) CDAC in sample mode. (**b**) CDAC at conversion end with C_{bit2} being ideal

might occur with 12 bit converters, which can be considered noise-free on one side but which still require trimming.

The result explained above should now be verified by calculation. The positive reference $REF+=REF$ is only sampled on the C_{bit2} capacitor and all other capacitors are connected to the negative reference potential, which for simplification should be ground ($REF-=0$ V). The sampled charge Q_S can be calculated to:

$$Q_S = C_{bit2} \cdot (0 \text{ V} - REF) \tag{3.47}$$

The charge is frozen and the conversion process starts with setting C_{bit1} and C_{bit2} to ground. It was estimated that C_{bit3}, C_{bit4} and C_{DEC}, are connected to the reference at the end of the conversion. So that the charge distribution Q_C at the end of the conversion can be described with

$$Q_C = (C_{bit1} + C_{bit2} + C_{bit5} + C_{GND}) \cdot (V_C - 0\ \text{V}) + (C_{bit3} + C_{bit4} + C_{DEC}) \cdot (V_C - REF) \tag{3.48}$$

Equations (3.47) and (3.48) will lead to Eq. (3.49) under the considerations of charge equity and that V_C has to converge to 0 V at the end of the conversion.

$$\begin{aligned} C_{bit2} \cdot (0\ \text{V} - REF) &= (C_{bit1} + C_{bit2} + C_{bit5} + C_{GND}) \cdot (0\ \text{V} - 0\ \text{V}) \\ &\quad + (C_{bit3} + C_{bit4} + C_{DEC}) \cdot (0\ \text{V} - REF) \\ C_{bit2} &= C_{bit3} + C_{bit4} + C_{DEC} \end{aligned} \tag{3.49}$$

Equation (3.49) therefore proves that the converter is comparing the capacitive matching. Ideally, the sum of $C_{bit3,}C_{bit4,}C_{DEC}$ equal C_{bit2}. Now if C_{bit2} would be too small by 1 LSB, then the comparator would decide to switch C_{DEC} back to ground and C_{bit5} to the reference instead (see the left side of Fig. 3.36). The converter result is 00111 instead of the ideal value of 01000. The error can be resolved by simply subtracting the two values.

The right side of Fig. 3.36 shows the conversion result, if C_{bit2} is too large by 1 LSB. Again, the converter result is 00111, but the value of 00010 has to be added as the error correction capacitor is switched to the reference additionally. The final result, which is 01001, ends up being 1 LSB larger than the ideal value 01000. Again subtracting the two results will provide the matching error.

It was explained above that the bit of interest has to be set actively to the negative reference potential during the trim conversion. This is particularly true for the 'dual CDAC architecture', where the conversion of the MSBs is even performed on another CDAC. The easiest implementation for the dual CDAC approach is to ignore the conversion result of $CDAC2$ and just set the first bits of $CDAC1$ as required by an additional digital state-machine.

For the typical CDAC approach, this additional logic can be minimized by a smart switching method. If the sampled charge can be reduced, then the bit of interest would automatically be set to ground in a normal conversion process and additional logic can be spared. This can be achieved by utilizing the dynamic error correction capacitor, which can lower the comparator input during the conversion. This capacitor was previously named as 'DEC down' and is symbolized with $C_{DEC\downarrow}$ below. This capacitor is initially connected to the reference and only switches to ground for the error compensation step. This capacitor can therefore be utilized to sample to ground first and switch it to the reference for the conversion process as shown in Fig. 3.37. This is generating an offset, which acts as if the sampled charge was reduced.

The correctness of the described offset generation can be calculated as following. At first, let's evaluate the comparator input voltage after sampling without switching the DEC↓ capacitor. The reference would be charged to the bit 2 capacitor, the charge is frozen, by opening the hold switch and then the charge of the bit 2 capacitor is redistributed to all capacitors, if the capacitor gets coupled to ground. The comparator input voltage V_c therefore settles to

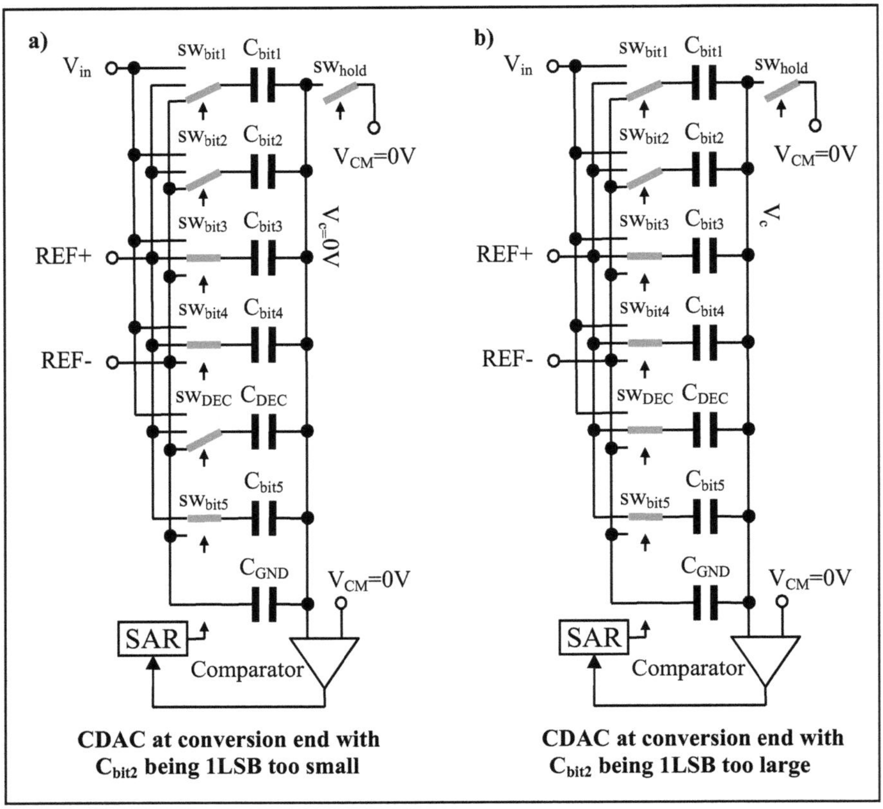

Fig. 3.36 Exemplary CDAC used for deriving a measurement algorithm for capacitive matching with non-ideal capacitors. (**a**) CDAC at conversion end with C_{bit2} being 1 LSB too small. (**b**) CDAC at conversion end with C_{bit2} being 1 LSB too large

$$V_c = -\frac{REF \cdot C_{bit2}}{C_{tot}} \tag{3.50}$$

If $C_{DEC\downarrow}$ is now switched back to reference prior to the conversion, then the comparator input voltage changes to

$$V_c = -\frac{REF \cdot C_{bit2} - REF \cdot C_{DEC\downarrow}}{C_{tot}} \tag{3.51}$$

C_{tot} is expressing the sum of all capacitors. Based on Eq. (3.51), an offset voltage is added to the sampled voltage, which will make sure that the bit 2 capacitor will switch back to ground during the conversion process, so that C_{bit2} gets always compared to the lower bit capacitors. The offset certainly has to be considered in the final error calculation.

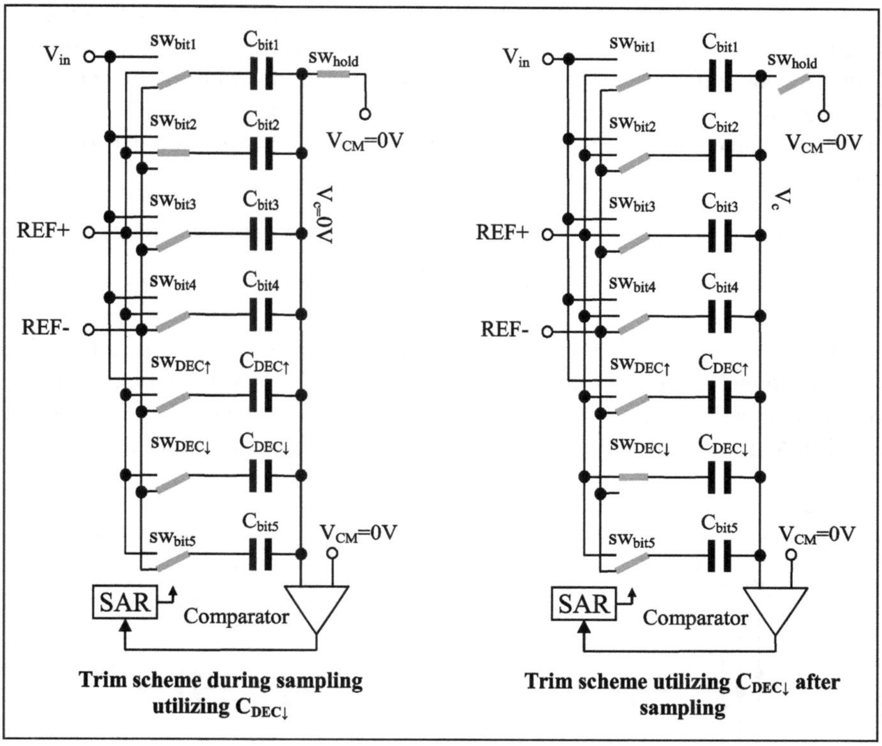

Fig. 3.37 Exemplary CDAC using $C_{DEC\downarrow}$ to generate an offset for the measurement algorithm

This circuitry was implemented on a test chip. It was generally functional, but generated an error of several LSB, where an accuracy of a fraction of an LSB is required. Also this error is highly dependent on the capacitor to be trimmed. Reason is the charge injection caused by the hold switch sw_{hold}, which is dependent on the impedance of the node V_c to ground. The impedance of the input switch, which couples the CDAC capacitors with the input voltage V_{in}, is of significantly lower impedance than the switches to the reference potentials. The input switches are dimensioned for low THD, while the reference switches are chosen for appropriate settling.

Therefore, a method is required to calibrate the charge injection. This can also be achieved with the SAR ADC itself. The idea is to sample a known voltage, which advantageously is the reference voltage again, on the capacitor of interest and then run a normal conversion similar to the trim conversion above. The delta between the digital representation of the sampled voltage and conversion result is caused by the charge injection and the capacitor mismatch. The trick to avoid the error caused by the capacitor mismatch is to sample on the erroneous capacitor of interest and to remain it connected to the reference voltage during the whole conversion.

An example is given in Fig. 3.38. The sampling, which is illustrated on the left side of Fig. 3.38 is identical to the sampling method during the capacitor comparison. The only difference is that now the charge injection is causing the node V_c to be different than 0 V after sampling. In Eq. (3.52), this voltage is called V_{inject} and it affects all capacitors. The sampled charge $Q_{S,cal}$ can be expressed the following way:

$$Q_{S,cal} = C_{bit2} \cdot \left(V_{inject} - REF\right) + \left(C_{tot} - C_{bit2}\right) \cdot \left(V_{inject} - 0\ \text{V}\right)$$
$$Q_{S,cal} = C_{tot} \cdot V_{inject} - C_{bit2} \cdot REF \tag{3.52}$$

During the conversion, the digital control circuitry needs to make sure that C_{bit2} remains connected to the reference as shown on the right side of Fig. 3.38. The control loop will still converge V_c to the common-mode potential V_{CM}, which here equals 0 V, so that the charge at the end of the conversion calculates to

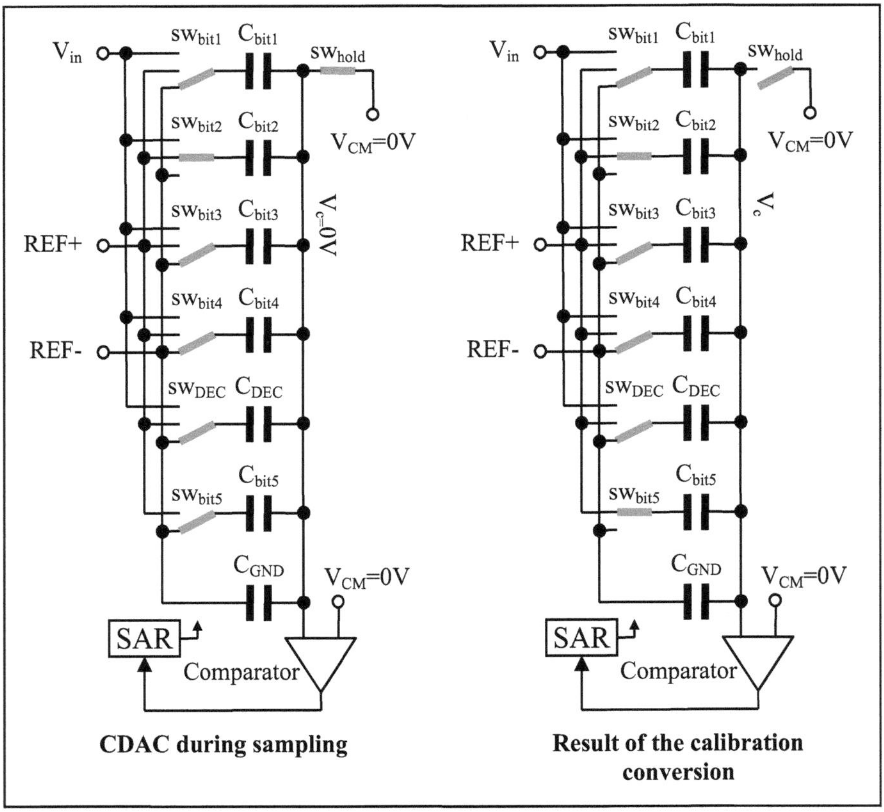

Fig. 3.38 Exemplary CDAC running a conversion to calibrate the charge injection

$$Q_{c,cal} = (C_{bit2} + C_{bit5}) \cdot (0 \text{ V} - REF) \tag{3.53}$$

In a last step, the offset caused by the charge injection can be calculated to

$$V_{inject} = -\frac{C_{bit5}}{C_{tot}} \cdot REF = -1 \text{ LSB} \tag{3.54}$$

Now after calibrating the charge injection, the capacitor matching can be evaluated with the capacitor of interest getting connected to negative reference potential during the conversion. The sampled charge would still follow Eq. (3.52). Let's assume the conversion ends as drawn on the left side of Fig. 3.36, then the charge redistribution $Q_{c,trim}$ after the conversion can be described with Eq. (3.55) considering that V_c converges to 0 V again.

$$\begin{aligned} Q_{c,trim} &= (C_{bit1} + C_{bit2} + C_{DEC} + C_{GND}) \cdot (V_c - 0 \text{ V}) + (C_{bit3} + C_{bit4} + C_{bit5}) \\ &\quad \cdot (V_c - REF) \\ Q_{c,trim} &= (C_{bit1} + C_{bit2} + C_{DEC} + C_{GND}) \cdot (0 \text{ V} - 0 \text{ V}) + (C_{bit3} + C_{bit4} + C_{bit5}) \\ &\quad \cdot (0 \text{ V} - REF) \\ Q_{c,trim} &= -(C_{bit3} + C_{bit4} + C_{bit5}) \cdot REF \end{aligned} \tag{3.55}$$

With the charge being frozen, the capacitive matching can finally be calculated as done in Eq. (3.56).

$$\begin{aligned} C_{tot} \cdot V_{inject} - C_{bit2} \cdot REF &= -(C_{bit3} + C_{bit4} + C_{bit5}) \cdot REF \\ C_{tot} \cdot \left(-\frac{C_{bit5}}{C_{tot}} \cdot REF\right) - C_{bit2} \cdot REF &= -(C_{bit3} + C_{bit4} + C_{bit5}) \cdot REF \\ C_{bit2} = C_{bit3} + C_{bit4} &= 3 \cdot C_{bit4} \end{aligned} \tag{3.56}$$

The capacitance of C_{bit2} should be 4 times the capacitance of C_{bit4} and therefore is too small by the amount of C_{bit4} in this example, which here is representing 2 LSB.

A trick can be utilized to simplify the digital logic in a similar manner as described above in Fig. 3.37. It must be ensured that the capacitor C_{bit2} remains connected to the reference during the conversion of the calibration. Therefore, this time the node V_c needs to be lowered by a known amount of voltage. This can be achieved by connecting $C_{DEC\uparrow}$ to the reference voltage during the sampling phase and switching it back to ground once the conversion starts. The additional sampled charge has to be included into the calculation.

It is expected that the time consumed for the trimming process is substantially reduced without any negative influence on the trim result. Even if noise will cause some variation on the trim result, the average trim result needs to remain identical if measured with the state-of-the-art method or with the novel method described above. The standard deviation should not increase with the novel method.

Table 3.1 Comparison of DNL trim times for a state-of-the-art ramp method and the novel method

	Single site	Dual site	Quad site
Partial ramp method	6.06 s	6.35 s	7.28 s
Novel trim method	0.30 s	0.34 s	0.45 s

Table 3.1 shows the comparison of trim times for the DNL for the method using a partial ramp over ±500 LSBs [13], which represents the state-of-the-art, and the novel trim method. Note that the trim times for the state-of-the-art version presented below in Table 3.1 already benefit from utilizing an ideal hardware including a real-time histogram, a high-speed data transfer and an ultra low noise signal source. Trim times for similar ADCs tested on previous hardware took longer by a factor of 3.

The trim time was reduced by a factor of 20 and is basically limited by setting up the supply and signal sources. The terms single, dual or quad site express that one, two or four integrated circuits are tested in parallel. Testing more parts in parallel will increase the total test time due to an increased amount of data transfer and required calculation. The remaining trim time of 0.30–0.45 s can be neglected if compared to the total ADC test time of roughly 18 s. Most of the time is now consumed by the linearity test of the complete ADC transfer function.

In the next step, the performance of the novel trim scheme was compared to the partial ramp test. Therefore, the matching of the bit 6 capacitor of the same device was measured one hundred times with the partial ramp method and one hundred times with the novel trim method. The distributions of the trim results are shown in Figs. 3.39 and 3.40.

The distribution of the state-of-the-art partial ramp method in Fig. 3.39 does not show the expected Gaussian distribution, which thermal noise would generate. The peak is shifted to the right. This can be explained by the fact that the error correction has to be turned off for the measurements of the capacitive matching. To reduce dynamic errors, the conversion typically runs at lower speed such as 250 kSPS instead of 1 MSPS. However, some error will remain. Nevertheless, the average DNL result is measured to -1.37 LSB and is used as baseline for the novel trim method. The value of -1.37 LSB expresses that the code of interest is missing and that additionally the code below it is still short by 0.37 LSB. Both codes are affected by the bit 6 capacitor matching.

The distribution of the DNL measured with the novel trim algorithm fits much better to the expected Gaussian distribution. The novel trim algorithm is still using the dynamic error correction and can correct for dynamic errors during the conversion process. Therefore, the average value is more central causing the average value to be shifted to -1.41 LSB. The correlation between the two methods is still excellent as the average values only differ by 0.04 LSBs. The standard deviation is actually better by 0.03 LSB if using the novel method.

Summarized, the novel trim algorithm works more reliable and is by a factor of 20 faster than the partial ramp method. This method was also implemented in the

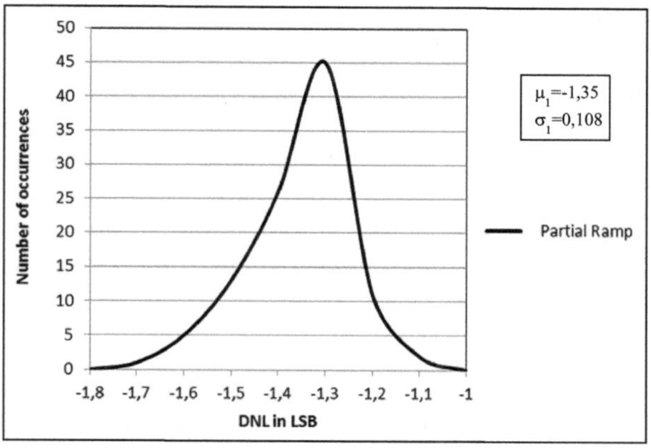

Fig. 3.39 Distribution of DNL measurement results with the partial ramp method [13]

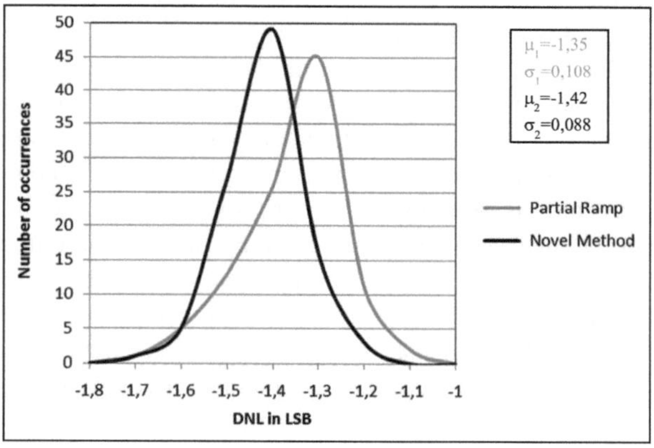

Fig. 3.40 Distribution of DNL measurement results with the novel trim method [13]

low voltage device presented in Sect. 3.8, but will not further be documented as it has no influence on the performance and on the design of the ADC. The only remaining feature that needs to get in place for this ultra-low power ADC is the zero-power sampling. The concept is explained in the next chapter.

3.6 Zero-Power Sampling [16, 17]

It was already mentioned in Chap. 1 that the power consumption is a key specification. Either mobile applications should have a long life time or an island application is controlled and supplied with a 4–20 mA current or the device density is so high that cooling issues exist.

The reduction of the bandwidth of the reference and the comparator helped to reduce the power dramatically from the second generation SAR ADCs (G2, year 2000) to the third (G3, year 2007) from 91 mW to 39 mW at a 5 V supply and even

Table 3.2 Power comparison of the G2 and the G3 SAR ADC generation [4]

Design	G2	G3	
Supply/Speed	5 V/0.5 MSPS	5 V/1 MSPS	2.7 V/1 MSPS
Comparator	7.3 mA/36.5 mW	4.2 mA/21 mW	3.1 mA/8.4 mW
Reference	8.8 mA/44.0 mW	1 mA/5.0 mW	1 mA/2.7 mW
VCM-Generation	1.5 mA/7.5 mW	1.7 mA/8.5 mW	1.4 mA/3.8 mW
Else (incl. Digital)	0.6 mA/3.0 mW	1.9 mA/9.5 mW	0.3 mA/0.8 mW
Total	18.2 mA/91 mW	8.8 mA/44.0 mW	**5.8 mA/15.7 mW**

The final result is shown in bold

15.5 mW at a 2.7 V supply while simultaneously doubling the conversion rate from 500 kSPS to 1 MSPS as is it demonstrated in Table 3.2 [4].

Table 3.2 shows the improvements in the comparator and the reference design. A further reduction in power might only be possible by going to faster processes and/or lowering the supply voltage.

Table 3.2 also shows that the digital power consumption is an essential contributor. The digital power consumption actually increased from the G2 to the G3 generation by a significant amount, as several new digital functions got added such as the digital error correction or digital register mapping. In the future, the digital gate count is expected to further grow. Features like digital post processing will be added such as oversampling filters, up-down counters or even microprocessors. Again, the power consumption can only decrease by moving to a faster process with smaller dimensions and by lowering the supply voltage.

Table 3.2 also lists a VCM-generation circuit. The hold switch connects the capacitors and the comparator input to a common-mode voltage during sampling and in this way is setting the operation point of the comparator. It is not the main contributor to the power consumption, but the VCM-Generation circuitry particularly has to be turned on during the sampling mode. This essentially affects the power saving modes that were added in the G3 generation. SAR ADCs are often not used at the maximum sample rate. Often a conversion is performed and the SAR ADC is placed in a power-saving mode until the next conversion starts. Different power saving modes exist that differ in number of components that remain biased.

The ADC should wake up quickly out of the NAP or SLEEP mode. Therefore, only circuits with large bandwidths are turned off, which are the comparator and the VCM-Generation. All circuits with slower settling such as the reference and the biasing remain turned on. The result in power scaling versus the conversion rate is shown in Fig. 3.41. In a deep power-down mode, all circuit blocks are turned off. The remaining power consumption falls to basically zero with the disadvantage of longer wake-up delays.

Now, the VCM-Generation forms a particular problem for the power saving modes. The input voltage cannot be stored across the sample capacitors until the common-mode voltage is stable. The ADC first needs to wake up and then a certain time needs to be dedicated for sampling, which is dependent on the bandwidth of the external circuitry driving the analog inputs. The higher the bandwidth, the

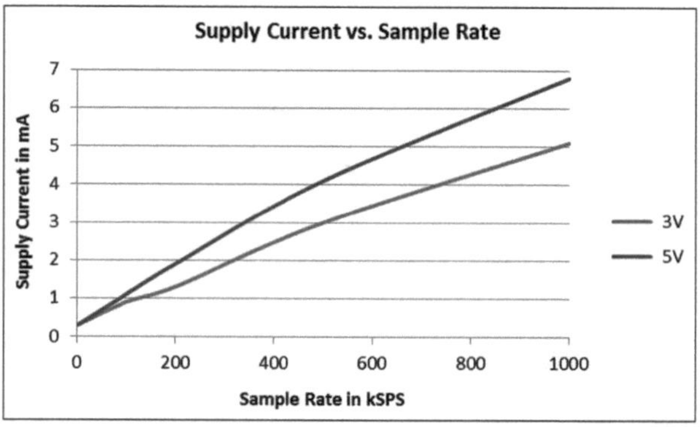

Fig. 3.41 Current consumption of a typical SAR ADC versus conversion rate in auto-nap mode [4]

shorter is the wake-up delay. The ADC can remain for a longer time in the power saving mode, but the higher bandwidth of the external circuitry typically comes with higher power consumption. The lower the bandwidth, the longer is the wake-up delay, but the current consumption of the driver is less.

The wake-up delay is also generating overhead for the application. A signal needs to be generated to wake the ADC up. A dedicated wait time needs to be added, which is dependent on the input bandwidth, and then the conversion start needs to be triggered. An asynchronous conversion start, which is triggered by an arbitrary signal, is not possible.

Therefore, it would be beneficial, if the ground could act as the common-mode voltage. This would avoid the wake-up issues above, the VCM-Generation would be unnecessary at all, so that the power consumption reduces together with the die area. If V_{CM} equals to ground, then the comparator input voltage V_c would normally be in the range of $\pm\frac{REF}{2}$. The transistors of the hold switch would start leaking, if the ADC operates from a single supply voltage (ground and supply). First of all the source-bulk diode would be forward biased and additionally, the gate-source voltage would be positive with gate being at 0 V and the source at $-\frac{REF}{2}$.

The solution, which is presented in [16], is based on the fact that ADCs are normally designed fully differential to minimize the charge injection of the hold switch, which generates the offset, offset drift and eventually non-linearities. In [16], the hold switches are coupling the comparator side of the capacitors to ground. After sampling, several capacitors of the CDAC switch to the reference potential *REF* on both the positive and the negative side to generate a valid common-mode input voltage at the comparator input V_{cp} and V_{cn}. This condition is called common-mode boost. The MSB decisions are then convered with the common-mode boost in place. The limited common-mode rejection of the CDAC and the comparator will generate an error, which is reflected in the transfer function as discontinuity as shown in Fig. 3.42.

It is known that the differential comparator input voltage converges to 0 V in the course of the conversion. Once it is small enough to avoid the leakage at the hold switch, the common-mode boost is reset and the error caused by the common-mode boost is eliminated. However, the error could already have affected the MSB

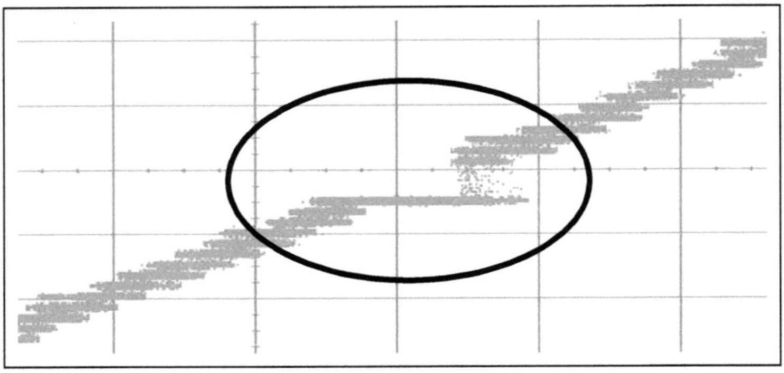

Fig. 3.42 Discontinuity in the ADC transfer function caused by common-mode shift during conversion

decisions prior to the common-mode reset. Therefore, an error-correction step has to follow the reset of the common-mode boost.

As an example, Fig. 3.43 shows a conversion with a simple common-mode boost architecture. For the MSB decision, the positive MSB capacitor switches as given to *REF* and the negative side to 0 V. The remaining capacitors on both the positive and the negative side, which are normally coupled to the ground potential, are now all switched to *REF* to generate the common-mode boost. As a consequence, the comparator input voltages are boosted by half the reference voltage, which is 2 V in the example. Please note that the positive comparator input voltage is input voltage dependent and is now in the range of $\pm\frac{REF}{2}$ around the common-mode boost voltage $\frac{REF}{2}$ or $V_{cp} = REF - V_{inp}$.

The MSB decision is 0 and the comparator input voltage has to be lowered by the quarter of the reference. Normally, the MSB capacitor would be connected to ground and the MSB-1 capacitor to *REF*. However, the MSB-1 capacitor (C/2) is already connected to *REF* for the common-mode boosting. In an alternative switching scheme, the MSB capacitor can remain at REF and MSB-1 capacitor is reset to 0 V. The MSB decision will cause the differential comparator input voltage to be less than a quarter of the reference, which in the example is less than ±1 V. Leakage is still possible.

In the example, the MSB-1 decision is 1 and V_{cp} needs to be increased by 1/8th of the reference. With the MSB-1 result, the differential voltage will be converging into the range of ±0.5 V, which is less than a diode drop. Let's assume that parasitic diodes mentions above have no leakage at voltages lower than 0.6 V. Then, the common-mode boost is not required anymore for the MSB-2 decision. The boost capacitors are switched back to ground and the conversion continues normally. Note that the error correction step was left out in the example above for simplicity.

Now that the common-mode potential is ground, the VCM generation is not required anymore. Furthermore, the comparator is a high bandwidth circuit with short wake-up times and might be turned off during sampling. Consequently, the power consumption can be turned off completely during sampling. This

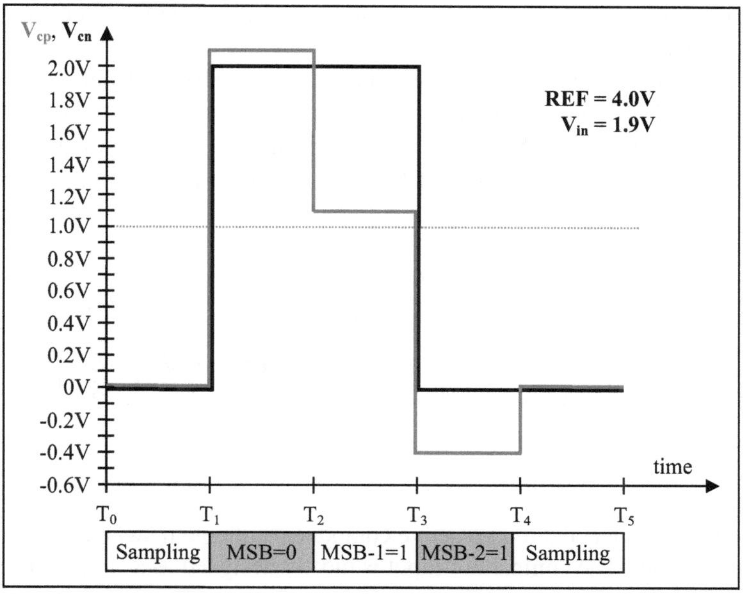

Fig. 3.43 Exemplary comparator input potentials V_{cp} and V_{cn} utilizing common-mode boosting

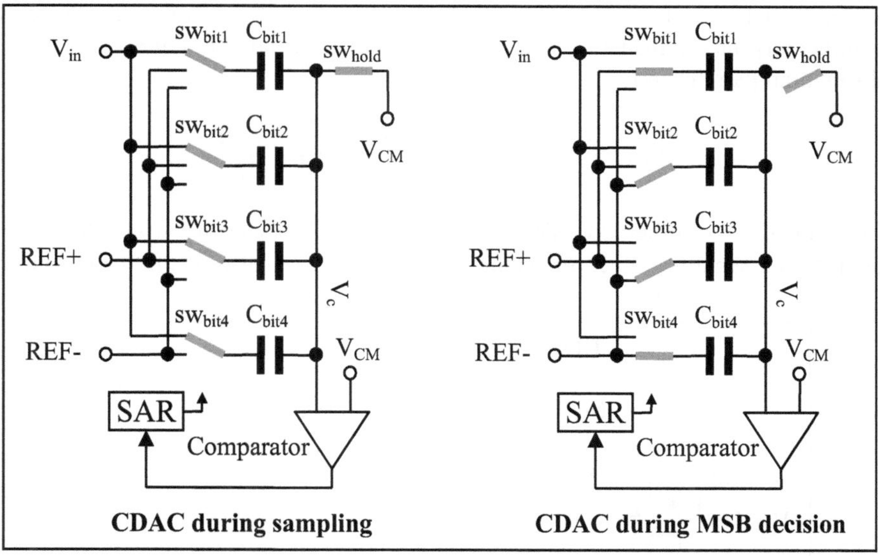

Fig. 3.44 Simplified CDAC used to calculate the maximum comparator input voltage

architecture is therefore called 'zero-power sampling'. This switching scheme mentioned above requires a comparator with large common-mode input range. This is a drawback as further power reduction should be achieved by decreasing the supply voltage, which will actually limit the common-mode range. The following will explain how zero-power sampling can be combined elegantly with the dual CDAC architecture (Fig. 3.44).

The comparator input voltage V_c has its highest possible peaks during the MSB decision, which was already discussed previously. For an input voltage V_{in} ranging from 0 V to the reference REF, the comparator input voltage can range from $V_{CM} - \frac{REF}{2}$ and $V_{CM} + \frac{REF}{2}$ as calculated below by estimating ideal binary weighted capacitors.

$$\begin{aligned} Q_S &= 2 \cdot C_{bit1} \cdot (V_{CM} - V_{in}) \\ Q_{c,MSB} &= C_{bit1} \cdot (V_{c,MSB} - REF) + C_{bit1} \cdot (V_{c,MSB} - 0\ \text{V}) \\ V_{c,MSB} &= V_{CM} + \left(\frac{REF}{2} - V_{in}\right) = V_{CM} + \left(\frac{REF}{2} - \frac{REF}{2} \pm \frac{REF}{2}\right) \\ &= V_{CM} \pm \frac{REF}{2} \end{aligned} \tag{3.57}$$

The node V_c is connected to two active components, the hold switch sw_{hold} and the comparator input. The comparator input is mostly designed as a simple differential input stage as drawn in Fig. 3.45. The common-mode voltage V_{CM} is constraint in two ways by the comparator input voltage.

First, the input transistors T_1 and T_2 should have a drain-source voltage V_{DS} larger than the saturation voltage $V_{DS,sat}$, so that the drain source resistance is significantly larger than the load resistance and does not limit the gain of the differential pair. Therefore

$$V_{CM} > V_{GS} - V_{DS,sat} + \frac{I \cdot R}{2} = V_T + \frac{I \cdot R}{2} \tag{3.58}$$

Equation (3.58) further considers that the gate-source voltage, which for PCH devices is a negative value, minus the saturation voltage between drain and source

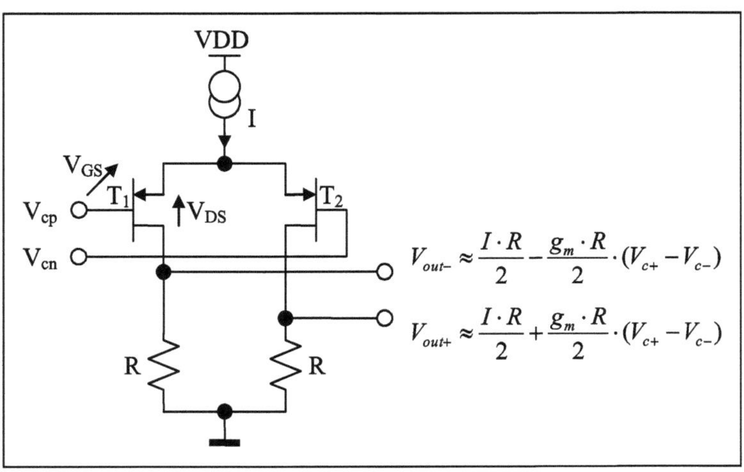

Fig. 3.45 Typical comparator input stage designed as differential pair

equal the threshold voltage V_T. The common-mode voltage can go all the way down to 0 V with selecting the current and/or the load resistance low enough. However, the current defines mostly the noise of the differential input pair. Keeping it low will result in higher noise. Lowering the resistance will affect the gain of the differential pair.

The common-mode voltage V_{CM} is also limited in its maximum value. The input transistors require a valid gate-source voltage and the current source, which in the simplest case is given by a transistor operating in saturation, requires an absolute value of the drain-source voltage larger than the saturation voltage. Note that V_{GS} and V_{DS} are negative voltages for the PCH transistor.

$$V_{CM} < AVDD + V_{GS} + V_{DS,sat} \tag{3.59}$$

The common-mode voltage is further limited by the hold switch, which is typically designed as a CMOS transmission gate (see Fig. 3.46). If source or drain voltages extend the ground or the supply, then the transistors start leaking through their parasitic diode to bulk. Another leakage path opens through the channel itself. As soon as charge is lost due to leakage, the conversion result is wrong and cannot be corrected anymore.

During sampling, all nodes are at defined values. As long as the common-mode voltage does not extend ground or supply, there are no leakage currents. However, the comparator input voltage V_c has a significant swing during the MSB decision. A trick needs to be invented, so that the common-mode voltage can equal ground and that the hold switch does not leak. The dual CDAC architecture is very helpful at this point. Sect. 3.2 explains how the ADC internal signal amplitude is attenuated by at least a factor of 4. A stronger attenuation is possible. In this way, the internal signal amplitude would be reduced from $V_{CM} \pm \frac{REF}{2}$ to $V_{CM} \pm \frac{REF}{8}$. If the common-mode voltage is 0 V and the reference is at 5 V, then the comparator input voltage is limited to ± 0.625 V.

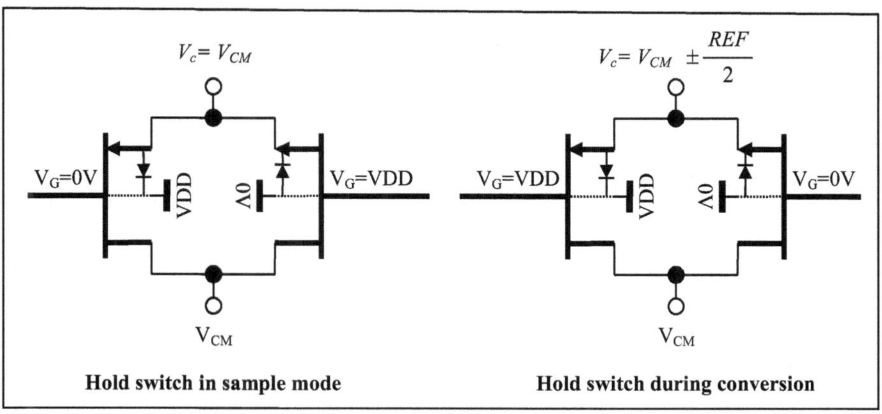

Fig. 3.46 Typical hold switch and its connectivity during sample and during conversion mode

A common-mode voltage at 0 V would simplify the hold switch to an NCH transistor, which only shows leakage when V_c gets more negative than the bulk and/or the gate voltage. During the conversion, both bulk and gate are normally connected to ground and both bulk and gate are high impedance nodes as long as the bulk does not start leaking to drain or source. A simple capacitor can be used to pull bulk and gate to a negative voltage such as $-\frac{VDD}{2}$. The only process requirement is that the bulk of the NCH transistor can be isolated from the common substrate.

A simplified circuit, which can perform the task of generating the negative voltage of $-\frac{VDD}{2}$ is shown in Fig. 3.47. A capacitor C_1 is charged to the supply voltage *VDD* during the sample phase, where the digital *Hold* signal is '0'. This connects the upper plate of $C_1 = C$ to *VDD*, while the lower plate is coupled to ground. The plates of a second capacitor C_2 with the same capacitance C are shorted to ground. The sample phase ends with *Hold* switching to '1'. This first opens the NCH transistor so that the output node, which couples to the bulk continuously and the gate of the hold switch in hold mode, is now in high impedance. In a next step, the upper plate of C_1 is pulled down from *VDD* to ground. The charge of C_1 therefore distributes to C_1 and C_2 and pulls the output node below ground. As they are of equal size, *VDD* is divided by the factor of 2.

The drain-bulk diode of NCH transistor would now start leaking. To prevent this from happening, the bulk of the NCH transistor is connected to the output node and is also pulled down to $-\frac{VDD}{2}$. The NCH transistor would still leak through its normal channel, as the drain would be pulled below the gate voltage of 0 V. Therefore, also the ground connection of the inverter needs to connect to the output node causing the gate of the NCH transistor to be at the same potential as the drain.

Summarized, the boost circuitry of Fig. 3.47 generates a negative potential during the hold phase, which connects to the gate and bulk of the NCH transistor of the hold switch. Together with the dual CDAC architecture, the source and drain voltages of the hold switch will never get below the bulk and gate potential and the transistor will not leak. Therefore, ground can be used as the common-mode voltage of the CDAC, which is also a valid operating point for the comparator as long as Eq. (3.58) is valid. The hold switch can therefore be simplified as shown in Fig. 3.48. The supply voltage can now be as low as 1.2 V.

Now, as ground can be used for the common-mode voltage, there is no further need for an internal common-mode amplifier. Power is no longer required during the sampling phase. Sampling can also be continuously performed. There is no need to wake up any circuitry before the end of the sample phase. The analog input voltage will not be disturbed by any wake up activity. External high speed or low impedance amplifiers for the analog input path are only required, if the ADC operates with its full conversion rate.

Chapter 3 presented novel architectures, which can be used for a significant improvement of SAR ADCs. The base is the floating CDAC architecture, which allows the conversion of signals with higher amplitude on a lower voltage core

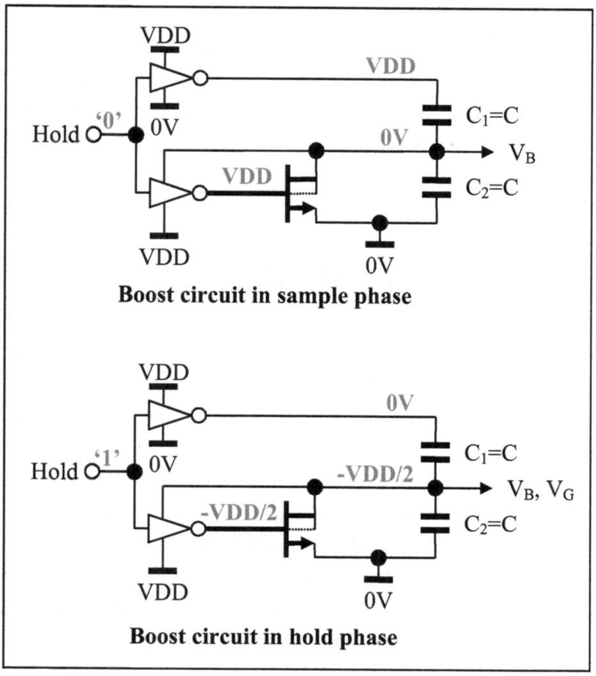

Fig. 3.47 Level-shift circuitry to generate negative potentials for the hold switch's gate and bulk

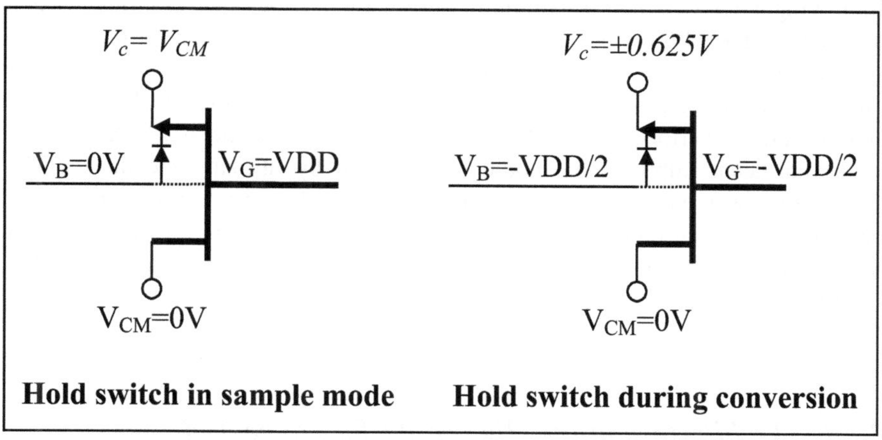

Fig. 3.48 NCH transistor used as hold switch for a 0 V common-mode voltage

supply. This architecture was improved in terms of robustness with the dual CDAC architecture. The novel ideas of scale-down trim, digital trim and INL trim will help to improve the performance of the ADCs in terms of gain, power, size as well as linearity. A method to measure the matching of capacitors was introduced, which reduces the test time significantly. This will make high performance SAR ADCs more affordable in many applications. And finally the concept of zero-power sampling was introduced to the dual CDAC architecture.

The following will now introduce the realization and results of these architectures on two examples. One is presenting a breakthrough in best SNR and resolution. The other one is presenting the lowest power 1 MSPS, 16 bit SAR ADC.

3.7 Highest Performance SAR ADC

State-of-the-art 16 bit SAR ADCs as described in [4] with 5 V input range show a typical SNR of 93 dB. The SNR can be improved by moving from pseudo differential inputs to fully differential inputs as shown in Fig. 3.49, which have twice the amplitude.

If the noise remains constant, the SNR would improve by 6 dB when doubling the signal range. This is only true, if it would not be limited by the quantization noise. Such fully differential SAR ADCs like the ADS8482 from Texas Instruments [18] are therefore designed as 18 bit converters. The SNR improves to 99 dB, which is reflected in a DC histogram as shown in Fig. 3.50, where the ADC inputs are shorted and converted 8192 times. The digital output has a code spread of 6 LSB.

Please note that such a fully differential signal path is difficult to maintain in an application. A single ended or pseudo differential input with a voltage range of ± 10 V, which is the standard for industrial applications, is much more convenient. Improving the SNR would be helpful in several applications such as programmable logic controller (PLC) or power protection applications. Some state-of-the-art SAR ADCs provide this ± 10 V input ranges, but divide the signal internally to a 0–5 V range. Such converters as the ADS8556 from Texas Instruments [5] therefore achieve a noise performance of 92 dB similar to the single ended 5 V ADCs.

A significant improvement could be achieved by utilizing the dual CDAC architecture, where the internal signal attenuation is not required. The ADC internal peak-to-peak signal range could increase from 5 V to 20 V by a factor of 4. The SNR would improve by 12 dB to 105 dB or in other words only 2 codes on an 18 bit resolution would be expected in a DC histogram.

3.7.1 Architecture of the Highest Performance ADC

Additional modes were implemented to push the performance of the dual CDAC architecture. Adding a fully differential mode regarding Fig. 3.49 with ± 10 V positive and negative signals would further increase the differential signal range to 40 V_{pp}, adding another 6 dB to the SNR. Again, the SNR would be limited by the quantization noise, so that a fully differential 18 bit and even 20 bit mode was implemented.

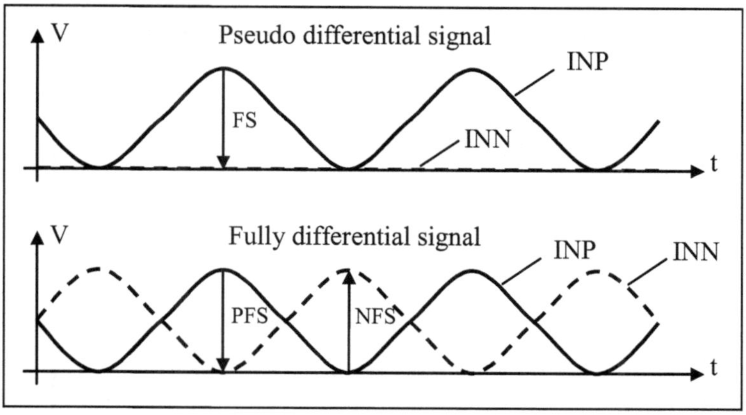

Fig. 3.49 Pseudo differential and fully differential signals used as ADC inputs

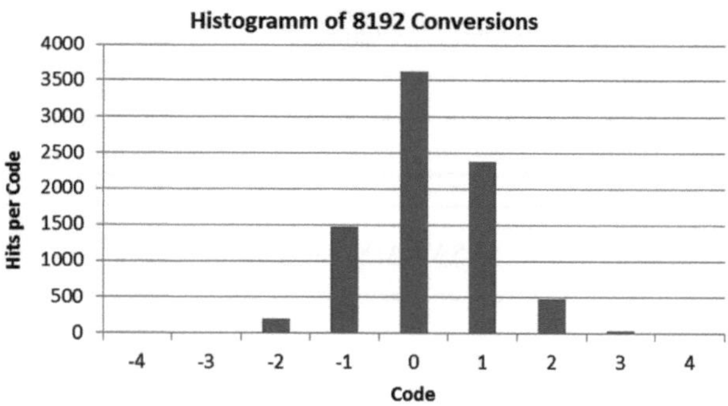

Fig. 3.50 DC Histogram of a state-of-the-art 18 bit SAR ADC showing 6 LSB of noise [18]

Figure 3.51 is showing the block diagram of the test chip. The positive and the negative input voltages INP and INN are sampled on a 4 bit coarse ADC and a 19 bit fine ADC. To simplify the schematic and the layout, each CDAC has its own comparator. The coarse CDAC reduces the internal signal amplitude by a factor of 8 to ± 1.25 V during the MSB decision with a simple capacitive divider. The coarse ADC converts the first 4 bits using the same reference potentials $REF+$ and $REF-$ as the fine ADC. This enables the best gain matching between the two ADCs. The reference potentials are typically at $REF+ = +10$ V and $REF- = -10$ V. However, the absolute value of the reference voltage can range from 5 V to 12 V. The sample capacitors of the fine ADC remain floating during the four MSB decisions and are coupled to their correct reference potential simultaneously after the first 4 bits are evaluated by the coarse ADC.

Zero-power sampling was partially implemented on the test chip. The CDACs are sampling to ground, just the power down of the comparator is not activated during sampling. The comparator input voltage during the bit 5 decision will remain in a ± 0.625 V range, which will already cause leakage at the hold transistors. Another risk is charge loss by voltage spikes, if the capacitors are not turning on

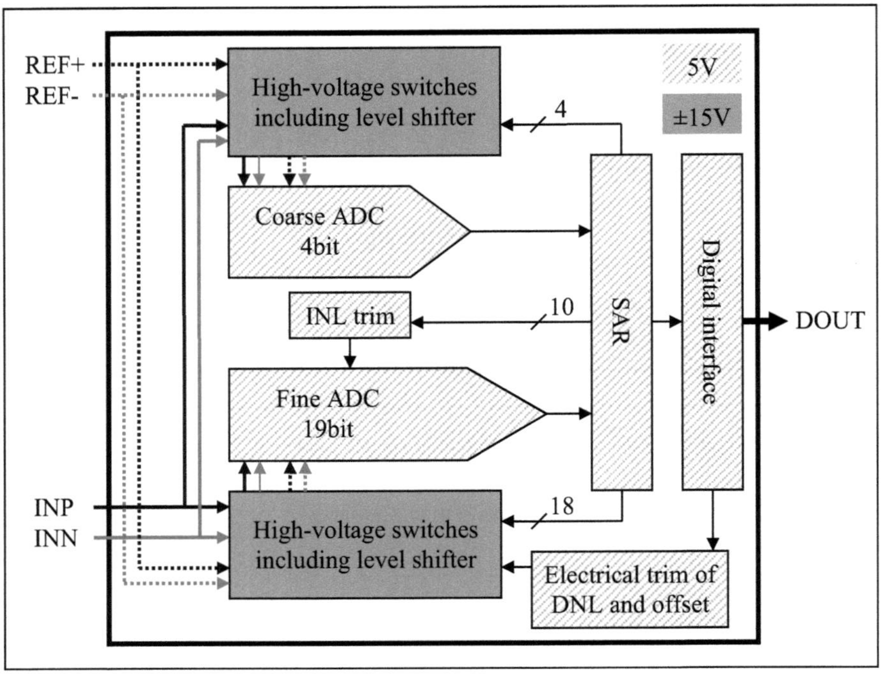

Fig. 3.51 Block diagram of the 20 bit SAR ADC using the dual CDAC architecture

exactly simultaneously or if the impedance of the switches to $REF+$ differs from the switch impedance to $REF-$.

Therefore, a positive common-mode shift is applied during the bit 5 decision by switching the lower bit capacitors on the positive and the negative CDAC side from $REF-$ to $REF+$. The common-mode shift will cause some error during the bit 5 decision due to limited common-mode rejection (see previous Fig. 3.42), which will later be compensated by the dynamic error correction. The common-mode shift is reset as soon as the bit 5 decision is finished.

Alternatively, one clock cycle could just be used to switch the floating capacitors to their reference potential based on the decisions of the coarse ADC with applying the common-mode shift. The common-mode shift could then be removed prior to the bit 5 decision.

More significant errors occur due to the mismatch of the coarse and the fine ADC. The mismatch is dominated by offset and gain errors. Based on experience on previous products, the offset can go all the way up to ± 2 mV and the gain error to 0.5 %. A 2 mV offset of the coarse ADC would equal to 0.04 % of the full scale range. The mismatch of the capacitors is another error source. The dynamic error correction should therefore provide a correction range of at least 0.75 % of the full scale range, which means that a first error correction step has to occur after bit 7. This equals an error correction range of 2048 LSB in 18 bit mode.

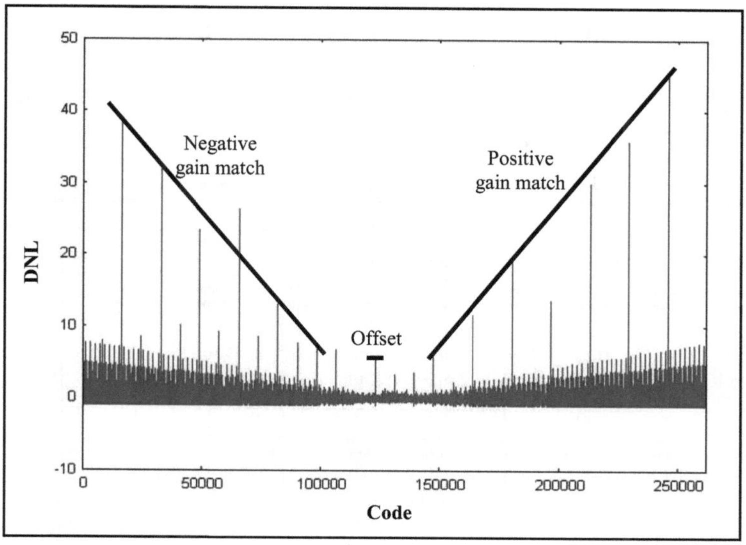

Fig. 3.52 SAR ADC transfer function with dual CDAC architecture and DEC being turned off

The matching on real silicon can actually be measured with an all-code DNL test as shown in Fig. 3.52 (also see Sect. 3.2), where the error correction is turned off. It shows that the matching between the ADCs is actually better than 50 LSB compared to the expected maximum of 2048 LSBs. The improvement was achieved with the implementation of the analog scale-down trim.

The capacitors of the dynamic error correction should be placed in the MSB array of the CDAC to avoid a significant increase in die size. The CDAC of the test chip, which is drawn in Fig. 3.53, therefore has the seven MSBs and the dynamic error correction ($DEC\uparrow 1$, $DEC\downarrow 1$) in the MSB array. It is followed by a first scale-down capacitor $SD1$, which couples bit 8 through bit 14 to the MSB array and includes a second dynamic error correction ($DEC\uparrow 2$, $DEC\downarrow 2$) and a second scale-down capacitor $SD2$. The LSB array following the second scale-down capacitor $SD2$ consists bit 15 through 18.

Note, that a bit 19 capacitor is actually not physically required. The 20 bit mode is a fully differential mode, where the MSB decision is evaluating, if the positive input INP is larger than the negative input. During the decision, both MSB capacitors from the positive and the negative CDAC array are coupled to the reference. They are switching accordingly to $REF+$ or $REF-$ for the MSB-1 decision.

Consequently, 19 bits can be evaluated with 18 bit capacitors in the positive and 18 in the negative CDAC. For the evaluation of the 20th bit, oversampling based on the patent described in [19, 20] is used (also see Sect. 2.6.1). The oversampling scheme is sampling the input voltage once and then is running a standard conversion from bit 1 through 19. As the noise is only a few LSBs, the next conversion for the matter of oversampling does not have to start from the beginning, but only needs to correct for the noise during the previous conversion with the last error correction

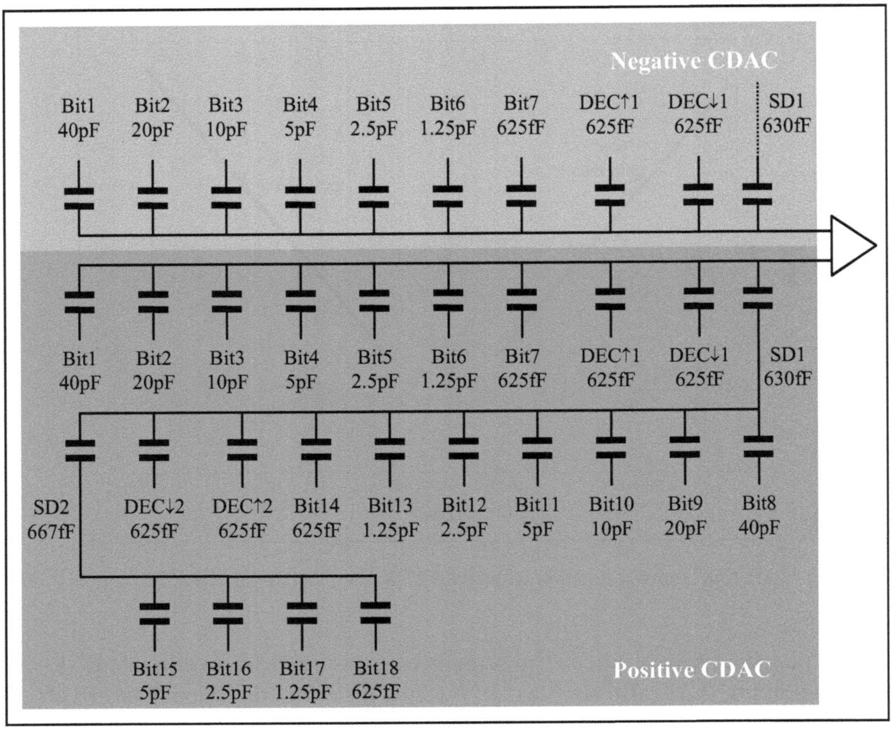

Fig. 3.53 Simplified CDAC for the high performance SAR ADC

stage and has to finish the LSB decisions. In this case, the conversions performed for oversampling would leave all settings from bit 1 through bit 14 and only convert the 2nd dynamic error correction and the bits 15 through 19 again. To improve the noise, the oversampling can be set to 2 or alternatively to 4.

The oversampling scheme is illustrated in Fig. 3.54. The first conversion follows the sampling and runs through a normal conversion cycle. Instead of sampling after the last conversion step, a second conversion is starting at the second dynamic error correction step and runs through to bit 19 again. In Fig. 3.54, the conversion process stops after the second conversion, but generally could be repeated several times. Note that only the conversion noise, which is basically caused by the comparator and the reference, will reduce by 3 dB with doubling the oversampling rate. The sample noise (kT/C) remains the same.

The capacitor representing the MSB is implemented with 40 pF vs. 20 pF that are typically used in state-of-the-art ADCs. This lowers the sampling noise by $\sqrt{2}$. The high reference voltages of ± 10 V together with the increased capacitance makes the reference settling challenging, which will be discussed later. This is particularly true as the bit 8 capacitor also starts with 40 pF again. This value was chosen for a good matching of the LSB capacitors.

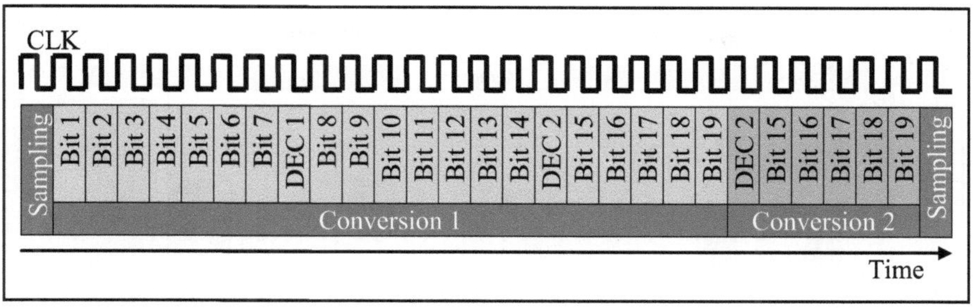

Fig. 3.54 Timing of SAR ADC with oversampling utilizing the dynamic error correction

Section 3.4 calculates that the expected INL of such a 20 bit SAR ADC approaches ±24 LSBs. The INL trim circuitry was therefore developed for this test chip and is implemented on silicon. The measurement results of this test chip were also demonstrated in Sect. 3.4. The INL of the ADC however, comes out significantly better than expected, so that the minimum amplitude of the correction signal was too large. The results shown in Sect. 3.7.3 are therefore measured with the INL trim circuitry turned off.

Finally, this test chip has all trim schemes implemented that were previously discussed. The scale-down trim is implemented to minimize the required amount of correction circuitry. The large amount of redundancy with two error correction steps makes the test chip eligible for the digital trim solution. However, the state-of-the-art analog trim solution is added as backup.

The layout is shown in Fig. 3.55. The most important blocks are marked. The die picture is provided in Fig. 3.56.

3.7.2 Process Selection

The dual CDAC architecture requires high voltage switches, which can withstand voltages above the maximum input range of ±12 V. In this particular implementation, the references can reach 12 V or −12 V, so that the maximum drain-source voltage must be larger than 24 V. Gate and bulk have to connect to the positive and negative high voltage supplies, which typically would be ±15 V ± 10 %. If a NCH input switch for example is turned on, then its gate could connect all the way up to 16.5 V, while the input voltage can go down to −12 V. The maximum gate-source voltage can therefore reach 28.5 V. Unfortunately, these high voltage transistors come with a significant on-resistance and a considerable parasitic capacitance, which change over the input voltage. These input voltage dependent parameters will affect total harmonic distortion (THD). The process, which was used for the test chip, has 36 V transistors available.

The comparator, the bias circuitry, the digital logic and interface, the memory and INL correction circuitry should run from a low voltage supply, which here is

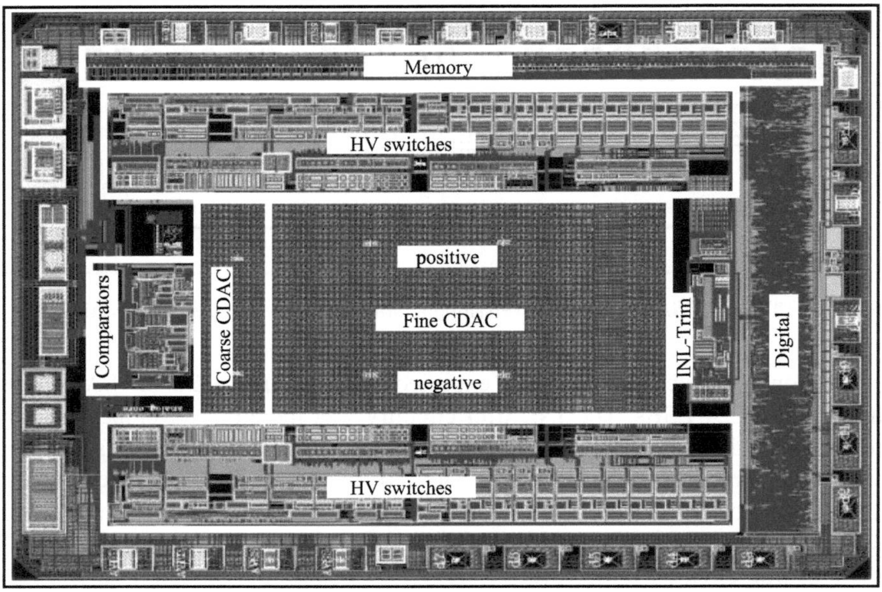

Fig. 3.55 Layout of a highest performance SAR ADC using dual CDAC architecture

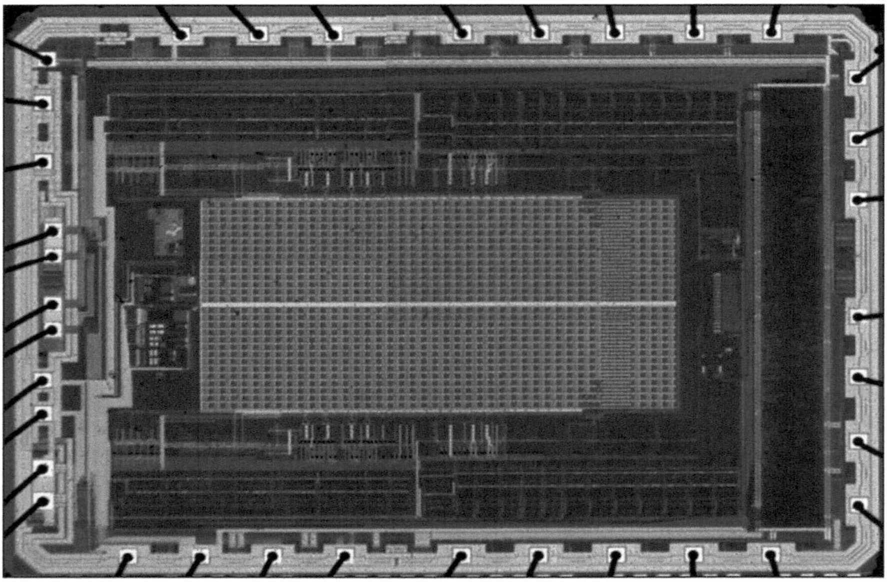

Fig. 3.56 Die picture of a highest performance SAR ADC using dual CDAC architecture

chosen to 5 V. Therefore, 5 V transistors must also to be available. A so called dual gate oxide process is required.

The capacitive DAC requires good matching capacitors. Trimming is possible, but the overhead is significant. Section 3.4 also discussed that voltage coefficients will affect the INL. The novel INL trim solution will correct for some amount, but is limited in its range. Small voltage coefficients are therefore welcome. These are typically provided by metal-metal or at least poly-poly capacitors. The total CDAC capacitance is above 320 pF. A high capacitance per area without neglecting matching is helpful to limit the die size. Additionally, they need to be able to withstand the maximum input voltage of 12 V.

The performance of high end SAR converters is often limited by oscillation on the internal supply voltage. Parasitic capacitors on die together with parasitic inductors from the bondwire, the package and wiring on the test board form an LC oscillation circuitry. Chip internal decoupling circuits can help to reduce the oscillation. Again, a large capacitance per area could help to limit the die size.

Finally, the devices need trimming. Memory and digital control circuitry are required. The process used for this device provides electrical poly fuses. A large digital gate density is advantageous.

3.7.3 Results

The most significant parameter of the test chip was noise. Anticipated was an 18 bit SAR ADC that only shows 1–2 codes at the digital output if a DC input voltage is continuously converted. Figure 3.57 shows a portion of the transfer function in the 18 bit mode. It is clearly visible that each code has some input voltage range V_{in}, where only that particular code occurs.

The DC histogram is shown in Fig. 3.58. It reflects the excellent noise performance, which was already presented in Fig. 3.57. Only 2 codes occur at the ADC output in the 18 bit mode compared to 6 codes peak-to-peak in state-of-the-art 18 bit SAR ADC. Note that this is achieved in single-ended mode, which is far easier to drive with external components compared to the fully differential input range of the state-of-the-art SAR ADCs.

Note that this noise performance is actually achieved with a low current consumption of 2 mA from the analog supply. Table 3.3 shows the relation between the power consumption and the conversion rate. The low power consumption is possible as the core utilizes zero-power sampling by coupling the sample capacitors to ground during the sample phase.

Table 3.3 also shows a dynamic portion, which is the change of the power consumption versus the conversion rate. This current is mostly caused by the power consumption of the digital circuitry. At 300 kSPS, the dynamic portion nearly reaches the quiescent current of the analog circuitry. This stresses the need to move towards smaller process dimensions and lower supply voltages for ultra-low power converters as it is proposed in Sect. 3.8.

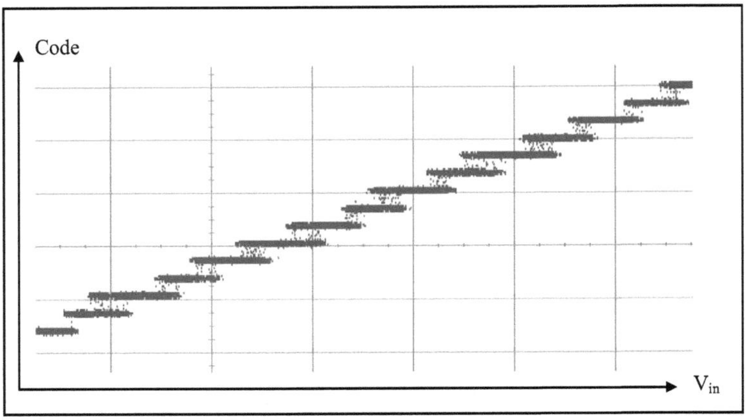

Fig. 3.57 Measured partial transfer function of the highest performance SAR ADC in 18 bit mode

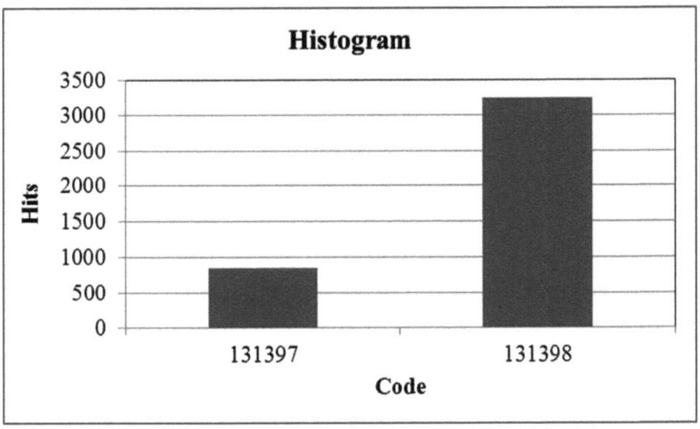

Fig. 3.58 DC Histogram of highest performance SAR ADC in 18 bit mode showing only 2 LSB of noise

Table 3.3 Power consumption of highest performance SAR ADC versus conversion rate

Conversion rate (kSPS)	0	75	150	225	300
Power consumption (mW)	10.2	11.6	13.9	16.2	18.5
Dynamic power portion (mW)	0	1.4	3.7	6	8.3

Probably the most important specification for applications using SAR converters is the linearity in form of DNL and INL. Figure 3.59 shows that the DNL clearly remains within the range of ± 1 LSB without any missing codes. The INL shown in Fig. 3.60 is slightly worse compared to the state-of-the-art converters caused by the voltage coefficients as higher voltages are directly sampled on the sampling capacitors.

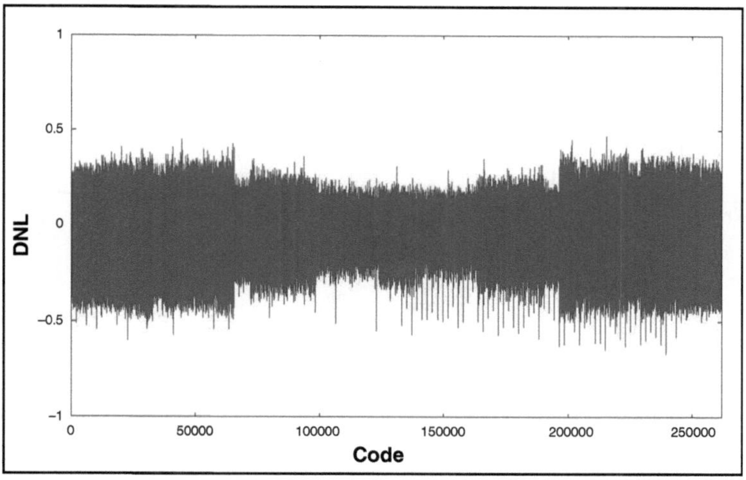

Fig. 3.59 DNL of highest performance SAR ADC in 18 bit mode

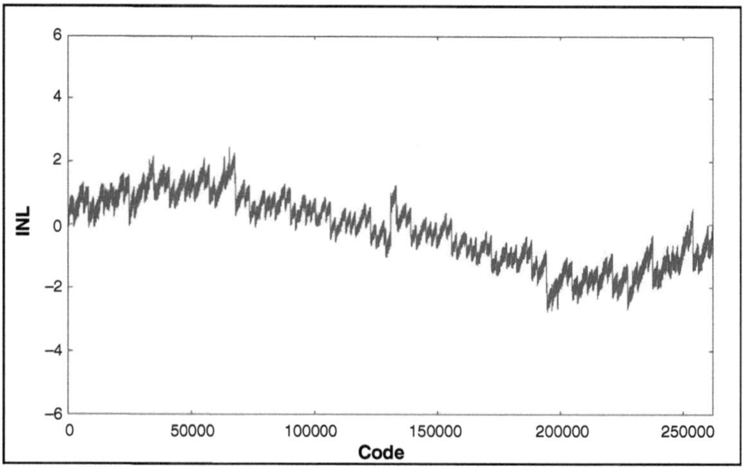

Fig. 3.60 INL of highest performance SAR ADC in 18 bit mode

The INL however has the expected S-shape and could be corrected with an INL trim scheme as introduced in Sect. 3.4. However, this INL performance was difficult to achieve, mainly because of the difficult load generated by the reference input. It would be beneficial to have optimized reference drivers on the die.

All state-of-the-art SAR ADCs work with a large external reference capacitor with a size so that the voltage across it will not drop by more than half an LSB during the conversion. Figure 3.53 showed that the CDAC is switching more than 320 pF between $REF+$ and $REF-$. Fortunately, the capacitors will always switch in series, so that the worst case load calculates to the quarter of that size

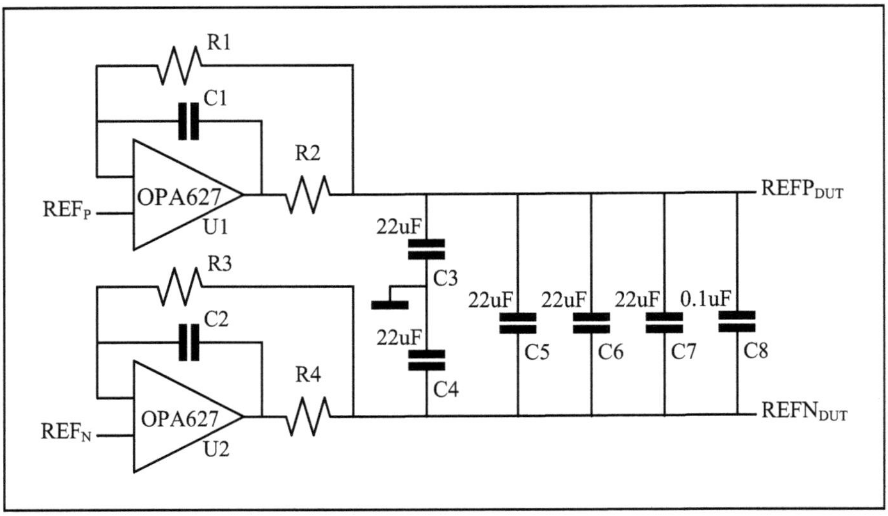

Fig. 3.61 Chip external driver circuitry for the positive and the negative reference voltage for the ADC

($C_{load} = 80$ pF). The external capacitor requires a size of approximately 2^n times C_{load} or roughly 80 μF. Such capacitors are very difficult to drive as most operational amplifiers get instable with capacitive loads. Figure 3.61 shows the driving circuitry on the test board.

$C5$, $C6$ and $C7$ each being 22 μF provide the capacitance to stabilize the reference voltage. $C8$ with only 0.1 μF is added to have a lower ESR capacitor available. $C3$ and $C4$ should further stabilize to reference voltages towards ground. Each reference potential is driven with an amplifier. Resistors ($R1, R2, R3, R4$) and capacitors ($C1, C2$) are added in the feedback path to lower the amplifiers bandwidth and therefore stabilize it.

This circuitry had the following issues. $C3$ and $C4$ coupled distortion from the ground asymmetrically into the reference voltages. Removing them improved the INL performance. Additionally, the low ESR of $C8$ enabled an oscillation circuitry with a high figure of merit together with the die internal capacitors and the parasitic inductors mentioned above. Capacitor $C8$ therefore had to be removed. The larger capacitors with higher ESR lower the figure of merit. Finally, the amplifiers $U1$ and $U2$ are also affected by the oscillation. Decoupling them from the capacitors would help. However, the reference input current is input voltage dependent. If any resistance is used for decoupling between the amplifiers and the capacitors, then an input dependent voltage drop would occur across the resistors and would generate INL.

Devices, which offer lowest DC resistance but higher impedance at frequencies of the oscillation, are ferrites. Therefore, $R2$ and $R4$ were replaced by ferrites. $R1$

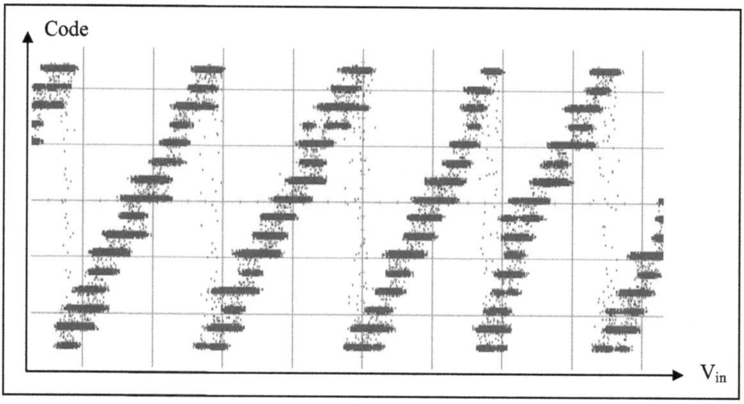

Fig. 3.62 Partial transfer function of highest performance SAR ADC in 20 bit mode

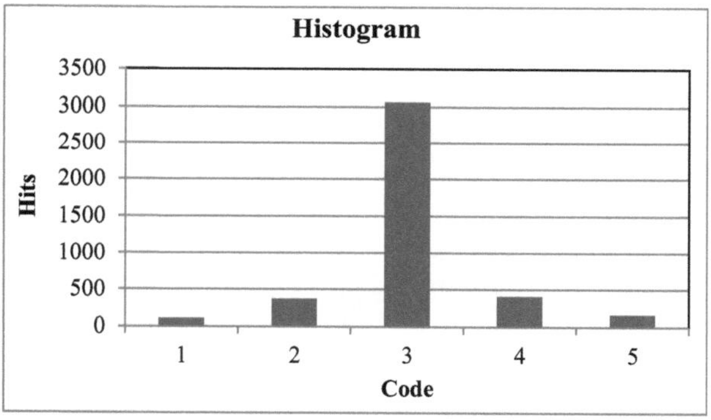

Fig. 3.63 DC Histogram of highest performance SAR ADC in 20 bit mode

and $R3$ were removed to keep the ferrites outside the amplifiers loop. The loop was then closed by replacing $C1$ and $C2$ with 49.9 Ω resistors.

Now as the 18 bit mode provides the expected performance with groundbreaking noise performance, data for the fully differential 20 bit mode will be presented starting with the noise performance in Figs. 3.62 and 3.63. Note, that only the last 4 bits of the transfer function are visualized. This is causing the jump after every 16th code.

The DC histogram in Fig. 3.63 proves that the test chip has less noise in terms of LSBs in the 20 bit mode than state-of-the-art SAR ADCs show with their 18 bits. The SNR therefore improves by more than 12 dB. The rms-value of the signal S can be expressed in LSBs by the total amount of codes 2^{20} divided the square root of 8. The rms-value of the noise N can be estimated to the peak-to-peak noise of 5 LSBs divided by 7.34, which would statistically reflect that one out of the 4096 samples is allowed to be outside the presented 5 LSB window. The rms-noise is

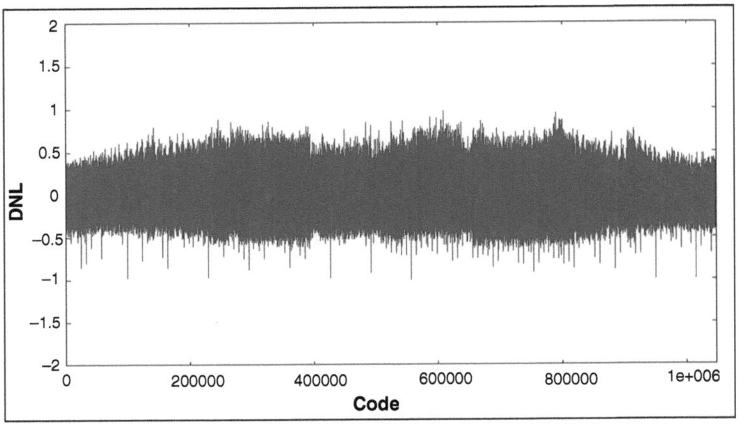

Fig. 3.64 DNL of highest performance SAR ADC in 20 bit mode

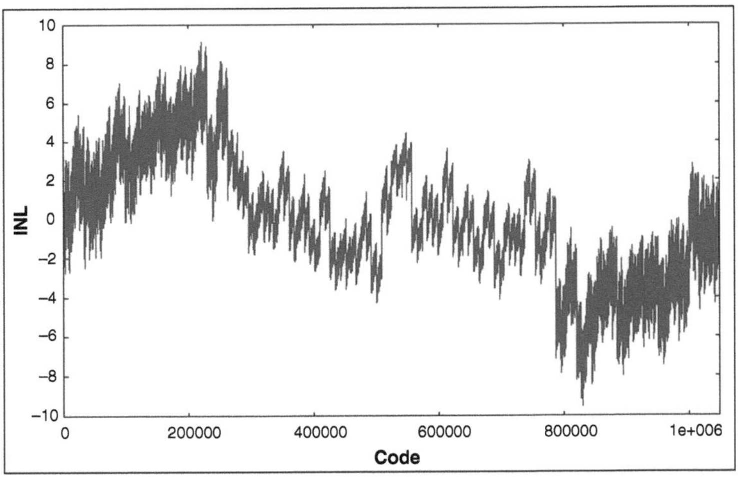

Fig. 3.65 INL of highest performance SAR ADC in 20 bit mode

then 0.685 LSB, which is very close to the exactly calculated value of 0.667 LSB. The SNR can then be estimated to

$$SNR = 20 \cdot \log\left(\frac{S}{N}\right) = 20 \cdot \log\left(\frac{370728}{0.685}\right) = 114.6 \text{ dB} \tag{3.60}$$

The DNL and INL performance are illustrated in Figs. 3.64 and 3.65. Basically all codes are in the DNL range of ± 1 LSB. Only 6 codes are on the edge to be missing.

Based on the calculations in Sect. 3.4, the INL was expected to be around ± 24 LSB, but came out a factor of 3.4 better than that. The quadratic voltage coefficient of this process is specified to 1 ppm/V^2. However, such small values are extremely difficult to measure accurately. This converter offers such accuracy. The general shape however is as expected.

Finally, the digital trim solution was implemented on the test chip. The functionality of the digital trim solution was verified. In a first step, the capacitor matching has to be measured. This could be done with the novel method to measure the capacitive matching presented in Sect. 3.5, which however was not implemented on this test chip. Section 3.5 also presents in Fig. 3.34 (also see Fig. 3.66) that the capacitor matching can also be measured via the DNL of the code prior to the capacitor value. If for example the matching M_{Cx} of the capacitor of bit x of an n bit ADC should be trimmed, then the trim value is equal to the DNL of the code below the respective transition:

$$M_{Cx} = DNL(code = 2^{n-x} - 1) \tag{3.61}$$

Such a code is called major bit transition. Distortions in the transfer function mostly occur at these major transitions. The DNL is not only caused by the capacitive mismatch, but also the settling requirements are most severe at these codes causing the already mentioned dynamic effects.

If for example bit 7 of the test chip in the 18 bit mode should be trimmed, then the DNL of code 2047 expresses the matching of the bit 7 capacitor to all smaller capacitors. Figure 3.66 also shows that it is easier to measure the matching, if the bit capacitor is too large. For that case, only the particular DNL needs to be measured. If the particular bit capacitor is too small, then several codes are missing as shown on the right picture of Fig. 3.66, which means that the DNL of several codes needs to be measured all the way to the first existing code that occurs prior to the bit transition.

The procedure of the digital trimming process is explained next. The DNLs of all major transitions are measured first. On the test chip, these are the transitions for all capacitors in the MSB array. It is assumed and proven that the matching of the capacitors below the first scale-down capacitor in Fig. 3.53 is sufficient. For the 7 capacitors in the MSB array of the test chip, these codes would be number 2047,

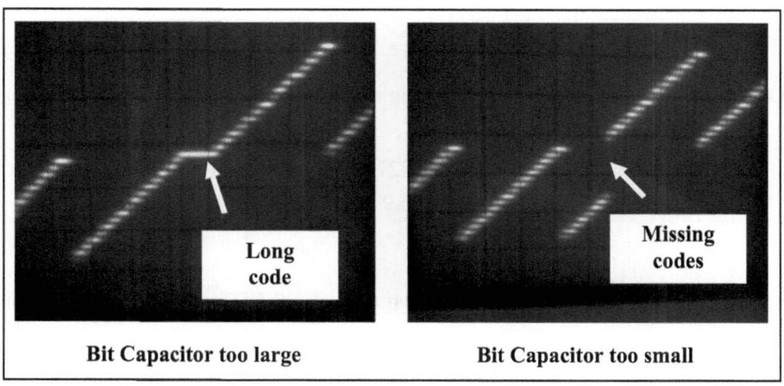

Fig. 3.66 Errors in the transfer function of an SAR ADC caused by capacitor mismatch

4095, 8191, 16383, 32767, 65535 and 131071. Note that the dynamic error correction would shift effects to other locations in the transfer function or would even smooth out capacitive mismatches. The dynamic error correction therefore has to be turned off. To avoid dynamic errors the conversion rate is typically reduced.

The capacitors of the first dynamic error correction are also placed prior to the first scale-down capacitor and therefore also require correction. Particular test modes were implemented to measure their matching, where the particular DEC capacitor to be measured is acting as the bit 7 capacitor.

The trim process is now explained for a particular device, where the DNL of the particular codes was measured as listed in Table 3.4. Let's assume that values can only be adjusted by full LSBs. Note that a minimum trim step of 1 LSB can result in a DNL better than ±0.5 LSB. The minimum trim step is a tradeoff between accuracy and trim overhead such as additional memory and logic.

The scale-down trim would be utilized for trimming the DNL of code 2047 to zero. This will minimize the gain error as explained in Sect. 2.2. The scale down trim would increase the weight of the lower bits, which affects all major transitions in the same way. Here the weight of the lower bits would be increased by 9 LSB, which will lower the DNL of the codes as shown in Table 3.5.

The DNL of code 2047 is −0.3 LSB and therefore within the range of ±0.5 LSB. The bit 7 capacitor therefore matches well enough to the lower bit capacitors. In the next step, the bit 6 capacitor will be evaluated. Code 4095 is too short by 2 LSB. This error value of +2 LSB has to be written to the correction register. The effect of positive correction values to the transfer function is shown in Fig. 3.67.

Shifting the transfer function down will generate an additional error of 2 LSB at all more significant transitions. In other words, the correction value of 2 LSB for the bit 6 capacitor will shorten the other major transitions of the upper capacitors from bit 5 through bit 1. Correcting the digital output code of bit 6 by 2 LSB will cause the DNL shown in Table 3.6.

Correspondingly, a negative correction value would shift the transfer function up respectively as shown in Fig. 3.68. For input voltages that generate long codes, the ADC will not achieve convergence. This gap will later be adjusted by the dynamic error correction. Moving the transfer function up by 2 LSB will make the next

Table 3.4 Initial DNL measurement for the digital capacitor trim of the high performance SAR ADC

Code	2047	4095	8191	16383	32767	65525	131071	DEC1↑	DEC1↓
DNL	8.7	7	6.9	11	6.9	5.2	12.5	0.5	1.4

Table 3.5 DNL after activating the scale-down trim of −9 LSB of the high performance SAR ADC

Code	2047	4095	8191	16383	32767	65525	131071	DEC1↑	DEC1↓
DNL	−0.3	−2	−2.1	2	−2.1	−3.8	3.5	−8.5	−7.6

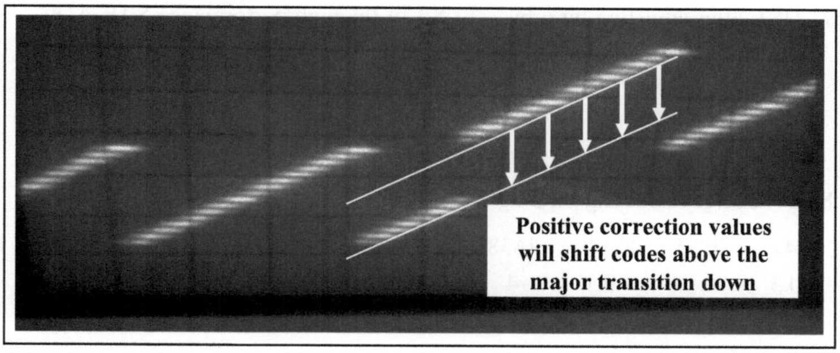

Fig. 3.67 The effect of positive trim correction values applying to the high performance SAR ADC

Table 3.6 DNL after correcting bit 6 by +2 LSB of the high performance SAR ADC

Code	2047	4095	8191	16383	32767	65525	131071	DEC1↑	DEC1↓
DNL	−0.3	0	−4.1	0	−4.1	−5.8	1.5	−8.5	−7.6

Fig. 3.68 The effect of negative trim correction values applying to the high performance SAR ADC

major transition longer by 2 LSB. If bit 6 would have been corrected with −2 LSB, then the DNL shown in Table 3.7 would be the consequence.

The correction steps based on Table 3.6 would continue the following way. The bit 5 capacitor, which is represented by code 8191, is now too small by 4.1 LSB and therefore has to be corrected with 4 LSB. This will cause the DNLs described in Table 3.8.

Based on Table 3.8, the capacitor representing bit 4 (code 16383) is now also short by 4 LSB and needs a 4 LSB correction. This will shorten code 32767 to −12.1 LSB, so that a correction of the bit 3 capacitor by 12 LSB is required. This will cause the intermediate result from Table 3.9.

Table 3.7 DNL after adjusting bit 6 by −2 LSB of the high performance SAR ADC

Code	2047	4095	8191	16383	32767	65525	131071	DEC1↑	DEC1↓
DNL	−0.3	−4	−0.1	4	−0.1	−1.8	5.5	−8.5	−7.6

Table 3.8 DNL after correcting bit 5 by +4 LSB of the high performance SAR ADC

Code	2047	4095	8191	16383	32767	65525	131071	DEC1↑	DEC1↓
DNL	−0.3	0	−0.1	−4	−8.1	−9.8	−2.5	−8.5	−7.6

Table 3.9 DNL after correcting bit 4 by +4 LSB and bit 3 by +12 LSB of the high performance SAR ADC

Code	2047	4095	8191	16383	32767	65525	131071	DEC1↑	DEC1↓
DNL	−0.3	0	−0.1	0	−0.1	−25.8	−18.5	−8.5	−7.6

Table 3.9 shows that bit 2 needs a correction value of +26 LSB, which will cause the bit 1 capacitor to be too small by 44.5 LSB. So that finally bit 1 needs correction of 44 or 45 LSB.

Also the dynamic error correction capacitors are within the MSB-array and therefore require trimming. Their influence on the comparator input voltage has to be less than the sum of all the bits below; otherwise the ADC might not reach convergence, which is the key requirement for the digital trimming. Therefore, the DEC capacitors are designed to be 5.25 LSB smaller compared to the bit 7 capacitors. The first dynamic error correction that can increase the comparator input voltage was measured to be −8.5 LSB shorter than the bit 7 capacitor and therefore needs correction by +3 LSBs. The first dynamic error correction that can lower the comparator input voltage was measured to be −7.6 LSB shorter than bit 7 and therefore needs correction by −2 LSB. Note that the sign changed, which is due to the inverting effect on the comparator input voltage.

Table 3.10 shows the overview of the correction values based on a scale-down adjustment of −9 LSB. Table 3.10 shows that particularly bit 1 requires a large correction range. The sum of all correction values is 92 LSB. These will also cause a gain error of 92 LSB.

The separate scale-down and bit 7 trims enable some options to optimize the required trim range. If for example the scale-down trim is only set to −8 LSB, then the correction values change to the values in Table 3.11. The sum of correction values reduces to −34 LSB.

Figures 3.69 and 3.70 prove the functionality of the digital trimming. The all-code DNL is illustrated in the 18 bit mode prior of the digital trimming (Fig. 3.69) and after the digital trimming (Fig. 3.70). The DNL after the digital trimming remains within the anticipated ±1 LSB range.

Note that the measurements in Figs. 3.69 and 3.70 were measured in a special 5 V trim mode. The signal range is 8 times lower. This enables the reference noise to enlarge the DNL towards the positive full scale.

Table 3.10 Required capacitor correction values, if the scale-down trim is set to −9 LSB

Bit	7	6	5	4	3	2	1	DEC1↑	DEC1↓
Correction value	0	2	4	4	12	26	44	3	−2

Table 3.11 Required capacitor correction values, if the scale-down trim is set to −8 LSB

Bit	7	6	5	4	3	2	1	DEC1↑	DEC1↓
Correction value	−1	0	0	−4	−4	−6	−19	2	−1

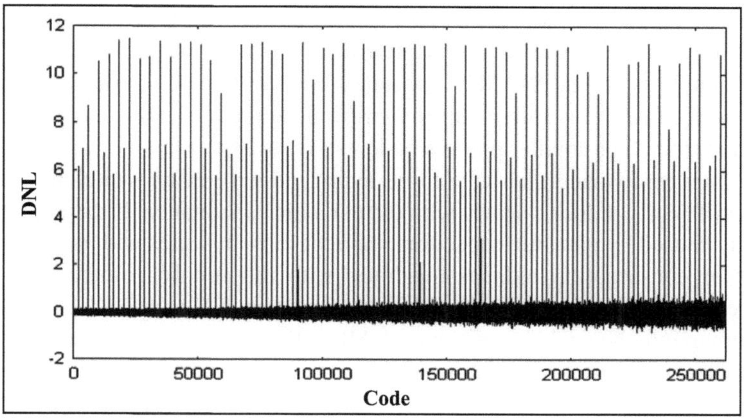

Fig. 3.69 DNL of the high performance SAR ADC prior to digital trimming

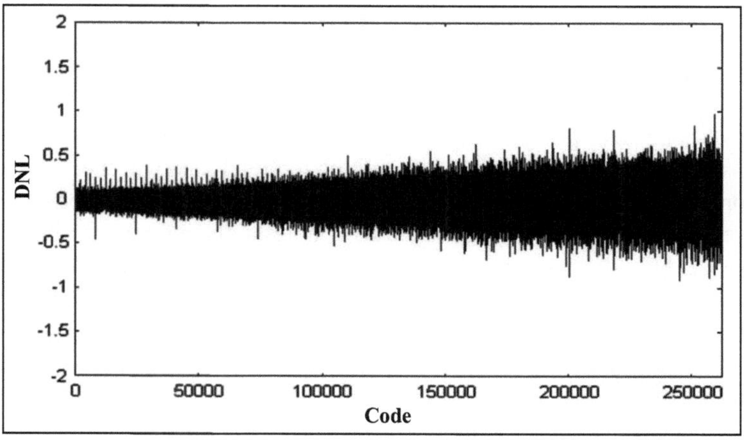

Fig. 3.70 DNL of the high performance SAR ADC after digital trimming

3.8 Lowest Power SAR ADC [21]

The architectures presented in Chap. 3 can also be used to move state-of-the-art 5 V SAR ADCs to smaller process nodes and lower supply voltages. In a first estimate, it could be assumed that the current consumption of the comparator remains about constant as the current is required to achieve good noise performance. Moving from a 5 V supply to a 1.2 V supply would then reduce the power consumption by a factor of 4.2.

Adding the zero-power sampling architecture enables the removal of the common-mode generation as the sampling occurs to ground. The power consumption of the common-mode generation was in the same range as the one of the comparator. Furthermore, the comparator can be placed into a power-down mode during sampling, so that the ADC consumes basically no power during sampling.

It was dicussed previously that a significant amount of power is burnt by the digital circuitry. The smaller process node will reduce the parasitic capacitors of the digital gates and wiring that have to be recharged with every switching step. The power will nearly reduce linearly with the parasitic capacitance. If the digital power is mostly caused by parasitic capacitors, then the current scales with the voltage and the power consumption reduces quadratic with the supply voltage. The digital power consumption can be neglected if moving from 5 V to 1.2 V.

Summarized, the goal is to achieve a SAR ADC with a power consumption in the range of 1–2 mW at a 1 MSPS conversion rate. Additionally, the new process node should help to reduce the die size dramatically by a factor of 2. The core circuitry of the state-of-the-art SAR ADC presented in [4] has a size of 1.62×2.05 mm.

Another aspect is the device sensitivity of state-of-the-art ADCs. As mentioned before, parasitic well capacitors and parasitic package and board inductors are forming oscillation circuitries for the die internal supply and reference voltages. Spikes on an internal 5 V supply can reach 1 V. The oscillation will couple into the analog signal path and will lower the analog performance in terms of offset, offset drift or DNL. The DNL of such a device is illustrated in Fig. 3.71.

The all-code DNL plot is slim in the middle, where the positive and the negative CDAC side are in similar states. Distortion on the reference will then only cause a common-mode voltage at the differential comparator inputs. At the positive and the negative full scale, the positive CDAC fully couples to the reference voltage and the negative CDAC side to ground and vice versa. Distortion on the reference will then directly couple into the signal path. As a result, the DNL in Fig. 3.71 widens up towards the full scales. The higher the integration level or the faster the clock rate is, the more issues are caused by the oscillation effect.

As an additional goal, decoupling capacitors should be added into the supply and the reference path to allow a more robust operation. This should enable higher integration such as adding the ADC core to a microcontroller or providing multichannel SAR ADCs with up to 16 ADCs integrated on one die.

Digital trimming and scale-down trimming further help to reduce the die size and to lower the reference load. They should therefore also be integrated on the die. Also the novel method to measure capacitive matching is implemented to keep the

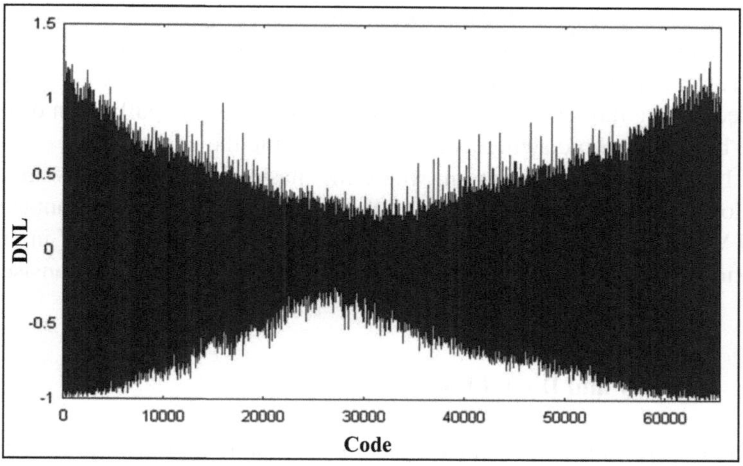

Fig. 3.71 DNL of a SAR ADC with distortion coupled into the signal by supply oscillation

production costs low. The INL trim was not implemented on this die. The 5 V input range together with a good capacitive voltage coefficient of the selected process should provide a good INL by itself.

A design investigation was implemented. Simulation results are provided below.

3.8.1 Process Selection

The required process components to achieve the goals above will be described in the following.

3.8.1.1 CDAC Capacitors

Most important are the CDAC capacitors as they have the largest influence on the performance such as DNL, INL and THD as well as on the die area. They need to provide

- Low voltage coefficients
- Large capacitance per area and good matching
- Metal-metal capacitors above metal 2 to enable the usage of the die area underneath
- A minimum of 4 metal layers for top plate shields of the CDAC capacitors

3.8.1.2 Dual Gate Oxide Process

The dual CDAC architecture should run from a 1.2 V supply with a 5 V input range and a 5 V reference voltage to optimize the SNR. A dual gate oxide process is required with 1.2 V transistors and 5 V transistors. The low voltage transistors should provide a good gain for the comparator input stage, which requires reasonable drain-source resistance and transconductance. It is advantageous if the bulk of the NCH transistors can be isolated from the substrate. This feature can be used for noise decoupling through substrate, but this also enables the pull-down of the gate and the bulk of the hold switch below ground during the conversion.

A 0.18 μm process was chosen, where the mask and wafer costs were significantly lower than of a comparable 0.15 μm process. It would be advantageous, if also 20 V transistors are available to enable a later extension to ±10 V inputs. The integration of internal reference circuits would benefit from bipolar transistors.

3.8.1.3 Memory and Digital Logic

Some kind of one time programmable memory is required for the trimming process. Electrical fuses or some kind of PROM are both usable. Electrical fuses are larger than PROM cells, but do not require programming overhead such as a charge pump.

The digital gate density is less of a concern as the gate count is typically in the order of 15 k. A good density on the other side will minimize the wiring and therefore the parasitic capacitance to provide low power consumption.

3.8.1.4 Decoupling

NWELL capacitors will be used for the supply and reference decoupling. They must be available for the 1.2 V supply with larger capacitance per area and for the 5 V reference. It would be preferable, if the 5 V NWELL capacitor can be placed underneath the metal-metal capacitors of the CDAC.

An overview of available processes is given in Table 3.12. Process 3 fits best for the requirements described above and was selected for this test chip.

Table 3.12 Exemplary process overview showing parameters relevant for SAR ADCs

Process	1	2	3	4	5	Unit
Minimum gate length	0.18	0.25	0.18	0.15	0.15	μm
Metal capacitor	Yes	Yes	Yes	Yes	Yes	
Linear voltage coefficient	–	12–50	12–50	0	0	ppm/V
Quadratic voltage coefficient	–	**5–30**	<5	5	5	ppm/V^2
Metal capacitance/area	1.12	1.18	1.18	1.3	1.24	fF/μm^2
Poly-NWELL capacitance/area	8.5	7	7.2	10	10.2	fF/μm^2
Number of metal layers	5	5	5	5	6	
HV transistors (maximum V_{gs})	5	5	5	**3.3**	**3.3**	V
Further HV transistors	–	–	22	–	–	V
Thin film resistors	No	No	Yes	Yes	No	
Memory	Yes	Yes	Yes	Yes	Yes	

Parameters of highest relevance are shown in bold

3.8.2 *Implementation and Design Challenges*

The core of the design is the dual CDAC architecture (see Fig. 3.72). Again, both CDAC were designed with their own comparator. The coarse ADC has 4 bit resolution. The fine ADC provides the full 16 bit. The reference can range between 2.5 V and 5 V. 5 V will provide the best results in respect of SNR. The SAR will control the conversion process of both ADCs and forward the raw data to the digital interface, which also includes the adjustments for the dynamic error correction and the state machine for the digital trimming. Analog trimming is not foreseen anymore beside the scale-down trim.

Both CDACs are sampling to ground, which is the requirement for zero-power sampling. Again, the bulk and the gate of the hold switch are pulled to half the supply below ground (−0.6 V) during the conversion. Comparator input voltages of ±500 mV will not cause charge leakage through the hold transistors. Both ADCs therefore need to attenuate the signal amplitude by a factor of 5. The coarse ADC will again attenuate the signal amplitude with a capacitive divider, the fine ADC will not activate the sampling capacitors until the coarse ADC has the first 3 bit evaluated.

The comparator of the coarse ADC has relaxed requirements in terms of noise and gain due to the error correction range of the fine ADC of nearly ±40 mV. It was therefore decided to have separate comparators for both ADCs. The coarse comparator will have smaller devices, less parasitic capacitance and a faster wake-up

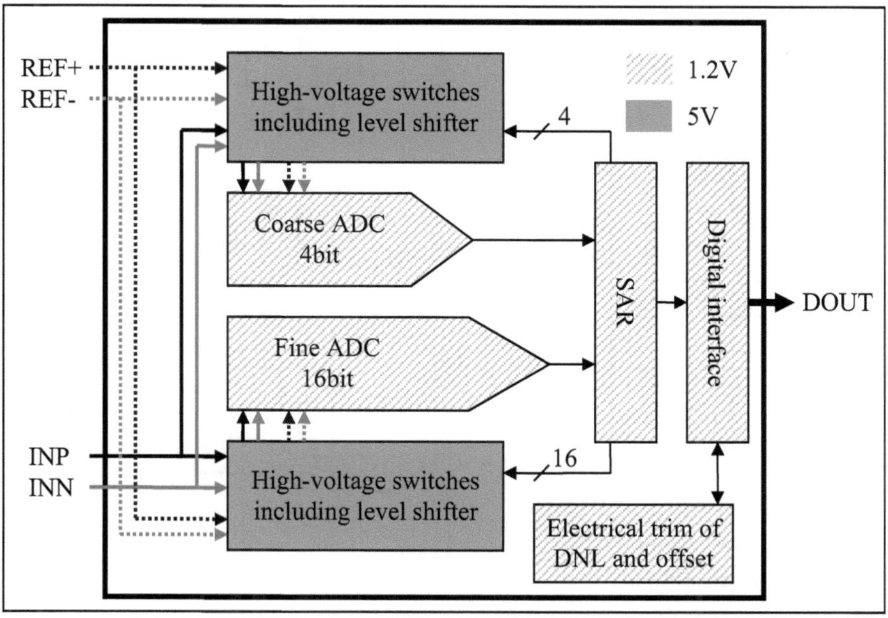

Fig. 3.72 Block diagram of the lowest power 16 bit ADC using dual CDAC architecture

time. The offset cancellation step of the coarse comparator can be performed within 2 clock cycle. During the next 4 clock cycles, it will convert the 4 MSBs and then turn off again. It is only active during 6 clock cycles per conversion. The fine comparator can use 5 of these clock cycles for its offset cancellation.

The switching process of the fine ADC is shown on the example of the MSB capacitor in Figs. 3.73 and 3.74. The bit capacitor is switched as always in the sample phase between the input voltage V_{in} and ground. The bold lines indicate the transistor state as off and on. After sampling, the capacitor is switched into the floating phase, while the coarse ADC is finding its decisions. The comparator input voltage V_c remains close to 0 V due to parasitic capacitors to ground.

After the evaluation of the first 3 bits, the fine ADC will couple the floating capacitors to the reference potentials in respect to the results of the coarse ADC. In this case, the MSB was evaluated to ‘1’. Section 3.6 discussed that the turn on process will generate glitches caused by timing or impedance mismatch of the reference switches. To keep the glitches low, a soft turn on was implemented. During this phase only a fraction of the reference switch closes, which is additionally followed by a resistor. The resistor value should be larger than the on-impedance of the switch. The resistor will smooth out the possible impedance mismatch of NCH switches to the negative reference (ground) and PCH switches to the positive reference (*REF*). The high impedance will further cause a slow settling and therefore avoids glitches caused by timing mismatches.

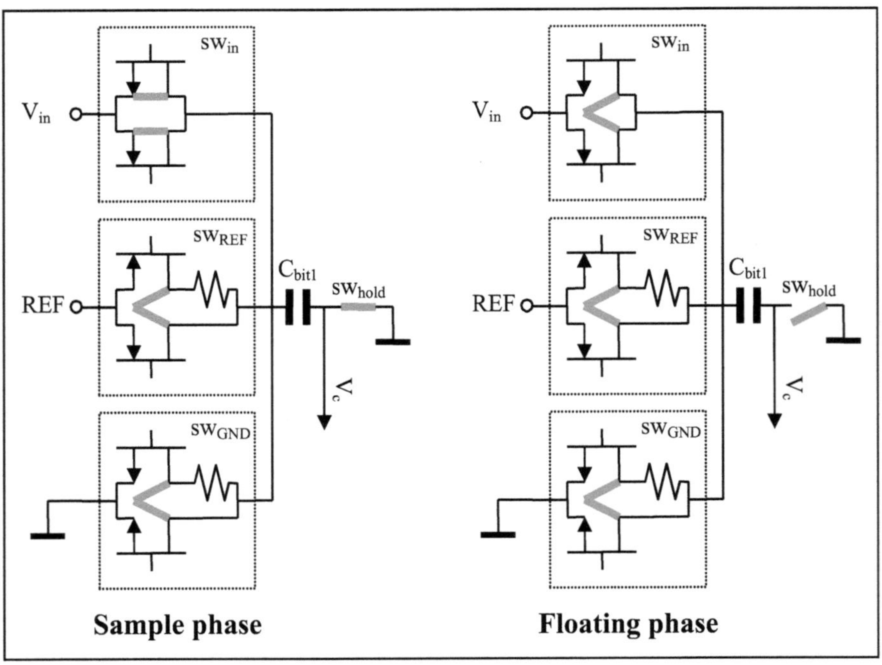

Fig. 3.73 Bit 1 capacitor of SAR ADC with dual CDAC architecture in sample and floating phase

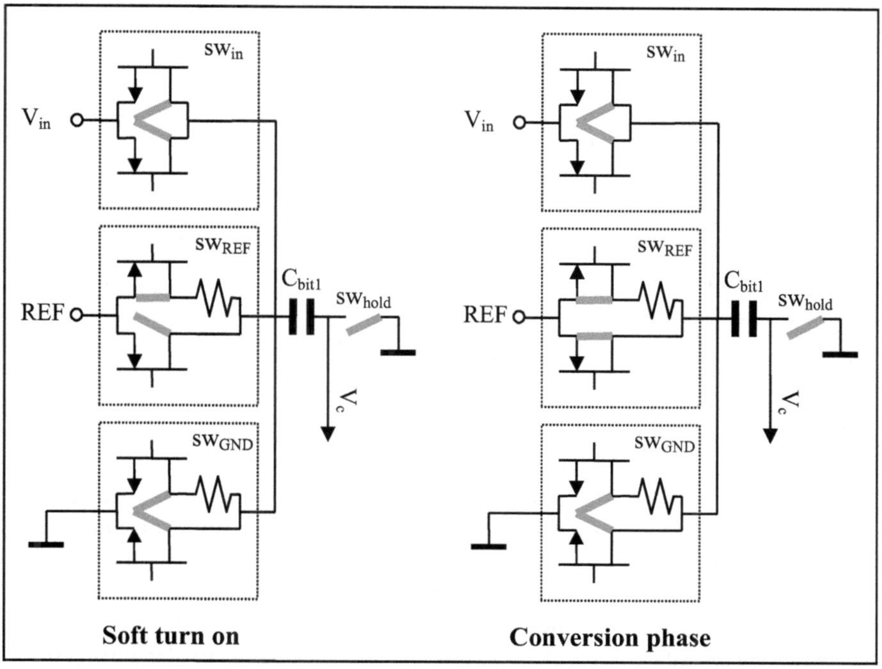

Fig. 3.74 Bit 1 capacitor of SAR ADC with dual CDAC in the soft turn on and conversion phase

Once the coarse CDAC finished the 4 MSB decisions, the fine ADC has to continue the conversion with full speed. Slow settling is not allowed anymore. Therefore, the larger fraction of the reference switch is turn on for the conversion phase as well.

The soft turn on of the fine ADC might cause distortion on the reference and influence the decisions of the coarse ADC. To minimize such effects, the reference of the MSB array REF_{MSB} has its own bondpad and also its own on-chip decoupling capacitor. The reference of the coarse ADC REF_c is shorted to the reference of the LSB array of the fine ADC. As long as the coarse ADC is evaluating bit decisions, the LSB capacitors are grounded. Also, when the LSB capacitors are active, the coarse ADC has finished its decisions and remains idle. Both references are shorted on package level with bondwires to the same pin. The negative reference, which here is ground, is treated the same way.

The area underneath the capacitor arrays is used for Poly-NWELL decoupling capacitors for the reference voltages REF_{MSB} and REF_c. The layer stack of the capacitor array is illustrated in Fig. 3.75. The linear capacitors of the CDAC are implemented in metal 2 and 3. Metal 4 can therefore be used to shield the capacitors from external influences like electric fields or inhomogeneous dielectrics of the package. The full area underneath the linear capacitors is used for poly-NWELL capacitors, which are connected with metal 1. It is ensured that the wiring is done symmetrical for all linear capacitors.

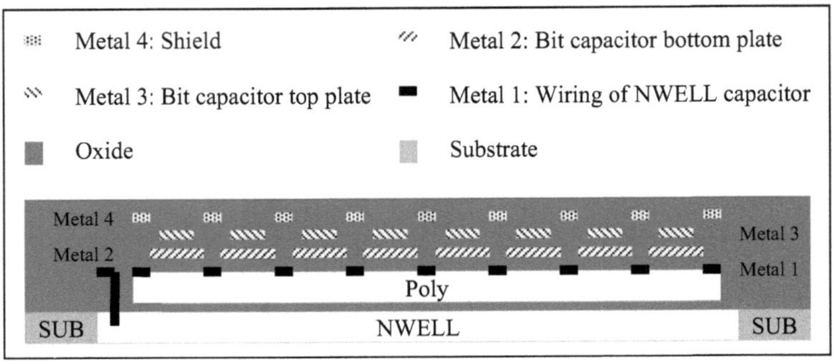

Fig. 3.75 Particular layer stack of the capacitor array in a CDAC with metal capacitors

345 pF are added for REF_{MSB} and 490 pF for REF_c. The more decoupling capacitance is used, the better the oscillation is attenuated. On the other side, the capacitors and the wiring also have around 6Ω resistance associated. The decoupling capacitors for REF_{MSB} of 345 pF would form a time constant of 2 ns together with the parasitic resistance. The largest reference load occurs, when the floating capacitors are connected to their reference potentials. Fortunately, the soft turn on will provide the reference a significant amount of settling time. This makes the switching process of the 1.25 pF of the bit 5 decision to the most critical reference load. If a discharged 1.25 pF capacitor connects to the 345 pF decoupling capacitor, then the chip internal REF_{MSB} will drop by 0.361 % or for a 16 bit converter by 237 LSB. If the internal reference settles back exponentially to the external reference voltage with at least half an LSB accuracy, then the settling time T_S would be around 18 ns based on Eq. (2.12).

This raises the question on how much settling time is available. A conversion rate of 1 MSPS is anticipated, which provides 1 μs for the sample and the conversion phase together. As a rule of thumb, the application should at least have 200 ns for the accurate sampling, otherwise the bandwidth together with the power consumption of the driving amplifiers is getting too large. This would leave 800 ns for the conversion. The conversion requires the 2 clock cycles for the offset cancellation of the coarse comparator. Then the coarse ADC will convert the 4 MSBs. The fine ADC will continue the conversion of bit 5 through 7 followed by a first dynamic error correction step. The conversion continues with the decisions of bit 8 through 13 and is then followed by the second dynamic error correction. The process is finished by the evaluation of the bits 14 through 16. Summarized, 20 clock cycles are required for the conversion and consequently one clock cycle will last 40 ns.

Typically, half a clock cycle is provided for the reference settling, the other half clock cycle for the comparator settling. This would leave 20 ns for the reference settling. The 345 pF internal decoupling capacitance is just small enough to allow accurate settling. Note that actually some error would be allowed as it can be corrected by the dynamic error correction.

The corresponding fine CDAC is shown in Fig. 3.76. Note that the largest capacitor in the LSB array, which is the bit 8 capacitor, was reduced to half the

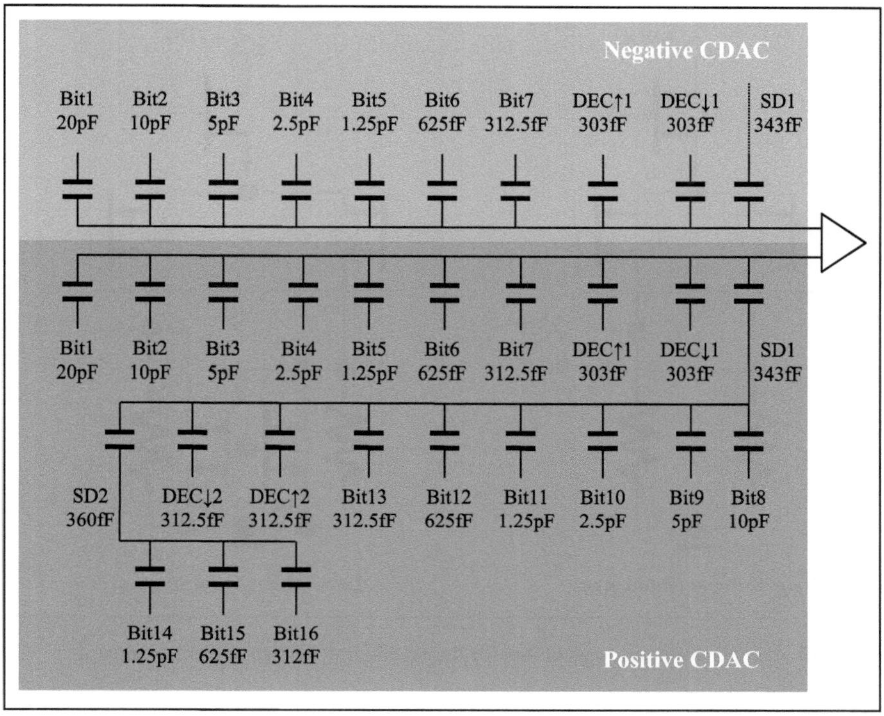

Fig. 3.76 Simplified CDAC for the lowest power SAR ADC

MSB capacitor to reduce the reference load. The minimum capacitor was chosen to 312.5fF. This seemed to be the smallest possible unity capacitance based on the process specifications providing sufficient capacitive matching. A second scale-down capacitor was added for the three LSB capacitors.

The low voltage poly-NWELL capacitor has an increased capacitance compared to the 5 V process used for state-of-the-art ADCs. This fact is used in this design to provide a significant amount of decoupling for every block on the chip. The different blocks like the digital logic of the SAR or the CDAC, the comparator or the biasing all have their particular supply capacitor of at least 100 pF, which further is decoupled with resistors from the supply pad. Crosstalk and oscillation are reduced to a minimum.

The comparator turned out to be particularly difficult to design. The comparator input stage is typically designed as differential pair as drawn in Fig. 3.77. The input voltages can drop below ground due to the zero-power sampling architecture. The ADC input and therefore the comparator input must allow some common-mode range of at least ± 100 mV. The voltage across the load resistors therefore has to be 0.1 V less than the threshold voltage of the transistors T_1 and T_2 and therefore has to remain below 200 mV. The current I of the comparator input stage has to be in the range of 1 mA to provide the required noise density, limiting the load resistors to 400 Ω. The low voltage transistors can further have very limited transconductance if simulated over models and temperature, so that the total gain can drop below a

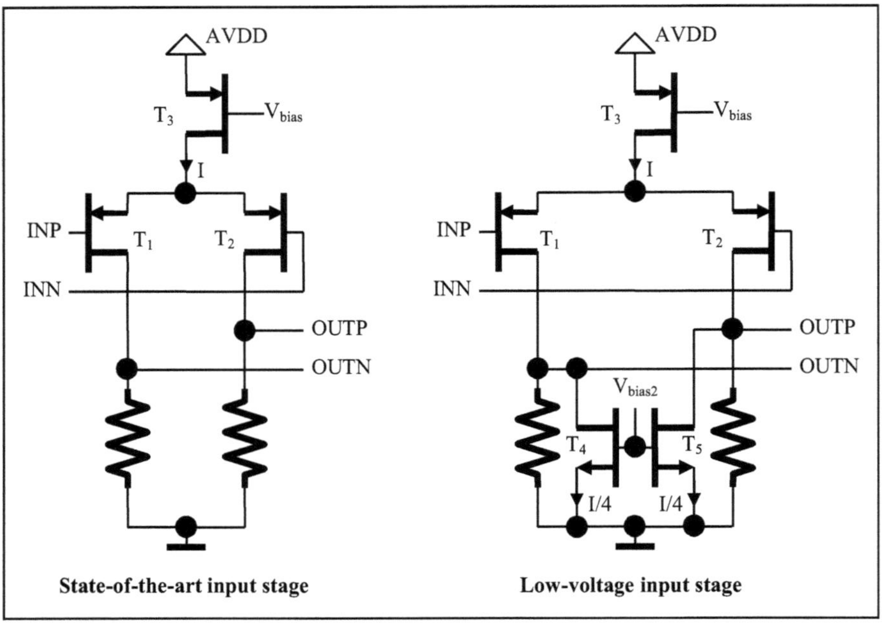

Fig. 3.77 Comparator input stage in state-of-the-art and low-voltage implementation

factor of 2. The comparator has to gain up signals with 20 μV up to 200 mV. The gain therefore has to reach the factor 10,000. 14 serial gain stages are required with a gain of 2 per differential stage. This is too much to achieve the required timing and power goals. The 200 mV output voltage is necessary for a possible offset of the output latch of the comparator.

Alternative 5 V transistors can reach a gate-source voltage above 1 V over process and temperature corners. Adding a small common-mode input voltage of 0.1 V and a saturation voltage of 0.2 V for the transistor T_3, which acts as the current source, causes the required supply voltage to exceed the anticipated goal of 1.2 V.

The only way to remain within the supply voltage limitations and still achieving reasonable gain is shown on the right side of Fig. 3.77. 1.8 V low leakage transistors were finally used as input transistors T_1 and T_2. Their threshold voltage is in the middle of the standard 1.8 V transistors and the 5 V transistors. The drain voltage is limited by reducing the current through the resistors with the additional transistors T_4 and T_5, which are connected as current sources. Note that their length must not be minimal as a limited drain-source resistance would again influence the gain over the simulation corners. The comparator consists of seven such stages connected in series followed by a differential latch.

The noise of the comparator adds to the ADC noise during the comparators offset cancelation and during critical bit decisions. The noise is dominated by the thermal noise of the input transistors T_1 and T_2 multiplied by the square root of the effective bandwidth. The bandwidth therefore has to be limited as much as possible. The first offset cancellation capacitors are placed after the second gain stage, where the offset including the noise is basically stored on capacitors (see Sect. 2.3). It is therefore important to limit the bandwidth of the second gain stage during the offset cancelation phase. An additional capacitor is switched between the outputs of the second differential stage.

This generated another idea. Noise during the conversion of the first bits is uncritical as it can be corrected with the second error correction stage. If the time is increased for the second error correction and the following bit decisions by for example allowing two clock cycles per decision, then the comparator bandwidth can be reduced as it is done with the second gain stage during the offset cancellation phase. A noise reduction mode was implemented that lowers the bandwidth by a factor of 3 and the effective comparator noise by a factor of $\sqrt{3}$ in trade off to four additional clock cycles per conversion.

3.8.3 Simulation Results

The design investigation concentrated on decreasing the power dissipation and reducing the die size by remaining state-of-the-art noise performance. Data is provided in this chapter to demonstrate the extraordinary performance.

Table 3.13 shows the power dissipation of the analog core supply. The power consumption of the digital circuitry was simulated with the extracted view including parasitic wire capacitance. The ADC consumes 1.5 mW at the full conversion rate of 1 MSPS. Best in class state-of-the-art converter achieve around 10 mW [22, 23]. Table 3.13 shows that the digital current still plays an important role. Please note that several additional features such as a programmable oversampling filter are already included. The ADC core itself consumes less than 1 mW. Also all power consuming blocks are fully scalable with the conversion rate f_s. Running the converter at 500 kSPS instead of 1 MSPS would reduce the power consumption by a factor of 2.

Until today, the power consumption of supporting circuitry was typically neglected. Such circuits are the power consumption of the ADC's reference input, a chip internal reference or the digital interface. All of them have to run from a

Table 3.13 Power consumption of the lowest power SAR ADC core

	Supply	Voltage (V)	Current (μA)	Power (μW)	Condition T = 27 °C	Scales with f_s
ADC	AVDD	1.2	800	960		Yes
Digital core	AVDD	1.2	340	408	Extracted layout	Yes
Interface	AVDD	1.2	75	90	Extracted layout	Yes
Sum	AVDD	1.2	1215	1458		100 %

Table 3.14 Power consumption of the supporting circuitry inside the lowest power SAR ADC

	Supply	Voltage (V)	Current (μA)	Power (μW)	Condition T = 27 °C	Scales with f_s
ADC	RVDD	3	185	555	V_{in} = 1.25 V	
Reference	RVDD	3	294	882		
Interface	BVDD	3	240	720	C_{load} = 20 pF	Yes
Sum		3	719	2157		

Table 3.15 Power consumption of the lowest power SAR ADC under worst case conditions

	Supply	Voltage (V)	Current (μA)	Power (μW)	Condition T = 125 °C	Scales with f_s
ADC	AVDD	1.26	1010	1272		Yes
Digital core	AVDD	1.26	460	552	Extracted layout	Yes
Interface	AVDD	1.26	33	40	Extracted layout	Yes
Sum	AVDD	1.26	1215	1864		100 %
ADC	RVDD	3	185	555	T = 27 °C	Yes
Reference	RVDD	3	400	1200		No
Interface	BVDD	3	175	525	C_{load} = 20 pF	Yes
Sum		3	719	2280		

higher voltage supply, which in Table 3.14 is 3 V. It can be seen that this circuitry is now consuming more power (2.16 mW) than the ADC core itself.

The power dissipation also changes over process and temperature corners. The worst case condition was found at 125 °C, where the PTAT bias current is high, and strong models, where digital gates have the highest cross current. Table 3.15 shows the summary of the worst case conditions. The ADC core still remains below the anticipated 2 mW. The support circuitry remains with 2.28 mW nearly constant.

The second significant goal was the ADC core size, which is the ADC without the IO cells as the ADC core is anticipated for higher integration such as microcontroller or multichannel ADCs. The core of the state-of-the-art ADC, which was presented in [4], measured 1.62 × 2.05 mm and therefore 3.3 mm^2.

The layout of this investigation is shown in Fig. 3.78. The core of the ADC measures 0.82 × 1.62 mm (1.46 mm^2) and is therefore by a factor of 2.3 smaller than the state-of-the-art ADC core. Note that the 1.46 mm^2 is already including a severe amount of decoupling capacitors to stabilize the operation of the core and to allow a higher integration level. The state-of-the-art ADC had no decoupling capacitors included. Here, nearly 1 nF of reference decoupling capacitance is buried underneath the two CDACs. The analog core supply is decoupled with another 1.2 nF. Decoupling capacitors are provided for each block such as the CDAC logic, the comparators, the biasing separately. Another 2.04 nF were placed within the IO ring.

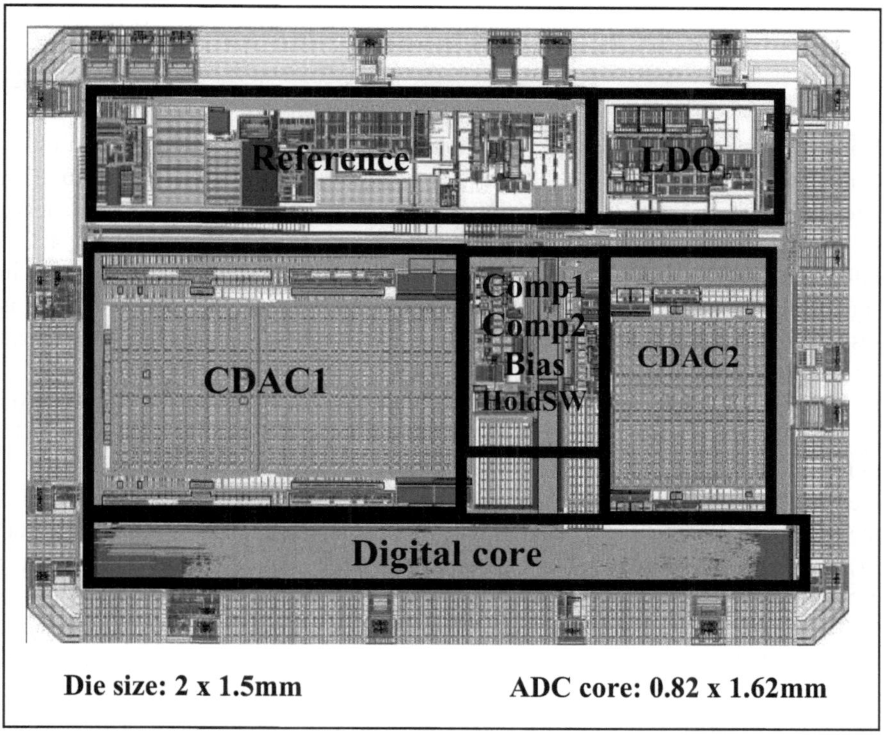

Fig. 3.78 Layout of the low power ADC

Figure 3.78 shows that the coarse CDAC ($CDAC2$) covers a significant amount of area. Measurement data regarding the matching between the fine CDAC ($CDAC1$) and the coarse CDAC needs to prove if the capacitors of $CDAC2$ can further be reduced. The same thing is true for the LSB array of $CDAC1$. Reducing the MSB array would cause less sample capacitance and therefore a higher sampling noise. Also the area of the digital core is larger than expected. However, the functionality exceeds the one in [4] significantly.

The reduction in power dissipation and die size is only acceptable, if the analog performance remains similar to the state-of-the-art ADC as presented in [4], which is achieving a typical of 92 dB in SNR at a 5 V input range.

Equation (3.62) is calculating the noise performance of this test chip. It considers the quantization noise V_q, twice the sampling noise for the positive and the negative sampling capacitor and the conversion noise given by the comparator V_{comp}, the comparator's offset compensation V_{AZ} and the reference voltage V_{REF}. The comparator noise and the reference noise are active at each critical bit decision. There are typically two critical bit decisions per conversion, where the comparator input is less than an LSB. The factor D_{CDAC} considers the attenuation of the internal signal amplitude.

$$V_{n\_rms} = \sqrt{{V_q}^2 + 2 \times {V_{kT/C}}^2 + \left(2 \times {V_{comp}}^2 + {V_{AZ}}^2 + 2 \times {V_{REF}}^2\right)/{D_{CDAC}}^2} \tag{3.62}$$

With 40 pF sample capacitance, a comparator noise of $V_{conv} = 13.5$ μV, an offset compensation noise of $V_{AZ} = 13.5$ μV, a reference noise of $V_{REF} = 9$ μV and an attenuation of $D_{CDAC} = 0.89$, the SNR calculates to 91.6 dB and 89.5 dB under worst case conditions at 125 °C. This data is simulated with an internal 4.1 V reference, while the ADC in [4] is driven with an external 5 V reference. The loss in signal amplitude by the factor of 1.22 due to the reduced reference voltage is responsible for a drop in SNR by 1.7 dB, so that the ADC would reach the typical 92–93 dB in SNR for 5 V single-ended inputs.

Summarized, the power and the die size were reduced in the design investigation by the expected amount. The power consumption was reduced by a factor of 5–10 in respect to comparable state-of-the-art SAR ADCs. The die size was reduced by a factor of 2.3 even after adding a significant amount of decoupling capacitance for the reference and the supply voltage. The internal signal amplitude remains similar to state-of-the-art approaches, so that the analog performance in terms of SNR stays around 92−93 dB.

These novel inventions carry the performance of SAR ADCs to a next level, but the research will not stop here. Improvement is for example required for the limited common-mode range of the dual CDAC ADC. Another wide field of research might be the integration of the sensitive SAR ADC together with power management circuitry, which generates all required internal supply voltages from the digital 3.3 V supply. Today, industrial ADCs might require a 15 V and −15 V supply to support the high input voltage range of ±10 V. It might further require a 5 V internal reference voltage and a 1.2 V supply for the ADC core.

Note that there are also further interesting developments ongoing. One is called hybrid SAR, which is basically combining the CDAC of the SAR ADC and with a flash ADC, which is replacing the comparator and converts around 4 bit at a time [23]. The other alternative is utilizing extensive redundancy with non-binary capacitor ratios. Redundancy will reduce the settling requirements on the cost of additional clock cycles per conversion. The optimum ratio regarding the amount of redundancy and extra clock cycles to achieve the fastest conversion rate is discussed in several papers and patents [24–26].

References

1. Frank Ohnhauser, Andreas Wickmann, *SAR analog-to-digital converter with large input range*, US Patent 7,773,024, Texas Instruments, 2010, URL: http://www.google.com.ar/patents/US7773024, status of 3/11/2015.
2. Frank Ohnhaeuser, Andreas Wickmann, *A floating CDAC architecture for high-resolution and low-power SAR A/D converter*, in the proceedings of the International IEEE Conference on

Semiconductors Dresden Grenoble (ISCDG2012), pp. 5-8, 2012, URL: http://ieeexplore.ieee.org/xpl/articleDetails.jsp?arnumber = 6360003&filter%3DAND%28p_IS_Number%3A6359978%29, status of 3/3/2015.

3. Frank Ohnhaeuser, Andreas Wickmann, *SAR ADC*, US Patent 7,965,218, Texas Instruments, 2011, URL: http://www.google.com.ar/patents/US7965218, status of 3/11/2015.
4. Frank Ohnhäuser, *Theory and realization of high-end analog-to-digital converters (ADC) based on the principle of successive approximation (SAR)*, PhD Thesis at the University of Erlangen-Nürnberg, Erlangen, 2008.
5. Product datasheet of the ADS8556, Texas Instruments, 2011, URL: http://focus.ti.com/lit/ds/symlink/ads8556.pdf, status of 03/11/2015.
6. Product datasheet of the AD7656, Analog Devices, 2006, URL: http://www.analog.com/static/imported-files/data_sheets/AD7656_7657_7658.pdf, status of 3/11/2015.
7. Michael Reinhold, Frank Ohnhäuser, Mikael Badenius, *Digital trimming of SAR ADCs*, US Patent 8,049,654, Texas Instruments, 2011, URL: http://www.google.com.ar/patents/US8049654, status of 3/11/2015.
8. Jan Bialek, Frank Ohnhäuser, Andreas Wickmann, Georg Fischer, Robert Weigel, Thomas Ussmüller, *Ein digitales Abgleichverfahren für SAR Analog-Digital-Umsetzer*, Kleinheubacher Tagung, 2012 (KHT2012-D.3-2).
9. Frank Ohnhäuser, Michael Reinhold, *SAR ADC and method with INL compensation*, US Patent 7944379, Texas Instruments, 2011, URL: http://www.google.com.ar/patents/US7944379, status of 3/11/2015.
10. Presentation on *Analog Process Technologies*, Texas Instruments, 2012, URL: www.ti.com/europe/downloads/Analog%20Process%20Technologies.pdf, status of 3/11/2015.
11. Abdullah Yilmaz, *LSB interpolation circuit and method for segmented digital-to-analog converter*, US Patent 6246351, Burr-Brown, 2001, URL: http://www.google.com/patents/US6246351, status of 3/11/2015.
12. Evert Seevinck, *Analysis and Synthesis of Translinear Integrated Circuits*, Booksurge Publishing, 2008.
13. Martin Allinger, *Development and implementation of a time efficient trim method for SAR analog-to-digital converters*, PhD thesis at the University of Erlangen, Erlangen, 2012.
14. Michael Reinhold, Martin Allinger, Frank Ohnhaeuser, *Electronic device and method for measureing dnl of an sar adc*, US Patent Application US 20130044015 A1, Texas Instruments, 2012, URL: http://www.google.com/patents/US20130044015, status of 3/5/2015.
15. Frank Ohnhaeuser, Michael Reinhold, Andreas Wickmann, Martin Allinger, *Novel Method to Measure the Capacitive Matching of A/D Converter Based on Successive Approximation*, IEEE Transactions on Instrumentation and Measurement, Volume 62, Issue 10, pp. 2652-2658, 2013, URL: http://ieeexplore.ieee.org/xpl/login.jsp?tp = &arnumber = 6523956&url = http%3A%2F%2Fieeexplore.ieee.org%2Fiel7%2F19%2F4407674%2F06523956.pdf%3Farnumber%3D6523956, status of 3/5/2015.
16. Michael Mueck, Michael Coln, *Analog-to-digital converter with the ability to asynchronously sample signals without bias or reference voltage power consumption*, US Patent 6667707, Analog Devices, 2003, URL: http://www.google.com/patents/US6667707, status of 3/11/2015.
17. Frank Ohnhaeuser, Andreas Wickmann, *ADC with low-power sampling*, US Patent 7,944,387, Texas Instruments, 2011, URL: http://www.google.com.ar/patents/US7944387, status of 3/11/2015.
18. Product datasheet of the ADS8482, Texas Instruments, 2006, URL: http://focus.ti.com/lit/ds/symlink/ads8482.pdf, status of 3/11/2015.
19. Seetharaman Janakiraman, Kiran M. Godbole, Surendranath Nagesh, *Increasing the SNR of successive approximation type ADCs without compromising throughput performance substantially*, US Patent 6894627, Texas Instruments, 2005, URL: http://www.google.com.ar/patents/US6894627, status of 3/11/2015.

20. Christopher Peter Hurrell, Gary Robert Carreau, *Analog-to-digital converter with signal-to-noise ratio enhancement*, US Patent 7218259, Analog Devices, 2007, URL: http://www.google.com.ar/patents/US7218259, status of 3/11/2015.
21. Frank Ohnhaeuser, Jan Bialek, *A 1.5mW 1MSPS 16bit SAR ADC with high performance*, in the proceedings of the International IEEE Conference on Semiconductors Dresden Grenoble (ISCDG2012), pp. 9-12, 2012, URL: http://ieeexplore.ieee.org/xpl/articleDetails.jsp?arnumber = 6360002&filter%3DAND%28p_IS_Number%3A6359978%29, status of 3/3/2015.
22. Product datasheet of the LTC2379-18, Linear Technology, 2011, URL: www.linear.com/docs/30158, status of 3/11/2015.
23. Yujendra Mitikiri, *High Speed High Resolution ADC Using Successive Approximation Technique*, US Patent 7796077, Texas Instruments, 2010, URL: http://www.google.com.ar/patents/US7796077, status of 3/11/2015.
24. Franz Kuttner, *A 1.2V 10b 20MSamples/s Non-Binary Successive Approximation ADC in 0.13μm CMOS*, in the proceedings of the International Solid-State Circuits Conference (ISSCC 2002), pp. 136-137, 2002.
25. Colin C. Price, Colin S. McIntosh, *Successive approximation analogue-to-digital converter*, US Patent 6351231, Analog Devices 2002, URL: http://www.google.de/patents/US6351231, status of 3/11/2015.
26. Yujendra Mitikiri, Visvesvaraya Pentakota, *Ternary Search SAR ADC*, India Patent Application 20110304493, Texas Instruments, 2011, URL: http://www.google.com.ar/patents/US20110304493, status of 3/11/2015.

Chapter 4
Basics on Delta-Sigma Converters

Industrial applications often require a high resolution and performance from an analog-to-digital converter at a moderate speed (see Chap. 1), so that SAR ADCs (Chaps. 2 and 3) and delta-sigma ADCs are most commonly used. This chapter will therefore concentrate on the basics of delta-sigma converters and then on the discrete-time approach utilizing the switched-capacitor technique, which is most commonly used. The alternative continuous-time architecture will be discussed in Chap. 5.

Delta-sigma converters were already introduced in Chap. 1. The analog signal A is coupled to a filter with feedback-loop as illustrated again in Fig. 4.1.

The quantizer, which in Fig. 4.1 is a comparator, is placed within a regulation-loop, resulting in the transfer function of Eq. (1.21):

$$Y_{\text{mod}} = A \cdot \frac{1}{1+s} + N \cdot \frac{s}{1+s}$$

The quantization noise N is suppressed within the low frequency signal band and the remaining high frequency noise is suppressed by a digital low-pass filter. This way, the delta-sigma converters achieve the best signal-to-noise ratios [1]. ΔΣ ADCs can convert signals with low signal frequency with high resolution. The oversampling and the digital low-pass filtering will even suppress higher frequency noise of the signal. Such signals occur often in industrial applications that use sensors. Examples are pressure or temperature sensors, flow or energy meters. Additionally, the analog filter provides a linear transfer function. Consequently, the differential non-linearity is extremely good. However, non-linear parameters of the electrical circuitry will still cause an integral non-linearity.

Further disadvantages are the low conversion rates caused by the required oversampling, the time delay caused by the digital filter and the fact that snapshots are not possible as the delta-sigma converter has to run continuously and synchronously with the clock.

F. Ohnhäuser, *Analog-Digital Converters for Industrial Applications Including an Introduction to Digital-Analog Converters*, DOI 10.1007/978-3-662-47020-6_4

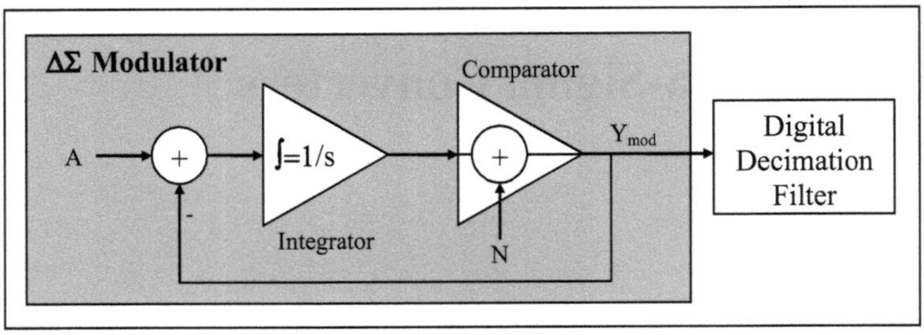

Fig. 4.1 Block diagram of a first order delta-sigma converter

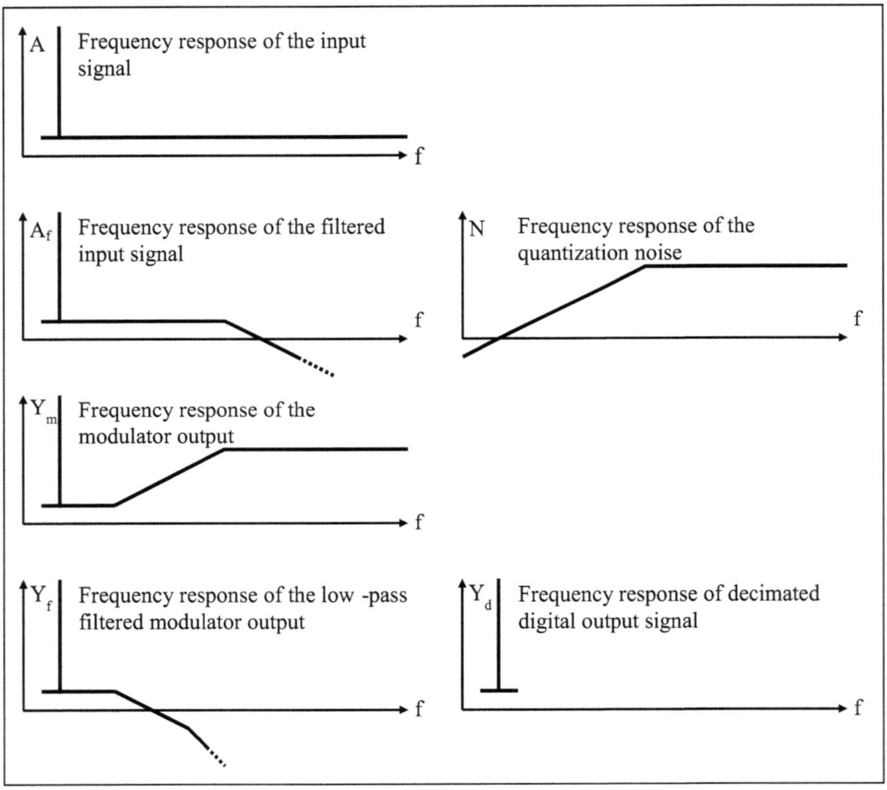

Fig. 4.2 Signal processing inside a delta-sigma converter

Figure 4.2 is showing the signal processing within the delta-sigma converter, assuming a signal A that has white noise associated. The signal A will get low-pass filtered (A_f) by the modulator causing the noise to decrease at higher frequencies. The quantization noise is added to the signal. The quantization noise was also considered white and will therefore be shaped as shown below after applying the high-pass transfer function (see signal N). The modulator output, which is named

Y_m in Fig. 4.2, combines the filtered signal and the shaped noise. The digital filter will suppress higher frequencies causing the signal Y_f. Finally, the frequency is decimated by reducing the data rate by the oversampling ratio. The signal Y_d will evolve.

The output data rate f_{data} can be calculated by the clock frequency f_{clk} divided by the oversampling ratio OSR. The maximum signal frequency f_{Nyq} is then expressed by the Nyquist frequency, which is half the data rate:

$$f_{data} = \frac{f_{clk}}{OSR} \tag{4.1}$$

$$f_{Nyq} = \frac{f_{clk}}{2 \cdot OSR} \tag{4.2}$$

The transfer function of Fig. 4.1 is described in the continuous-time domain and can be realized with the circuitry in Fig. 4.3.

The amplifier with the resistors at the input R_{in} and the feedback R_F together with the capacitor C form the integrator. Its output is inverted and therefore connects to the negative input of the comparator. The bitstream at the output is marked with Y_{mod} again. For the inverse feedback F, Y_{mod} is controlling a switch, which connects a positive reference voltage $REFP$ to the feedback resistor if $Y_{mod} = 0$ and a negative reference voltage $REFN$, if $Y_{mod} = 1$. The calculation below proves that it generates the anticipated transfer function. As a start, the feedback F can be expressed with Eq. (4.3), where $REFP = REF$, $REFN = 0$ V and $R_F = R_{in} = R$.

$$\begin{aligned} F(Y_{\text{mod}}) &= -Y_{\text{mod}} \\ F(Y_{\text{mod}} = 0) &= REFP \\ F(Y_{\text{mod}} = 1) &= REFN \end{aligned} \tag{4.3}$$

The negative input of the integrator is at virtual ground, so that a current flow I_{int} is generated from A and F through the resistors R, which has to flow to the integrator

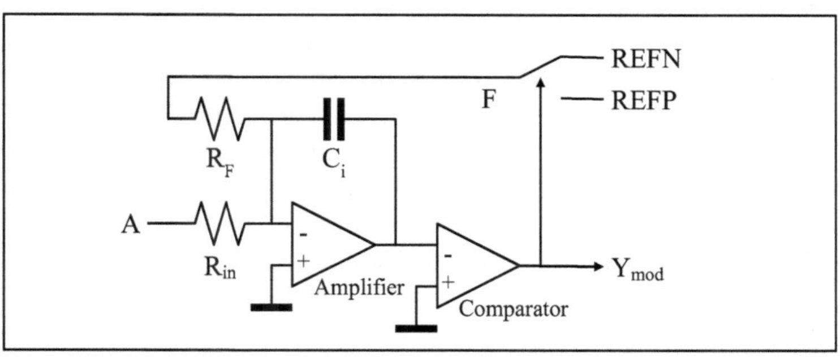

Fig. 4.3 Continuous-time, first order delta-sigma modulator

capacitor C. The current is expressed in Eq. (4.4), the voltage V_{int} across the integrator capacitor with Eq. (4.5).

$$I_{\text{int}} = \frac{A}{R} + \frac{F}{R} = \frac{A}{R} - \frac{Y_{\text{mod}}}{R} \tag{4.4}$$

$$V_{\text{int}} = -I_{\text{int}} \cdot \frac{1}{sC} = -\frac{A - Y_{\text{mod}}}{sRC} \tag{4.5}$$

The comparator is finally inverting V_{int} and adds the quantization noise N.

$$Y_{\text{mod}} = \frac{A - Y_{\text{mod}}}{sRC} + N \tag{4.6}$$

Equation (4.6) can finally be solved for Y_{mod} to receive the transfer function.

$$Y_{\text{mod}} = A \cdot \frac{1}{1 + sRC} + N \cdot \frac{sRC}{1 + sRC} = A \cdot STF + N \cdot NTF \tag{4.7}$$

SFT in Eq. (4.7) represents the 'signal transfer function' and *NFT* the 'noise transfer function'. Equation (4.4) shows that the sign of I_{int} cannot be influenced anymore, if A > REFP or A < REFN. Consequently, the integrator would not be regulated anymore and would converge to the supply rail. Y_{mod} would continuously be 1 or 0 respectively. A therefore has to remain within

$$A \in [REFN; REFP] \tag{4.8}$$

With REFN = 0 V, the input voltage range is unipolar. Bipolar input ranges could be achieved by either providing a negative reference voltage $REFN = -REFP$ or by implementing the modulator fully differentially.

Now the time-continuous delta-sigma modulators come with some technical disadvantages. The most significant disadvantage is the performance loss generated by clock jitter, which was discussed in Chap. 1. The effect of the clock jitter on a normal ADC has the effect that the input voltage can change during the statistical variation of the charge freezing clock edge caused by noise. A wrong input voltage is sampled and reduces the noise performance.

The effect to time-continuous delta-sigma converters is more severe, as the clock edge also triggers the comparator. The comparator output, which has some time delay associated that can jitter additionally, will set the position of the feedback switch. The feedback F erroneously connects to the wrong reference potential for the time Δt. The feedback is directly subtracted from the input A, so that an erroneous feedback acts identical as an erroneous or in other words noisy input voltage. The allowed total jitter Δt can be estimated.

The charge that flows from the input A to the integrator during one clock cycle T_{clk} can be calculated to

$$Q_{\text{int}} = I \cdot T_{clk} = \frac{A}{R} \cdot T_{clk} \tag{4.9}$$

The charge representing 1 LSB Q_{LSB} of a converter with n-bit resolution can be calculated by setting A in Eq. (4.9) to

$$\begin{aligned} A_{LSB} &= \frac{REF}{2^n} \\ Q_{LSB} &= \frac{REF}{2^n \cdot R} \cdot T_{clk} \end{aligned} \tag{4.10}$$

The error in charge, which is generated by the jitter, is expressed with

$$Q_{\Delta t} = \frac{REF}{R} \cdot \Delta t \tag{4.11}$$

$Q_{\Delta t}$ should be less than half of Q_{LSB}. So that the allowed jitter can finally be estimated to

$$\begin{aligned} \frac{REF}{R} \cdot \Delta t &\leq \frac{REF}{2 \cdot 2^n \cdot R} \cdot T_{clk} \\ \Delta t &\leq T_{clk} \cdot 2^{-(n+1)} \end{aligned} \tag{4.12}$$

With a 16 bit resolution and a clock frequency of 10 MHz, which correlates to $T_{clk} = 100$ ns, the maximum jitter should not exceed 0.7 ps. Fortunately, oversampling will reduce the noise and therefore the jitter requirements. However, former processes were too slow to keep the jitter in such a low range. Continuous-time modulators therefore do not play a significant role in commercially available delta-sigma ADCs. Modern processes, which are significantly faster, can support the jitter requirements. Continuous-time modulators are a significant topic in today's research. They are therefore discussed in detail in Chap. 5. The technique used today are time-discrete modulators, which solve this problem and are discussed in Sect. 4.5.

4.1 Delta-Sigma ADCs with Higher Order

Both approaches, continuous time or discrete time, can increase the resolution by using higher order systems and by replacing the comparator with a multi-bit converter. Higher order modulators increase the number of integrators in the system. Each additional integrator will cause the noise to decrease with additional

20 dB per decade towards smaller frequencies. Unfortunately, each integrator causes a phase shift of 90°, so that the stability and frequency responds of the STF needs to be evaluated for systems with an order higher than one. This should be exercised for a second order modulator as shown in Fig. 4.4.

For the phase analysis, the open loop system needs to be evaluated. The ADC and DAC do not influence the stability, if a zero-delay is assumed. Such an open-loop circuitry is drawn in Fig. 4.5. Note that this assumption of the zero-delay is inadequate for continuous-time modulators and is further investigated in Chap. 5. Time-discrete integrators include a time delay of one clock cycle, which already takes the delay of the ADC and DAC into account.

The open loop transfer function $F(s)$ from Fig. 4.5 is expressed by

$$F(s) = \frac{Y_{\text{mod}}}{A} = \frac{1}{s \cdot (s + k)} \tag{4.13}$$

For the stability analysis, the phase is calculated for $|F(s)| = 1$, which means that the unity circle is evaluated. s will be replaced by $j\omega/\omega_0$, $F(j\omega/\omega_0)$ is solved for the real and imaginary part (see Eq. 4.14), the real part expresses the cosine of the anticipated phase (Eq. 4.15) and the imaginary part equals the sine of the phase (Eq. 4.16). A good value for the phase margin might be 60° to achieve fast settling at good stability. Accordingly, the phase is 120°.

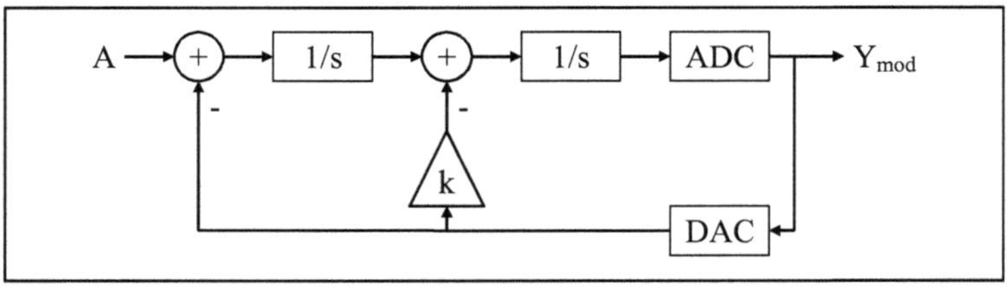

Fig. 4.4 Block diagram of a second order delta-sigma modulator

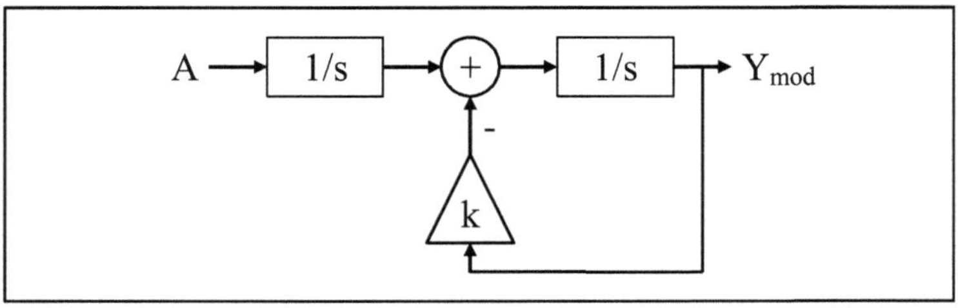

Fig. 4.5 Open-loop block diagram of the second order modulator used for stability calculation

$$F\left(\frac{j\omega}{\omega_0}\right) = \frac{1}{\frac{j\omega}{\omega_0}\cdot\left(\frac{j\omega}{\omega_0}+k\right)} = \frac{1}{-\left(\frac{\omega}{\omega_0}\right)^2+\frac{j\omega}{\omega_0}k} \tag{4.14}$$

$$\cos\left(120^\circ\right) = -\left(\frac{\omega}{\omega_0}\right)^2 = -0.5 \Leftrightarrow \frac{\omega}{\omega_0} = \frac{1}{\sqrt{2}} \tag{4.15}$$

$$\sin\left(120^\circ\right) = \frac{\omega}{\omega_0}k = \sqrt{\frac{3}{4}} \tag{4.16}$$

k can finally be calculated by replacing ω/ω_0 in Eq. (4.16) with the result from Eq. (4.15).

$$k = \sqrt{\frac{3}{4}}\cdot\frac{\omega_0}{\omega} = \sqrt{\frac{3}{2}} \tag{4.17}$$

Often, k is selected to 2 [2], generating a 76° phase margin. Finding the best coefficients for systems with order higher than three is difficult and the resulting SFT might still be unsatisfying. This problem is therefore solved by cascading modulators with lower order [3–6] as illustrated in Fig. 4.6.

The quantization error of the first delta-sigma modulator with order m_1 is connected to a second delta-sigma modulator with order m_2. This architecture is often used with $m_1 = m_2 = 2$. The quantization error is expressed by the output of the second modulator Y_2 and will be subtracted within the digital filter from the first modulator result Y_1. Such cascaded modulators act as a modulator with order $m_1 + m_2$.

4.2 Multi-bit Architectures

The noise performance can also be improved by utilizing a multi-bit quantizer instead of a comparator. Typical are 3 bit or 4 bit ADCs. With every additional bit, the quantization noise N decreases by 6 dB. Equation (1.21) proves that a lower

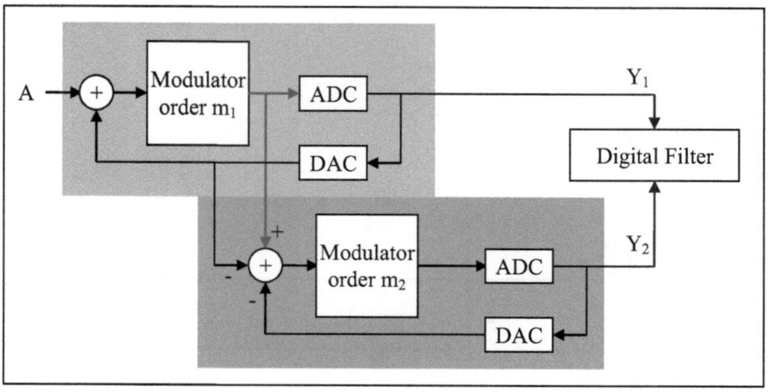

Fig. 4.6 Cascaded architecture to achieve higher order delta-sigma modulators

quantization noise N will linearly affect the remaining noise in the bitstream Y_{mod}. The total SNR will therefore also improve by 6 dB with every additional bit.

The digital Y_{mod} needs to be fed back to the input, which is done by a DAC (see Figs. 4.4 or 4.6). It needs to have the same resolution as the ADC. If the resolution would be less, then another quantization error would be applied inside the feedback path. Noise in this path directly adds to the input signal A and is not suppressed by the loop.

This multi-bit feedback also causes linearity problems. Even if the DAC in the feedback only requires a low resolution, it must provide the full DC accuracy of the delta-sigma modulator. If the modulator should for example have an 18 bit linearity, then the DAC requires an 18 bit linearity as well. This is caused as the DAC provides supporting points in the modulators transfer function, which is shown in the following.

Figure 4.7 shows the required DAC transfer function for a unipolar modulator with input range from 0 V to REF. F is representing the feedback voltage of the modulator, which equals the DAC output voltage. For the evaluation of the negative full-sale, Fig. 4.7 illustrates that the feedback voltage should be 0 V. If the analog input is less than the 0 V, then the difference between A and F would be negative and the integrator would continuously add up this negative difference until it reaches the supply rail. Consequently, the modulator would output a stream of '0' for analog input voltages smaller than the lowest feedback voltage from the DAC.

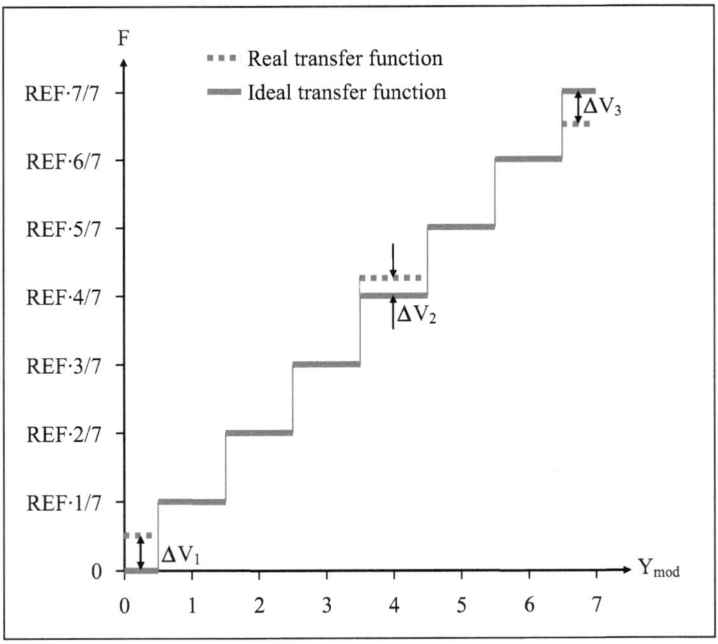

Fig. 4.7 Exemplary DAC transfer function used in the feedback loop of a delta-sigma modulator

As soon as A rises slightly above $F(Y_{mod}=0)$, the integrator sums a positive voltage. As soon as the integrator output reaches the first transition voltage of the ADC, the DAC feeds $F(Y_{mod}=1)$ back, the integrator makes a negative step, so that the integrator output is below the first ADC transition voltage again. The procedure starts from the beginning. It can be seen that the modulator is now in a correct feedback operation. $F(Y_{mod}=0)$ is the negative full-scale of the modulator. As a result, if the DAC has an offset ΔV_1 or in more general words a negative full-scale error, then the delta-sigma ADC has a negative full-scale error.

In a next step, let's evaluate the reaction of the modulator on linearity errors of the DAC such as marked with ΔV_2 in Fig. 4.7. The situation is very similar to the offset. As long as A is less than $F(Y_{mod}=4)$, the integrator will sum a negative voltage difference, until the integrator output reaches the lower transition voltage of the ADC. As a result for this example, the average of Y_{mod} will be less than 4 if $A=4/7 \cdot REF$. The average of $Y_{mod}=4$ is achieved, if $A=F(Y_{mod}=4)$. The DAC is setting supporting points in the transfer function of the ΔΣ ADC. Any linearity error of the DAC will cause a non-linearity in the ΔΣ ADC's transfer function.

An example for a time-discrete delta-sigma modulator with an order of 1 is given in Table 4.1. For simple calculation, the reference *REF* was chosen to 7 V, so that an ideal LSB of the DAC is 1 V. The input of the modulator A is chosen to the ideal 4 V. Similar to Fig. 4.7, $F(Y_{mod}=3)$ is set to the ideal 3 V and $F(Y_{mod}=4)$ is 4.25 V and therefore above the ideal value. The integrator input is named INT_{in} and the integrator output with INT_{out}. The output connects to the ADC, which is also non-ideal. However, an erroneous ADC transfer function acts as additional quantization noise and is therefore suppressed by the filter. In the example, the ADC outputs $Y_{mod}=4$ for $INT_{out}>3.4$ V and $Y_{mod}=3$ if INT_{out} is less than 3.4 V.

Table 4.1 Example of the operation of a delta-sigma modulator with 3 bit quantizer

Conditions	Time	A	F	INT_{in}	INT_{out}	Y_{mod}
REF = 7 V	T0	4	4.25	−0.25	4	4
A = 4 V	T1	4	4.25	−0.25	3.75	4
$F(Y_{mod}=4)=4.25$ V	T2	4	4.25	−0.25	3.5	4
$F(Y_{mod}=3)=3$ V	T3	4	3	1	3.25	3
$ADC_{in}(Y_{mod}=4)=3.4$–4.4 V	T4	4	4.25	−0.25	4.25	4
$ADC_{in}(Y_{mod}=3)=2.5$–3.4 V	T5	4	4.25	−0.25	4	4
	T6	4	4.25	−0.25	3.75	4
	T7	4	4.25	−0.25	3.5	4
	T8	4	3	1	3.25	3
	T9	4	4.25	−0.25	4.25	4
	T10	4	4.25	−0.25	4	4
	T11	4	4.25	−0.25	3.75	4
	T12	4	4.25	−0.25	3.5	4
	T13	4	3	1	3.25	3
	Average					3.79

Note, that Table 4.1 represents a time-discrete modulator of first order based on the state-of-the-art switched-capacitor technique that is explained in Sect. 4.5. Y_{mod} is the ADC decision based on the ADC's input INT_{out}. This result will set the DAC output F and therefore the integrator input INT_{in} for the next clock period.

Table 4.1 shows that the average of Y_{mod} is around of 3.8 instead of the ideal 4. In first order, this value is caused by the non-linearity of the DAC:

$$\overline{Y_{\text{mod}}} = A \cdot \frac{F_{ideal}(Y_{\text{mod}} = 4)}{F_{real}(Y_{\text{mod}} = 4)} = 4 \cdot \frac{4}{4.25} = 3.76 \tag{4.18}$$

Note that only the INL is affected. The linear transfer function of the modulator will interpolate that range between $F(Y_{mod}=3)$ and $F(Y_{mod}=4)$. If this range is too large as shown in Fig. 4.7, then the DNL of the codes of the delta-sigma converter within this range will be slightly too large.

In the following, the DNL is estimated for the example of an INL error of 0.25 V within a 7 V input range for a 10 bit delta-sigma converter with 3 bit quantizer. An LSB is 7 V divided by 1024 and therefore roughly 6.834 mV. The INL of 250 mV can also be expressed in 36.5 LSB. The error is caused because $F(Y_{mod}=4)$ is wrong. The DNL will therefore be off in the range of $F(Y_{mod}=3)$ and $F(Y_{mod}=4)$. This range has 1024 LSB divided by $2^3-1=7$ as a 3 bit quantizer is used. The error of 36.5 LSB distributes linearly over 146 codes. The DNL will therefore be around 36.5 LSB/146 = 0.25 LSB.

Finally, the full-scale error of the DAC in Fig. 4.7 ΔV_3 will also cause a full-scale error in the delta-sigma converter's transfer function. If the analog input is above $F(Y_{mod}=7)$, the integrator input will always be positive and the integrator will converge to the supply rail again. Consequently, if A is larger than $F(Y_{mod}=7)$, the modulator output Y_{mod} will be '1' continuously.

Gain and offset are typically calibrated in industrial applications, as there are several such errors in the signal path outside the delta-sigma ADC. The output signal of a sensor might for example be amplified with an operational amplifier. The amplifier adds an offset error. The resistors around the amplifier, which set the transfer function of the amplifier configuration, are mostly limited to an accuracy of 1 %. However, the loss in linearity is a severe disadvantage and needs further evaluation. Note that a 1 bit DAC does not have linearity errors, because it only feeds back the positive and the negative full-scale voltage.

4.2.1 Calibration Techniques

There are several papers regarding calibration and self-calibration techniques [7–11]. They basically have a lookup table implemented that converts the bitstream into a more accurate value. Y_{mod} has to represent the feedback value F with the accuracy of the delta-sigma converter. An interesting self-calibration method is shown in Fig. 4.8.

Fig. 4.8 Delta-Sigma modulator with n-bit feedback and self-calibration using a lookup table

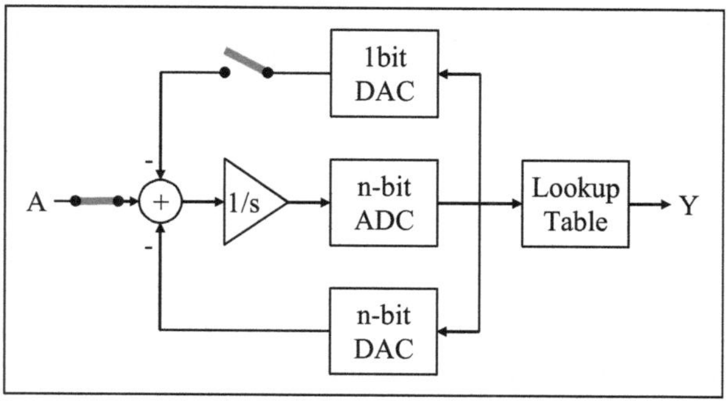

The delta-sigma converter is implemented as given, however a second 1 bit DAC is connected in parallel to the n-bit DAC. The 1 bit DAC is disconnected in normal operation and is realized by a switch that is feeding back a positive or a negative reference potential. In normal mode, the analog input *A* is connecting to the modulator.

The same modulator is used to calibrate the n-bit DAC, where the n-bit DAC acts as an input to the modulator. For this reason, the input *A* is disconnected and the feedback is done through the 1 bit DAC, which now gets connected to the adding node. This modified modulator generates a bitstream, which represents the inverted DAC output. Only the MSB of the ADC is utilized during calibration so that the 1 bit DAC can represent it accurately. The output needs filtering. The most simple filter implementation is an accumulator. The high resolution result for each DAC output voltage *F* is stored in the lookup table. In normal operation, the n-bit ADC is addressing the lookup table and the value in the lookup table is forwarded to the digital filter.

The method, however, which is most frequently used to compensate the linearity error is called dynamic element matching, which is presented in the next chapter.

4.2.2 Dynamic Element Matching

The DAC should be evaluated to find an alternative cancellation scheme for mismatch. A possible resistive implementation for a 3 bit DAC is shown in Fig. 4.9. Note from Fig. 4.3 that in the particular case *F* is at virtual ground so that a current is generated through each resistor that is the voltage *REFP* or *REFN* divided by the resistor value. This current flows on the integrator and generates a change in the output voltage of the integrator that is linearly dependent on the current. The required negation is implemented by switching $7 - Y_{mod}$ resistors to the *REFP* and Y_{mod} resistors to *REFN*.

Non-linearity is generated, if the resistors do not match each other. The typical matching performance of resistors on a semiconductor process is 0.1 %. The idea of

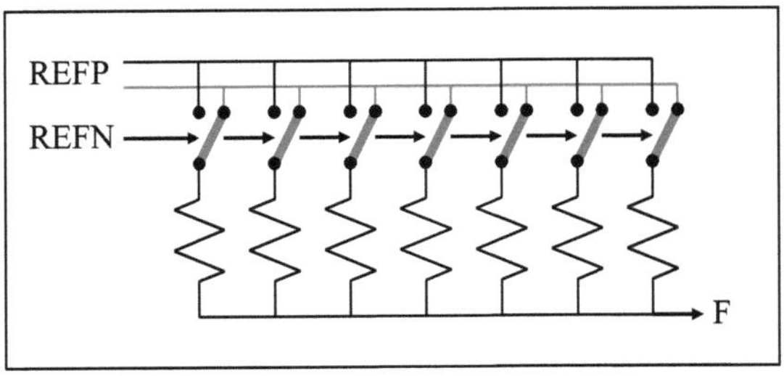

Fig. 4.9 Exemplary resistive DAC to illustrate dynamic element matching

the dynamic element matching is that each resistor is dynamically switching between *REFP* and *REFN* in a way, so that each resistor is Y_{mod} times connected to *REFN* and $7 - Y_{mod}$ times to *REFP* during 7 clock cycles. Then the linearity error caused by mismatch is canceled. This should be proven in the following by calculation. To simplify the calculation, *REFN* is assumed to be 0 V, six resistors should have the value R and one resistor the value $R + \Delta R$.

The current through the node F for the case that $7 - Y_{mod}$ resistors with the value R connect to *REFP* can be expressed by

$$I_{F1} = \frac{REFP}{\frac{R}{7-Y_{\text{mod}}}} \tag{4.19}$$

This feedback current will flow Y_{mod} times within 7 clock cycles. For $7 - Y_{mod}$ times during 7 clock cycles, the resistor with the value $R + \Delta R$ together with $6 - Y_{mod}$ resistors with the value R are connected to *REFP*, so that the current is

$$I_{F2} = \frac{REFP}{\frac{R}{6-Y_{\text{mod}}}} + \frac{REFP}{R + \Delta R} \tag{4.20}$$

Finally the average current is expressed in Eq. (4.21).

$$\begin{aligned}
\overline{I_F} &= \frac{1}{7} \cdot [Y_{\text{mod}} \cdot I_{F1} + (7 - Y_{\text{mod}}) \cdot I_{F2}] = \\
&= \frac{REFP}{7} \cdot \left[Y_{\text{mod}} \frac{7 - Y_{\text{mod}}}{R} + (7 - Y_{\text{mod}}) \frac{6 - Y_{\text{mod}}}{R} + (7 - Y_{\text{mod}}) \frac{1}{R + \Delta R} \right] = \\
&= \frac{(7 - Y_{\text{mod}}) \cdot REFP}{7} \cdot \left[\frac{6}{R} + \frac{1}{R + \Delta R} \right] = \frac{(7 - Y_{\text{mod}})}{7} \cdot \frac{REFP}{\left[\frac{R}{6} \middle\| (R + \Delta R) \right]}
\end{aligned} \tag{4.21}$$

Equation (4.21) shows that the current is now inverse linear to Y_{mod}, which was the goal of the dynamic element matching. The maximum current for $Y_{mod} = 0$ is the reference voltage *REFP* divided by the parallel resistance of six resistors with the value R and one resistor with the value $R + \Delta R$. Consequently, the linearity is not affected by the mismatch anymore.

Time-discrete implantations mostly use a capacitive feedback. However, the principle of the dynamic element matching works identically [12–14].

4.3 Dithering

Table 4.1 shows the interesting fact that the exemplary modulator outputs four times a '4' followed by a '3' (Y_{mod} = 44434444344443…). Such a repeated pattern can occur for DC input signals A, if the system noise is too low. The averaged output still describes the right input voltage, but the periodic pattern after every fifth bit will add a frequency portion in the output signal at $f_{data}/5$ for the example. Such frequency portions in the output signal are called idle tones. A real example of a measured idle tone can be seen in Fig. 4.10.

In the example of Fig. 4.10, a DC voltage was applied to a second order delta-sigma modulator. An idle tone can clearly be seen around 15 kHz after applying a FFT to the converters output data.

To avoid such idle tones [15], the repeated pattern needs to be avoided. This can be done by adding a sufficient amount of noise to the modulator. Unfortunately, adding white noise to the modulators input would derate the noise performance of the total delta-sigma converter. Adding high frequency noise on the other side will not derate the system performance, as it will be suppressed together with the quantization noise by the digital low-pass filter. Two examples are shown in Figs. 4.11 and 4.12.

In Fig. 4.11, white noise is generated. Before applying the noise to the modulator input, it is high-pass filtered, so that only high frequency noise remains. The high frequency noise is later suppressed by the digital low-pass filter. This architecture can be used in combination with an offset cancellation technique presented later in this chapter.

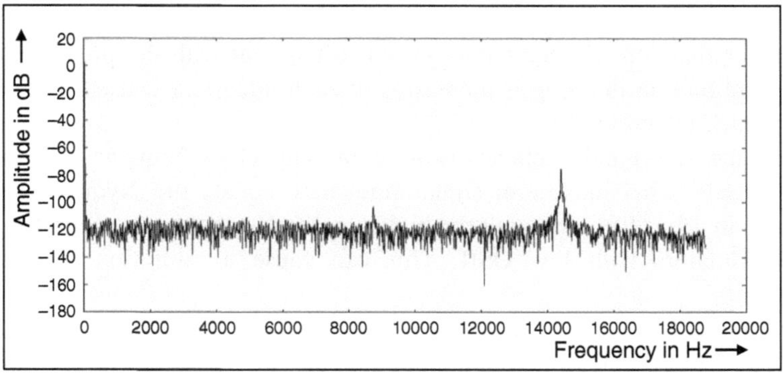

Fig. 4.10 Measured idle tones in the spectrum of the output of a delta-sigma modulator with DC input

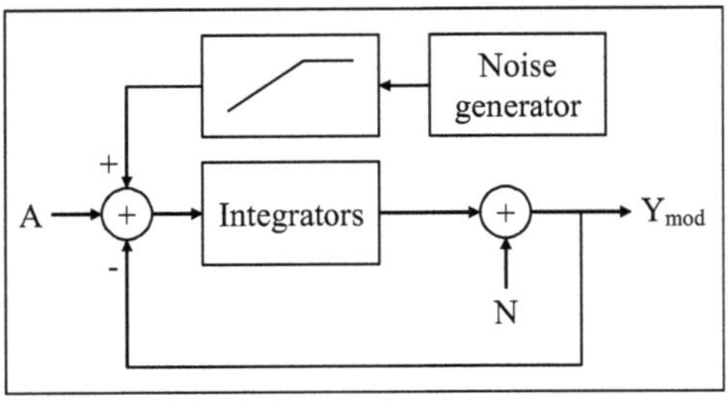

Fig. 4.11 Shaped dither noise added to the input of a delta-sigma modulator to avoid idle tones

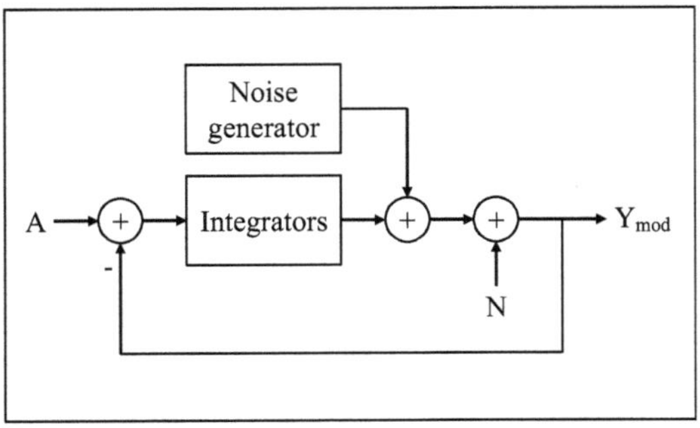

Fig. 4.12 Dither noise applied prior to the quantizer of a delta-sigma modulator to avoid idle tones

In Fig. 4.12, the dither noise is applied prior to the quantizer. Low frequency noise that would derate the system noise performance is already suppressed by the modulator's transfer function. The disadvantage is that the amplitude of the applied noise has to be significantly higher compared with the solution in Fig. 4.11.

4.4 Feed-Forward Architectures

The structures, which were proposed, had a general transfer function as shown in Fig. 4.13. Such an architecture is called feedback architecture, as Y_{mod} is fed back in the transfer function. Higher order systems often deal with the problem that the signal amplitude in the loop is increasing at each integrator caused by the signal delay of each integrator.

The higher the signal frequency is in ration to the clock frequency, the higher is the amplitude. This maximum signal frequency equals the Nyquist frequency described in Eq. (1.9). Particularly high amplitudes therefore occur for higher order modulators with low OSR. This can cause a saturation of the upper integrators.

The transfer function can also be realized with the architecture in Fig. 4.14. The input signal A as well as the integrator outputs are forwarded to the quantizer [16–18]. This provides two positive effects. First, a portion a_0 of the input signal

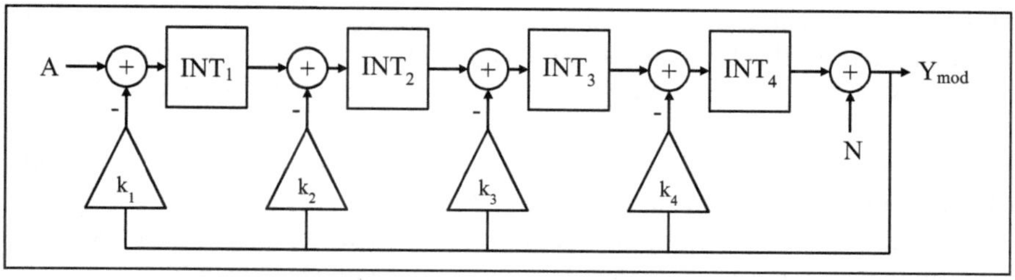

Fig. 4.13 Block diagram of a delta-sigma modulator with feedback architecture

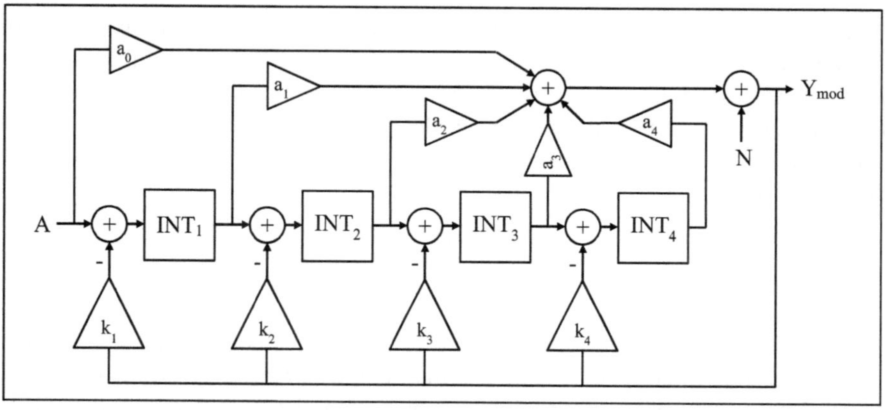

Fig. 4.14 Block diagram of a delta-sigma modulator with feed-forward architecture

is directly forwarded to the output. This will reduce the signal delay until A reaches the output Y_{mod} and will therefore reduce the amplitudes through the integrators.

Additionally, it can be seen in Fig. 4.13 that only the output of the last integrator connects to the quantizer. If the last integrator saturates due to high slopes of the input signal, then the delta-sigma converter will output wrong data. If the last integrator saturates within the feed-forward architecture of Fig. 4.14, then the previous integrators can still operate normally and feed their value to the output Y_{mod}. The saturated last integrator will add a constant voltage to the quantizer. Such an offset will be compensated by the feedback loop. Consequently the modulator will still output valid results, but acts as a modulator, where the order is reduced by one.

4.5 Switched-Capacitor Approach

It was mentioned previously that continuous-time delta-sigma modulators are very sensitive to clock jitter as it does not only affect the point in time, when the input signal is sampled, but also adds a significant integration error of the feedback path.

This can be avoided, if only a fixed amount of charge is integrated per clock cycle. Therefore, the feedback voltage as well as the input voltage are sampled with one clock phase and then transferred to the integrator on the other clock phase. Now the system is operating at discrete points of time, so that we talk about discrete-time modulators. They are typically realized with the switched-capacitor technique, which is presented in this chapter [19].

The idea is illustrated in Fig. 4.15. During phase 1, the input voltage A is sampled on a sample capacitor C_s. The sampled charge Q_s can be calculated to

$$Q_s = C_s \cdot A \tag{4.22}$$

After the phase 1, the switch between A and C_s is opened first, before the switch to the amplifier closes for phase 2. The potential of the inverted input of the amplifier is at virtual ground. Consequently, C_s has to discharge. As the switch is open and the amplifier's input is high impedance, the charge has to flow on the integrator capacitor C_i. A voltage will build up at the integrator output V_i that can be calculated to

$$\Delta V_i = -\frac{Q_s}{C_i} = -A \cdot \frac{C_s}{C_i} \tag{4.23}$$

The negative sign is added as the current flow is inverse compared to the voltage between V_i and virtual ground. Note that this charge is transferred during one clock cycle. The switched-capacitor circuit should now be compared to the time-continuous approach from Fig. 4.16.

The change of the integrator voltage V_i after one clock cycle should be identical to the time-discrete approach. The current flowing on the integrator capacitor is $I_i = A/R$. The current will cause a change in the output voltage, which for a constant voltage A can be calculated to

$$\Delta V_i = \frac{\Delta Q_i}{C_i} = \frac{-I_i \cdot T_{clk}}{C_i} = -\frac{A}{f \cdot RC_i} \tag{4.24}$$

Comparing Eqs. (4.23) with (4.24) results in

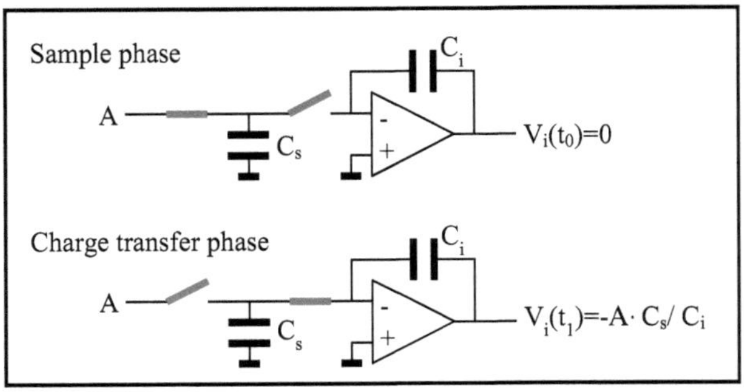

Fig. 4.15 Integrator in switched-capacitor technology in sample and charge transfer phase

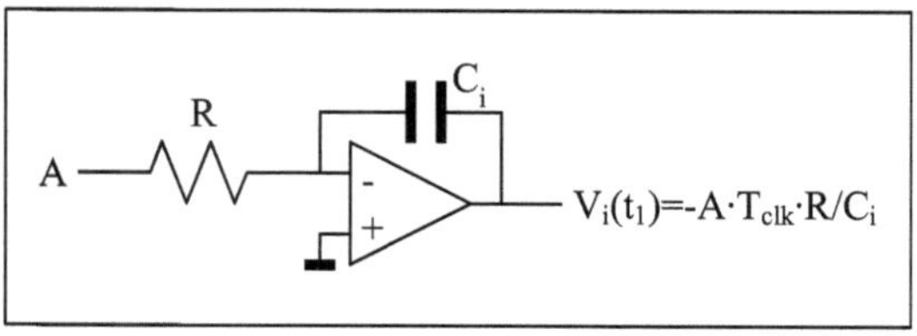

Fig. 4.16 Standard continuous-time integrator

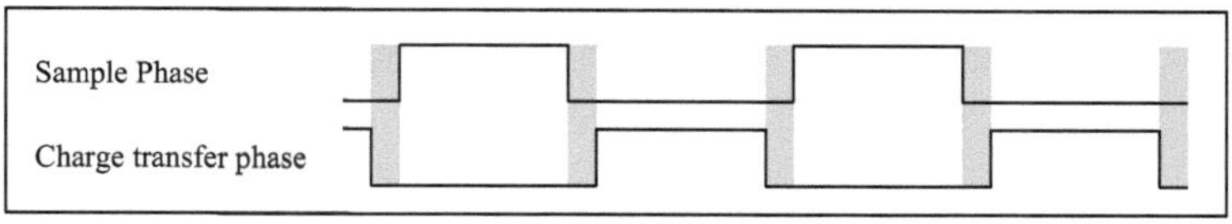

Fig. 4.17 Non-overlapping clocks must be used in switched capacitor technologies

$$C_s = \frac{1}{f \cdot R} \tag{4.25}$$

Equation (4.25) proves that the resistor can be replaced by a capacitor C_s that is switched between the input and the amplifier at a frequency *f*. Again, it is crucial that no charge is lost. Non-overlapping clocks as shown in Fig. 4.17 are required.

Furthermore, it was mentioned above that Eqs. (4.24) and (4.25) are only equal, if *A* is constant. The switched-capacitor approach is sampling the input voltage *A* at a certain point of time. The transfer function can therefore not be expressed by the continuous-time Laplace transfer function. The discrete-time description is required.

Hereby, the switched-capacitor integrator is adding a delay of one clock cycle. In the z-transformation, this is described with z^{-1}. With every clock cycle, the new input value is added to the existing value. Taking these two considerations into account, the integrator can be described with the block diagram in Fig. 4.18. The integrator output is called Y_i.

Based on the integrator in Fig. 4.18, the time-discrete version of the delta-sigma converter from Fig. 4.4 is shown in Fig. 4.19. Finally, Fig. 4.20 illustrates this modulator in switched-capacitor technology assuming the feedback coefficient *k* to be 2. The illustrated modulator is currently in sample mode with $Y_{mod} = 0$.

4.6 Differential Approach

In Chap. 2, the sample and hold circuitry was discussed. Most critical was the charge injection caused by the input switches, which will cause offset that will significantly drift over parameters like supply voltage and temperature. The charge injection will further be input voltage dependent, so that a significant non-linearity

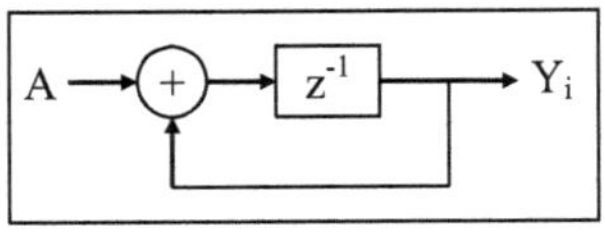

Fig. 4.18 Integrator described in time-discrete z-domain

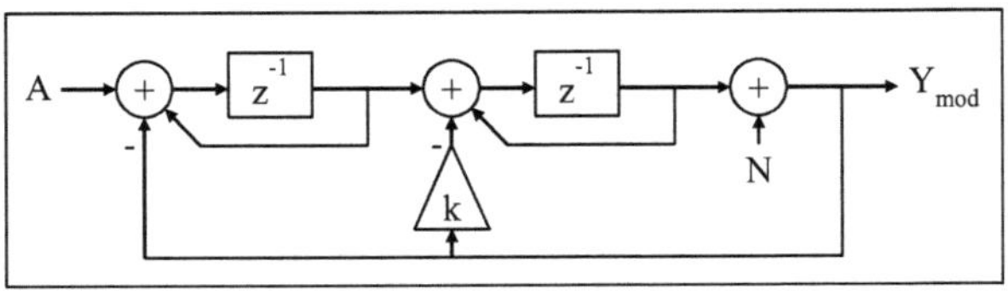

Fig. 4.19 Block diagram of a second order discrete-time delta-sigma modulator

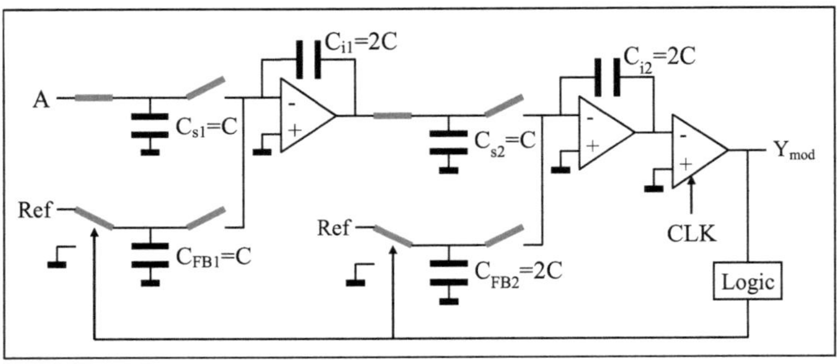

Fig. 4.20 Simplified circuit of a second order, 1 bit delta-sigma modulator in switched-capacitor technology

is added. One technique to reduce the charge injection uses shorted dummy transistors with an inverse clock. Such a circuitry is illustrated in Fig. 4.21.

The dummy transistors T_{dA} and T_{dY} should theoretically compensate the source-gate capacitance of the switch transistor T_{sw} and the drain-gate capacitance accordingly. The charge injection however is dependent on the impedance of input node *A* and output node *Y* and the clock edge, so that the charge injection of the dummy devices will not match the charge injection of the switch transistor. Furthermore, the transistor T_{sw} that is turned on has charge in the channel, when the switch is turned on. This charge will also contribute to the charge injection. The dummy devices during on-mode are turned-off and the channel does not exist. Consequently, this compensation technique only works to some extent. It must be used, if a switched-capacitor circuitry is realized single-ended.

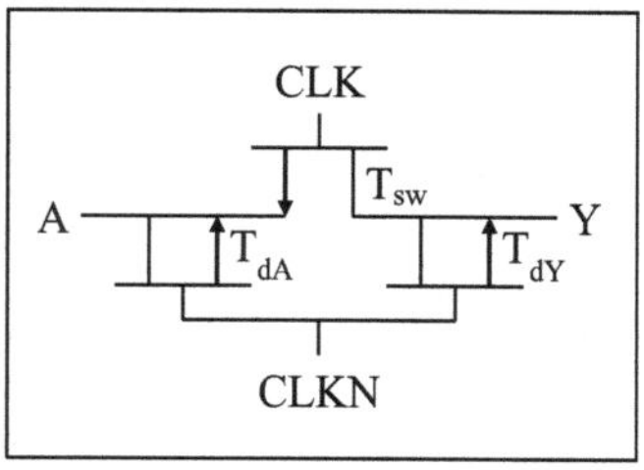

Fig. 4.21 Switch with dummy transistors T_{dA} and T_{dY} to compensate charge injection

However, similar to the SAR ADC, it is always better to realize the circuitry fully differential. In single-ended mode, the negative input can still be connected ground. An example of a fully differential integrator approach is illustrated in Fig. 4.22.

If the clock is high, which is the case in phase (a) of Fig. 4.22, the sample capacitors C_{sp} and C_{sn} are charged to the differential input voltage between nodes A_p and A_n. The integrator is disconnected.

With the falling clock edge, which is illustrated below in phase (b), the small switch between C_{sp} and C_{sn} is opened first. The small switch is common, so that the charge injection only causes a common-mode voltage change and does not affect the signal performance. The charge is now frozen.

Once the charge is frozen, the large input voltage switches to A_p and A_n can open as well, which is illustrated in phase (c). The time delay between phase (b) and (c) will be realized with an analog delay circuitry. The charge injection of the input switches is significantly larger and input voltage dependent, but it will not affect the ADC performance as the charge is already frozen.

After the inputs are completely disconnected, the capacitors can be connected in series on one side and to the integrator on the other sides in phase (d). The differential voltage will be transferred to the integrator.

Care must also be taken that the charge injection of the integrator switches does not add charge onto the integrator capacitors. Fortunately, they will be at same potentials after the integration phase, which is the common-mode input potential of the differential amplifier. Note that the amplifier requires a common-mode regulation. The common-mode regulation can be realized with a switched-capacitor circuitry as well. When disconnecting the integrator, it might be preferable to open the switches to the integrator first as shown as phase (e) and then the common switch (phase f). The modulator can now continue with the sample phase again.

The integrator and the input circuitry have requirements regarding their bandwidth and noise, which are discussed in the next chapter.

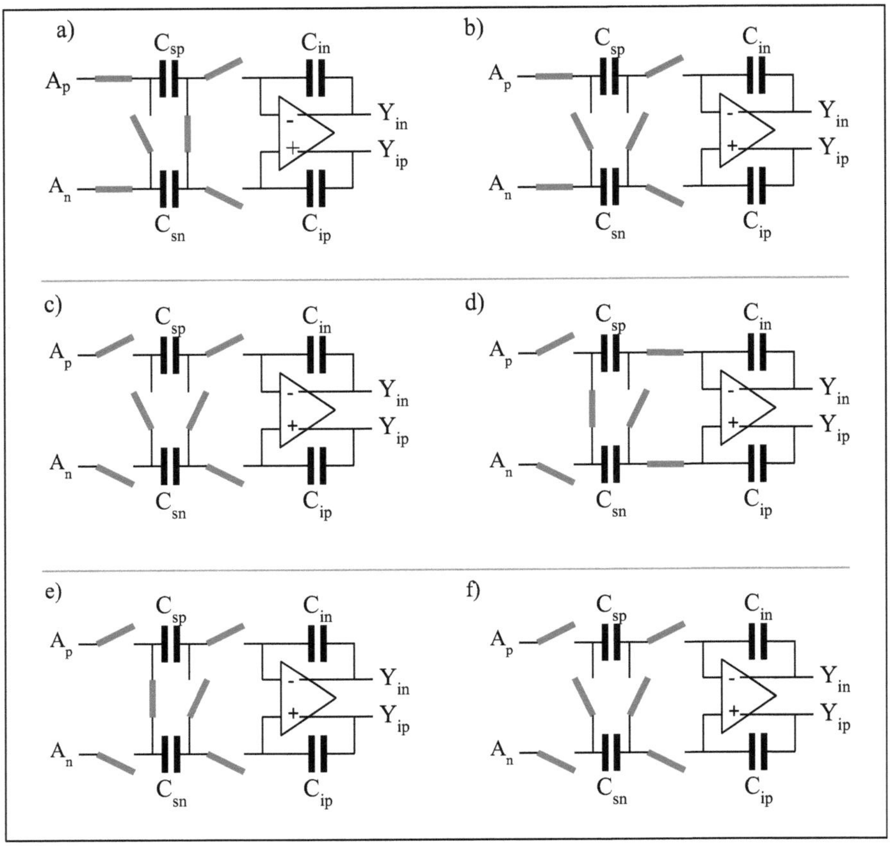

Fig. 4.22 Timing phases of the charge transfer in a differential delta-sigma modulator approach. (**a**) Sample phase. (**b**) Freezing sampled charge. (**c**) Disconnecting the inputs. (**d**) Integration phase. (**e**) Disconnecting the integrator, step 1. (**f**) Disconnecting the integrator, step 2

4.7 Bandwidth and Noise

It was discussed before that two settling phases are required for a switched-capacitor delta-sigma modulator. First the input and feedback capacitors need to be charged and then the charge has to be transferred to the integrator. Both steps need to be performed to the accuracy of the modulator. Half a clock cycle is available for each settling process, if neglecting the time for the switching process and neglecting any duty cycle of the clock.

The required bandwidth can then be calculated based on Eq. 2.13

$$f_{-3\ \mathrm{dB}} > 2 \cdot f_{clk} \cdot \frac{\ln(2) \cdot (n+1)}{2\pi} \tag{4.26}$$

Note that the bandwidth of the input circuitry and integrator need to be based on the clock frequency and not on the maximum signal frequency. This is significantly higher compared to the requirements of a time-continuous approach. Consequently the power consumption of the time-discrete solution is significantly higher and the possible sample rates are lower.

Please note that the sample capacitors are discharged, when they connect back to the inputs. If assuming that this settling is exponential, then the voltage at the sample capacitor can be described with Eq. (4.27). The time, where the sampling starts was set to t = 0 s.

$$V_s = V_{in} \cdot \left(1 - e^{-\frac{t}{RC}}\right) \tag{4.27}$$

The sampled voltage V_s at $t = \frac{T_{clk}}{2}$ equals

$$V_s\left(\frac{T_{clk}}{2}\right) = V_{in} \cdot \left(1 - e^{-\frac{T_{clk}}{2RC}}\right) = a \cdot V_{in} \tag{4.28}$$

Equation (4.28) expresses that the exponential term is constant for fixed settling time. If the settling is insufficient, then primarily a gain error is the consequence. Unfortunately, the exponential settling is only an assumption, so insufficient sampling will to some extend also affect the linearity. This effect needs to be simulated carefully.

Also the settling of the reference path is relaxed. For differential modulators, the reference will be fed back to either the positive or the negative integrator with each clock cycle. Insufficient settling will not affect the linearity, as long as the feedback charge is constant for each clock cycle. A wrong feedback value will again mainly affect the gain. As an advantage to the SAR ADC, the average reference current is also code independent. This makes the reference design easier. Particular the tight load rejection as discussed for SAR ADCs in Chap. 6 is not required.

High bandwidths always affect the noise and will be discussed in the following. It was discussed previously that the noise is filtered for low frequencies. However, this is only true for noise that is generated within the loop such as the quantization noise. Noise that is generated at the input such as thermal noise of the first integrator or the kT/C noise of the sampling capacitors for the inputs are directly added to the input signal and will accordingly show up at the output. However, the oversampling will reduce the input noise by −3 dB per doubling of the OSR.

Differential modulators have four capacitors that connect to the input, which normally have an equal value C_s. Each capacitor adds the sampling noise, so that the rms-noise V_{ns} can be calculated to

$$V_{ns} = \sqrt{\frac{4kT}{C_s \cdot OSR}} \tag{4.29}$$

The integrator and the reference will also add their noise. Beside the thermal noise, which was already discussed for SAR ADCs in Chap. 2, the low frequency noise, mostly called 1/f-noise (named $V_{1/f}$ below), might be a significant noise contributor. Only the thermal noise will be reduced by the oversampling. Also the reference will always be fully fed back to the input, so that it will not contribute its noise code dependently as the SAR ADC does. The noise caused by the reference V_{nref} and the integrator V_{ni} are described by Eqs. (4.30) and (4.31).

$$V_{nref} = \sqrt{V_{1/f,ref}^2 + \frac{n_{ref}^2 \cdot f_{eff,ref}}{OSR}} \tag{4.30}$$

$$V_{ni} = \sqrt{V_{1/f,i}^2 + \frac{n_i^2 \cdot f_{eff,i}}{OSR}} \tag{4.31}$$

n in both equations is expressing the thermal noise density of the reference and the integrator respectively. $f_{eff,ref}$ and $f_{eff,i}$ are the effective bandwidth, which for first order systems equals π/2-times the −3 dB-bandwidth.

The influence of the noise sources above is illustrated in Fig. 4.23. For an ideal modulator, the noise would continuously decrease towards lower frequencies. The thermal noise from the noise sources above will cause the noise to flatten out, once

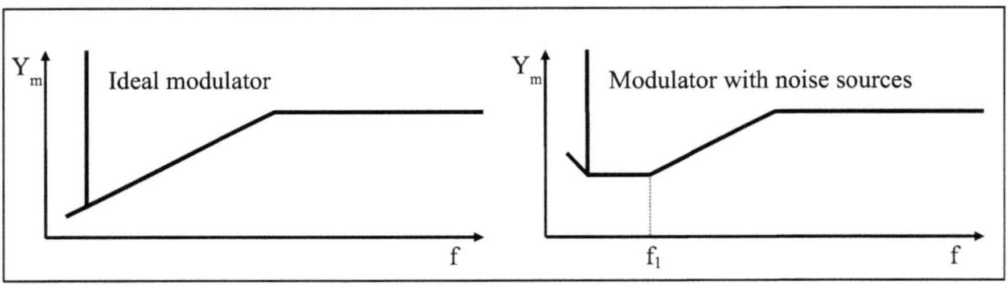

Fig. 4.23 Schematical frequency responds of an ideal modulator and a modulator with noise sources

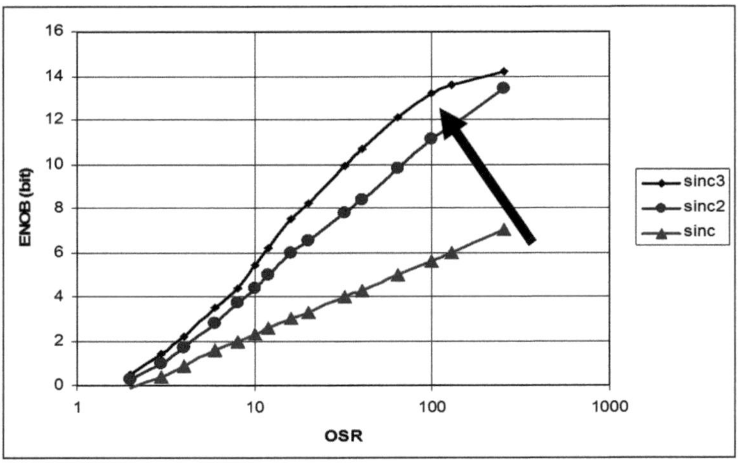

Fig. 4.24 Thermal noise affecting ENOBs of a delta-sigma modulator at high OSRs [20]

the noise sources dominate the quantization noise. The 1/f-noise might cause the noise floor to increase again for lower frequencies.

The noise sources above will affect the SNR. The noise-shaping of the modulator has no effect anymore, once the OSR of the digital filter is so high that frequencies above f_1 in Fig. 4.23 are already suppressed. From there on, additional oversampling will only improve the SNR by 3 dB and the ENOB by 0.5 bits for doubling the OSR. This effect is shown in Fig. 4.24.

4.8 Offset Cancellation

Gain and offset are the two DC specifications, which are generally less critical than linearity or noise, because they are calibrated in the application. However, it is required that the drift over temperature remains low. This is particularly not the case for CMOS amplifiers as used in the integrator. This amplifier offset as illustrated in Fig. 4.25 during the integration phase will directly be reflected in an offset of the modulator. Note that only the offset of the first integrator is essential. The regulation loop will eliminate any offset of following stages.

The sample capacitors are precharged to the input voltage and the feedback voltage. The offset will cause that not the full charge is transferred to the integrator capacitor, but that a residual charge remains. The charge Q_A from the input sample capacitors C_{sA}, which is transferred to the integrator, can be calculated with the sampled charge Q_{sA} reduced by the residual charge Q_{sR} after the integration phase, which is caused by the offset voltage.

$$Q_A = Q_{sA} - Q_{sR} = C_{sA}A - C_{sA}V_{off} = C_{sA}\left(A - V_{off}\right) \tag{4.32}$$

Equation (4.32) clearly shows that the offset of the integrator directly affects the modulators offset performance.

Methods are available to cancel the offset similar to the auto-zero function in SAR converters. These techniques utilize the differential approach of the integrator, which is already in use to eliminate charge injection. The gain of an ideal fully

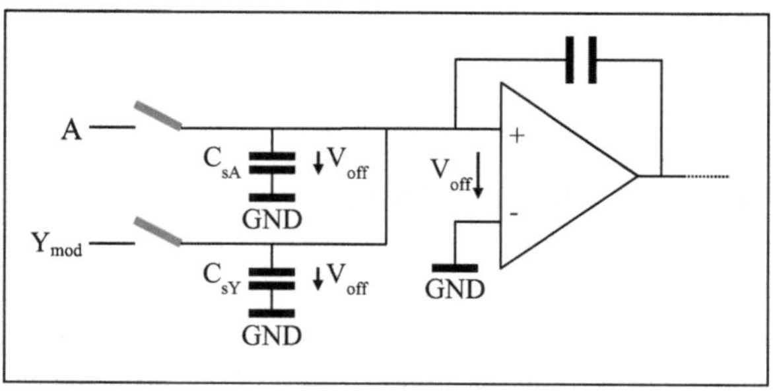

Fig. 4.25 Charge transfer phase of an integrator with amplifier offset V_{off}

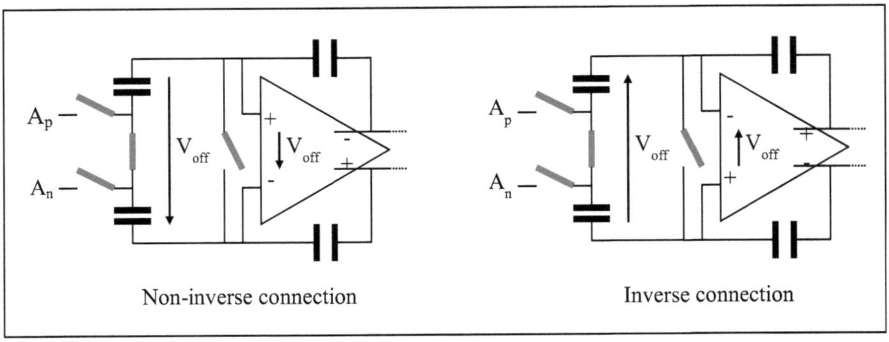

Fig. 4.26 Differential integrator with amplifier offset in non-inverse and inverse amplifier connection

differential amplifier and the feedback force the differential input voltage to be zero. It does not matter, if the amplifier is connected inverse or non-inverse. If the amplifier has an offset, then the input will be forced to this offset voltage V_{off} instead of zero. Figure 4.26 shows a simplified integrator in inverse and non-inverse connection including the offset voltage.

According to Fig. 4.26, the remaining charge on the capacitor would be the positive offset voltage in the non-inverse configuration and the negative offset voltage in inverse configuration. The idea of the offset cancellation technique is to swap the amplifier in consecutive conversion steps. If two following integration steps have the amplifier first connected non-inverse configuration, where the transferred charge is called $Q_{A,ni}$, and then in inverse configuration with transferred charge $Q_{A,i}$, then the average transferred charge Q_A can be calculated with Eq. (4.33).

$$\overline{Q}_A = \frac{Q_{A,ni}}{2} + \frac{Q_{A,i}}{2} = \frac{C_{sA}}{2}(A - V_{off}) + \frac{C_{sA}}{2}(A + V_{off}) = C_{sA}A \tag{4.33}$$

Equation (4.33) shows that the offset is cancelled. Two offset cancellation techniques, which are the correlated double sampling and the chopper technique, are most frequently used in modulators. Both utilize the method of swapping the differential amplifier.

4.8.1 Correlated Double Sampling

The correlated double sampling technique [21, 22] always has two integration steps before the ADC digitizes the next output Y_{mod}. In the first step, the analog input is sampled and integrated as shown in Fig. 4.27. The amplifier is connected in non-inverse operation, so that the offset is causing the according error.

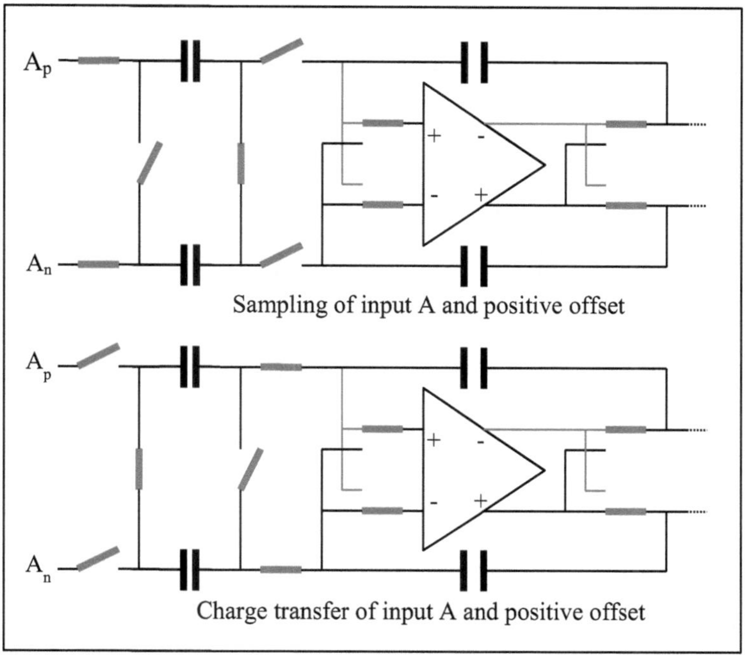

Fig. 4.27 Sampling and charge transfer of the input of a modulator with correlated double sampling

In a second sampling step, the capacitors might be shorted during sampling, so that no charge is stored. The amplifier is then connected inversely. Accordingly, the inverse offset would be integrated as illustrated in Fig. 4.28. After the integration of the inverse offset, the ADC can digitize its input voltage again.

The essential disadvantage of this technique is that two integration processes are required for one modulator result. Consequently, the power consumption is unnecessarily high or the conversion rate is reduced by a factor of two. The chopper technique is avoiding this disadvantage.

4.8.2 *Chopper Techniques*

As an obvious next step, the offset can be canceled not as an additional integration step, but the input voltage can be sampled also during the second sample process and integrated with the inverse amplifier configuration. This way, the first integration step is identical to the one in Fig. 4.27. The second integration step is then illustrated in Fig. 4.29. Shorting the inputs is not required. This architecture is called chopper technique [23, 24]. The digitizer is converting after each integration process.

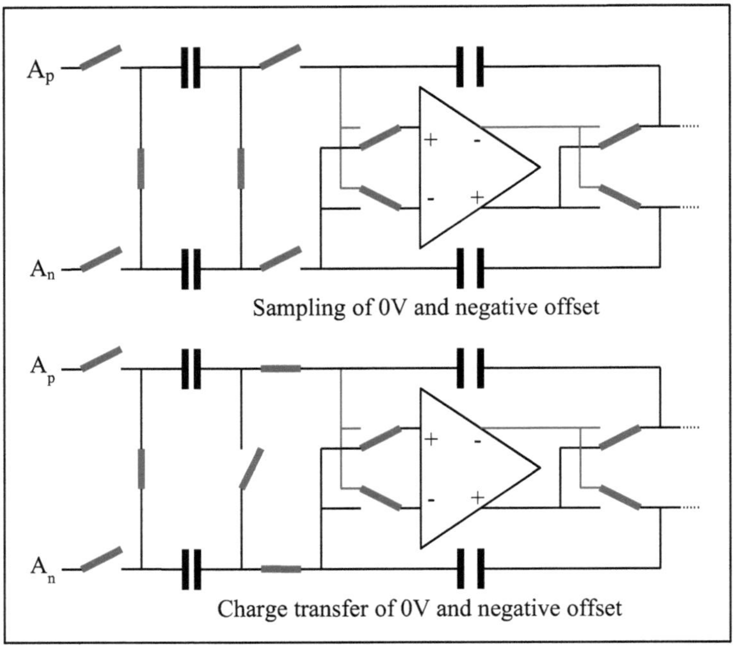

Fig. 4.28 Sampling and charge transfer of 0 V in a modulator with correlated double sampling

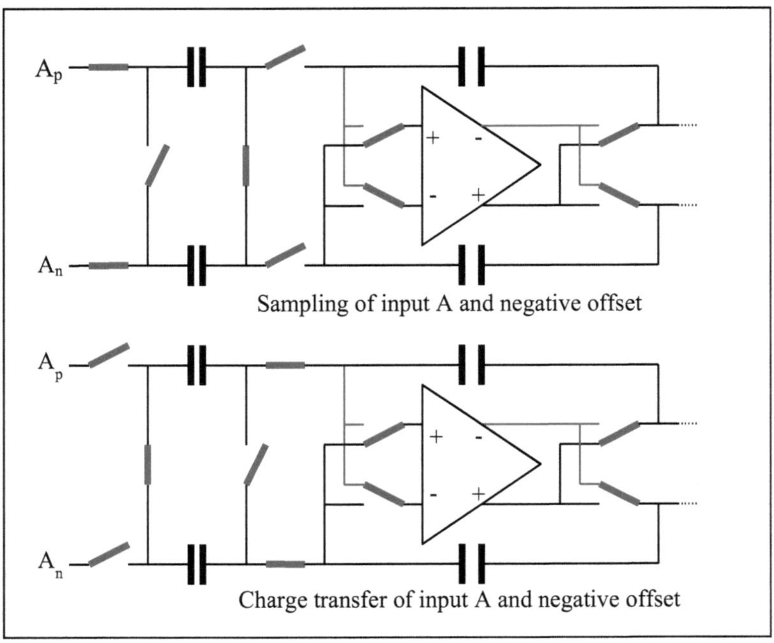

Fig. 4.29 Second charge transfer step of a modulator with chopper technique

It is also obvious that an error is caused if the OSR has an odd number. If this error would be generated randomly, then it would act as noise. Consequently, the noise can be used as the required dither noise mentioned above. If the noise is not only random, but also provides a high-pass characteristic then it would further be suppressed by the following digital low-pass filter.

Special care must be taken that charge on of the integrator capacitors is not lost during the swapping process of the amplifier. This effect is called integrator leakage. It will cause errors particularly in the middle of the transfer function, where Y_{mod} has the highest frequency.

4.9 Layout

From a layout perspective, the input path and the first integrator are most critical. Offset, noise and distortion at that point will directly couple into the input signal. Therefore the following rules are advisable.

All analog signals to the input of the first integrator should be wired differentially, so that noise and distortion from the substrate or surrounding signals act as common-mode noise, which is suppressed by the common-mode rejection of the integrator. The input signals should further be shielded with wires in the same metal layer that are connected to a current free ground. The shields should remain some distances to the signal wires to reduce parasitic capacitance. Noise and distortion of surrounding signals will then be suppressed.

The sample capacitors of the input signal and the capacitors from the feedback of the first integrator need to match well. Mismatch will cause a gain error. Fortunately, this gain error will not drift over a varying supply voltage or temperature, if the settling times are sufficient. All rules for capacitive matching, which were introduced for the SAR ADC, also apply here.

The switches to all capacitors particularly in the input path and the first integrator must match well. The switching process will cause charge injection that adds to the input related charge. This is particularly important for the switches that freeze the charge on the sample capacitor and the integrator. Charge injection, which differentially adds to a positive and negative input, is common and will be rejected by the differential architecture. Note that dummy switches can be used to reduce the total injected charge. The dummy devices are injecting the inverse charge of the switches. For differential signals, the total value is uncritical as long as it is common. However, the charge injection of dummy devices will add further mismatch, so that they will increase the mismatched charge injection, which is critical for differential architectures. Dummy switches are therefore not advisable.

Similar to the CDAC layout of the SAR ADC, a clean cut between the digital circuitry and the analog circuitry should be done to avoid digital coupling into sensitive analog signals. If digital signals have to cross analog signals, then shields should be used and the crossing should occur differentially. Close wiring of digital signals in parallel to analog signals must be avoided.

The integrator input is a floating node during the integration phase. Similar to the SAR ADC, all floating nodes are sensitive to capacitive noise coupling. All parasitic capacitance particularly at the integrator inputs should therefore be minimized.

It was mentioned above that the offset of the integrator will directly influence the offset of the modulator. The offset cancellation techniques will suppress the offset, but it is still advisable to remain it as low as possible by utilizing interleaving layout techniques with dummy transistors on the sides. The rules that were discussed for the comparator input of the SAR ADC apply here as well. Note that the random chopper technique will convert the offset into dither noise, which will increase the noise floor.

References

1. Fernando Medeiro, Belen Pérez-Verdú, Angel Rodríguez-Vázquez, *Top-Down Design of High-Performance Sigma-Delta Modulators*, Kluwer Academic Publishers, Bosten/Dordrecht/London, 2010.
2. R. Schreier, *The delta-sigma toolbox*, URL: http://www.mathworks.com/matlabcentral/fileexchange/19-delta-sigma-toolbox, status of 7/13/2013.
3. Rocío Río Fernández, Fernando Medeiro Hidalgo, Belén Pérez Verdú, José Manuel Rosa Utrera, Ángel Rodríguez-Vázquez, *CMOS Cascade Sigma-Delta Modulators for Sensors and Telecom*, Springer, New York/Dordrecht/Heidelberg/London, 2006.
4. Guangming Yin, Frederic Stubbe, Willy Sansen, *A 16-b 320-kHz CMOS converter using two-stage third-order ΔΣ noise shaping*, in the IEEE journal of Solid-State Circuits, Volume 28, Issue 6, pp. 640–646, 1993.
5. Tapani Ritoniemi, Eero Pajarre, Seppo Ingalsuo, Timo Husu, Ville Eerola, Tapio Saramäki, *A stereo audio sigma-delta A/D converter*, in the IEEE journal of Solid-State Circuits, Volume 29, Issue 12, pp. 1514–1522, 1994.
6. Guangming Yin, Will Sansen, *A high-frequency and high-resolution fourth-order ΔΣ A/D converter in BiCMOS technology*, in the IEEE journal of Solid-State Circuits, Volume 29, Issue 8, pp. 857–865, 1994.
7. Frank Ohnhaeuser, *Untersuchung von Sigma-Delta-Modulatoren höherer Ordnung mit Multibit-Rückführung*, diploma thesis at the Friedrich-Alexander-University, Erlangen 1996.
8. Yves Geerts, Michiel Steyaert, Willy M Sansen, *Design of Multi-Bit Delta-Sigma A/D Converters*, Kluwer Academic Publishers, Bosten/Dordrecht/London, 2002.
9. Mohammed Sarhang-Nejad, Gobor C. Temes, *A high-resolution multibit ΔΣ ADC with digital correction and relaxed amplifier requirements*, in the IEEE journal of Solid-State Circuits, Volume 28, Issue 6, pp. 648–660, 1993.
10. T. Cataltepe, A.R. Kramer, L.E. Larson, G.C. Temes, R.H. Walden, *Digitally corrected multi-bit ΔΣ data converters*, in the proceedings of the IEEE International Symposium on Circuits and Systems (ISCAS 1989), pp. 647–650, 1989.
11. J.W. Fattaruso, S. Kiriaki, M. de Wit, G. Warwar, Self-calibration techniques for a second-order multibit sigma-delta modulator, in the IEEE journal of Solid-State Circuits, Volume 28, Issue 12, pp. 1216–1223, 1993.
12. Bosco H. Leung, *Architectures for multi-bit oversampling A/D converter employing dynamic element matching techniques*, in the proceedings of the IEEE Conference on Circuits and Systems (ISCAS 1991), pp. 1657–1660, 1991

13. Kofi Makinwa, Johan Huijsing, *Dynamic Offset Compensated CMOS Amplifiers*, Springer, New York/Dordrecht/Heidelberg/London, 2009.
14. Yongjian Tang, Hans Hegt, Arthur van Roermund, *Dynamic-Mismatch Mapping for Digitally-Assisted DACs*, Springer, New York/Dordrecht/Heidelberg/London, 2013.
15. Kaveh Hosseini, Michael Peter Kennedy, *Minimizing Spurious Tones in Digital Delta-Sigma Modulators*, Springer, New York/Dordrecht/Heidelberg/London, 2011.
16. Konstantin Schmid, *Entwurf und Layout eines Delta-Sigma-Modulators in Feed-Forward-Architektur*, diploma thesis at the Friedrich-Alexander-University, Erlangen, 2010
17. Qiang Lu, Xiaowei Liu, *A fourth-order cascaded full feed-forward sigma-delta modulator*, in the proceedings of the International Conference on Optoelectronics and Microelectronics (ICOM 2012), pp. 466–469, 2012.
18. Kirk C.-H. Chao, Shujaat Nadeem, Wai L. Lee, Charles G. Sodini, *A higher order topology for interpolative modulators for oversampling A/D converters*, in the IEEE journal of Transactions on Circuits and Systems, Volume 37, Issue 3 pp. 309–318, 1990
19. Patrick J. Quinn, Arthur H.M. van Roermund, *Switched-capacitor techniques for high-accuracy filter and ADC design*, Springer, New York/Dordrecht/Heidelberg/London, 2007.
20. Product datasheet of the ADS1202, Texas Instruments, 2004, URL: http://www.ti.com/lit/ds/symlink/ads1202.pdf, status of 3/11/2015.
21. J.E. Johnston, *A 24-bit delta-sigma ADC with an ultra-low noise chopper-stabilized programmable gain instrumentation amplifier, in the proceedings of the third conference on Advanced A/D and D/A Conversion Techniques and Their Applications*, pp. 179–182, 1999.
22. Chang Ying-Hwi, Wu Chung-Yu, T.-C. Yu, *Chopper-stabilized sigma-delta modulator, in the proceedings of the International Symposium on Circuits and System* (ISCAS 1993), pp. 1286–1289, 1993.
23. C.C. Enz, G.C. Temes, *Circuit techniques for reducing the effects of Op-AMP imperfections: autozeroing*, in the proceedings of the IEEE, Volume 84, Issue 11, pp. 1584–1614, 1996.
24. N. Saputra, M.A.P Pertijs, K.A.A. Makinwa, J.H. Huijsing, *Sigma delta ADC with a dynamic reference for accurate temperature and voltage sensing*, in the proceedings of the International Symposium on Circuits and System (ISCAS 2008), pp. 1208–1211, 2008.

Chapter 5
Continuous-Time Delta-Sigma ADCs

Chapter 4 introduces the principle of delta-sigma ADCs. It discussed the switched-capacitor technique together with its advantages. The precision of the transfer function is only dependent on capacitive matching, which is typically 0.1 % accurate. The clock frequency can further vary in a wide range without affecting the stability. This is due as only the sampled charge is transferred to the integrator, which therefore will not saturate when changing the clock frequency. Furthermore, the clock jitter only affects the system noise hardly. The input signal will be sampled at a wrong point of time. However, due to the oversampling, the input frequencies are low anyhow, where noise based on clock jitter can normally be neglected.

As a severe disadvantage, there are two settling phases within one clock period. First the sample capacitor needs to be charged accurately to the input voltage. The sampled charge then needs to be transferred to the integrator. It is also important that the sample capacitor is fully discharged. The two settling phases require a minimum bandwidth, which can be estimated with Eq. (4.26).

$$f_{-3\text{dB}} => \frac{\ln(2) \cdot (n+1)}{2\pi \cdot \frac{T_{CLK}}{2}}$$

A second order modulator to monitor electrical motor currents typically has to run on a 20 MHz clock frequency ($T_{CLK} = 50$ ns) so that failure conditions of a motor can be detected quickly. If used for the normal motor control loop, the resolution n has to achieve 12 bit. The bandwidth of the input stage and the integrator therefore needs to be above 57 MHz. This high bandwidth together with the noise requirements of the first integrator causes a high power consumption. Lower settling requirements, would enable higher speed or lower power dissipation. This is why an essential amount of research is currently performed on the topic of continuous-time delta-sigma modulators [1–4] that are presented in this chapter.

Figure 5.1 is showing the principle of a second order continuous-time modulator to illustrate the anticipated power and bandwidth requirements. The first integrator

F. Ohnhäuser, *Analog-Digital Converters for Industrial Applications Including an Introduction to Digital-Analog Converters*, DOI 10.1007/978-3-662-47020-6_5

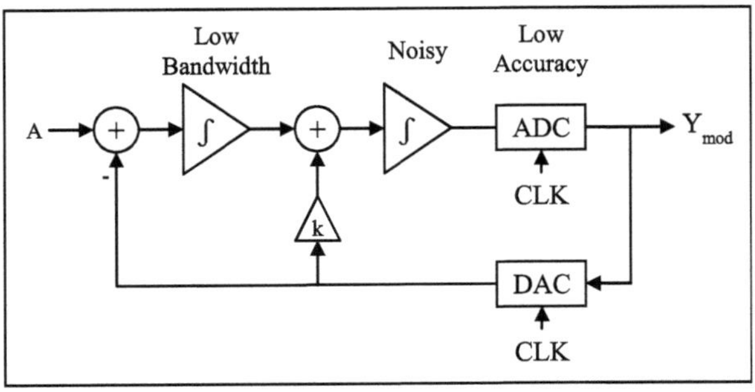

Fig. 5.1 Block diagram of a second order delta-sigma modulator

needs to follow the input signal A and can be of low bandwidth as it operates continuously. The second integrator can additionally be noisy. It is already inside the loop, so that the input referred noise will already be shaped with first order. The ADC is fully in the feedback path. Its resolution sets the quantization error. Simple latched comparators can be used, which have no DC current. Unfortunately, the real requirements of the circuitry are harder and will be discussed in Sect. 5.3.

The combination of low power consumption and/or high speed makes it interesting for many applications [5]. In industrial applications, this might be battery driven systems such as flow meter to measure the water or the gas flow in a pipe. Low power is also required for electrical equipment in island operation that is controlled with a 4–20 mA current. Electrical motor drives need to measure the current in the phases of the motor, where galvanic isolated delta sigma modulators are in use. These devices also require an isolated power supply, which is difficult to generate. The lower the power consumption, the more techniques are available to integrate the isolated power supply. A similar current measurement is required in power supplies or in electrical power meter.

Unfortunately, the continuous-time approach shows some disadvantages, which make the design difficult and sensitive. Particularly critical in continuous-time modulators is the DAC in the feedback path. The DAC output voltage adds to the input voltage A and any error that is generated by the DAC or at the input of the first integrator will cause a non-linear transfer function. This will be discussed in the following chapters.

The various effects are illustrated with simulated results in form of FFTs for a third order modulator with 32 MHz clock. The spectrums of a 15 kHz signal is illustrated in Fig. 5.2, where in a first case the input signal had no thermal noise associated. The spectrum illustrates the pure quantization noise. In a second case a noise density of $7.5\frac{\text{nV}}{\sqrt{\text{Hz}}}$ was added to the input signal. The bandwidth will be reduced by the connected digital filter based on the oversampling ratio. The reduced bandwidths are illustrated with the vertical lines for OSRs of 32 and 128.

The following chapters will show spectrums that are derated by various effects. Keep in mind that only effects are essential, whose spectrum exceeds the illustrated spectrum with thermal noise. Base on that, effects below the -130 dB are uncritical.

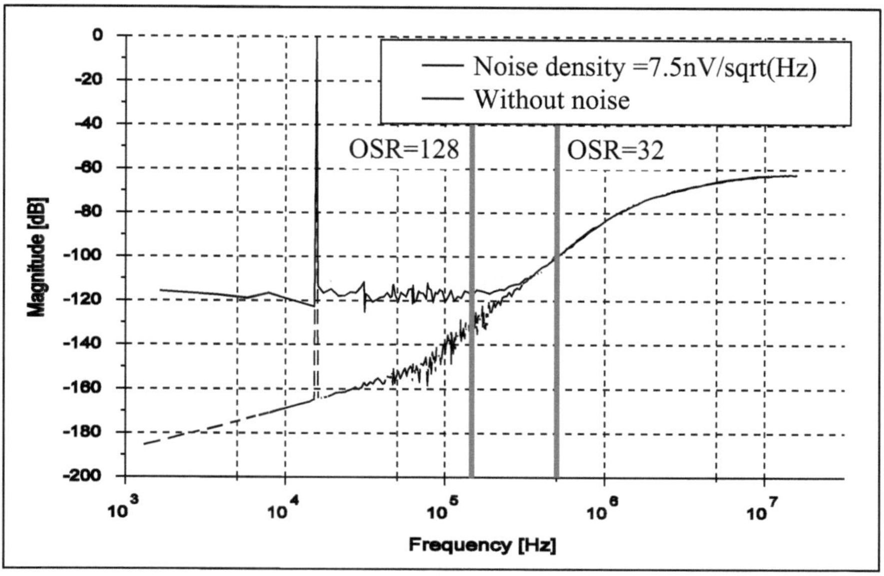

Fig. 5.2 Simulated spectrum of a delta-sigma modulator with and without input noise

5.1 Clock Jitter

The clock jitter is the first severe issue. It has no effect on the analog input voltage *A* that it is continuously integrated. For the example in Fig. 5.3, the analog input voltage was therefore set to 0. It was further estimated that a positive reference voltage REF or a negative reference voltage −REF will be fed back. It can easily be seen that the clock jitter Δt generates an erroneous integrator output voltage of ΔV_{int1}.

Note that this integration error does not occur in switched-capacitor approaches as the feedback charge on sample capacitors is fix over time as soon as voltages are settled. The integration error $\varepsilon_{\Delta t}$ of Fig. 5.3 can be calculated. Per clock cycle, the feedback voltage is connected to *–REF* instead of *REF* for the duration Δt with $2 \cdot REF = FSR$. Consequently, the error is the full-scale range times the ratio of the jitter to one clock cycle. The error $\varepsilon_{\Delta t}$ in ratio to the full-scale *FSR* can therefore be expressed with Eq. (5.1)

$$\frac{\varepsilon_{\Delta t}}{FSR} = \frac{\Delta V_{\text{int1}}}{FSR} = \frac{\Delta t}{T_{CLK}} \tag{5.1}$$

If the error $\varepsilon_{\Delta t}$ should be less than half an LSB, then

$$\frac{LSB}{2 \cdot 2^n LSB} > \frac{\Delta t}{T_{CLK}} \Leftrightarrow \Delta t < \frac{T_{CLK}}{2^{n+1}} \tag{5.2}$$

In the example of the electrical motor drive, the modulator has to operate at 20 MHz ($T_{CLK} = 50$ ns) and a resolution of $n = 12$ bit is required. If furthermore 12 bits

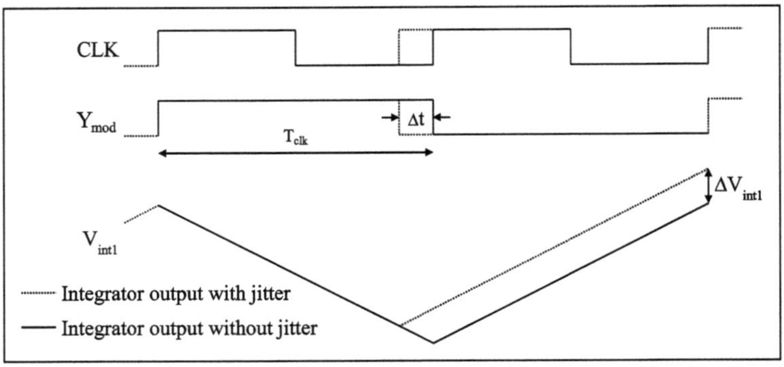

Fig. 5.3 Effect of the clock jitter on an integrator output voltage

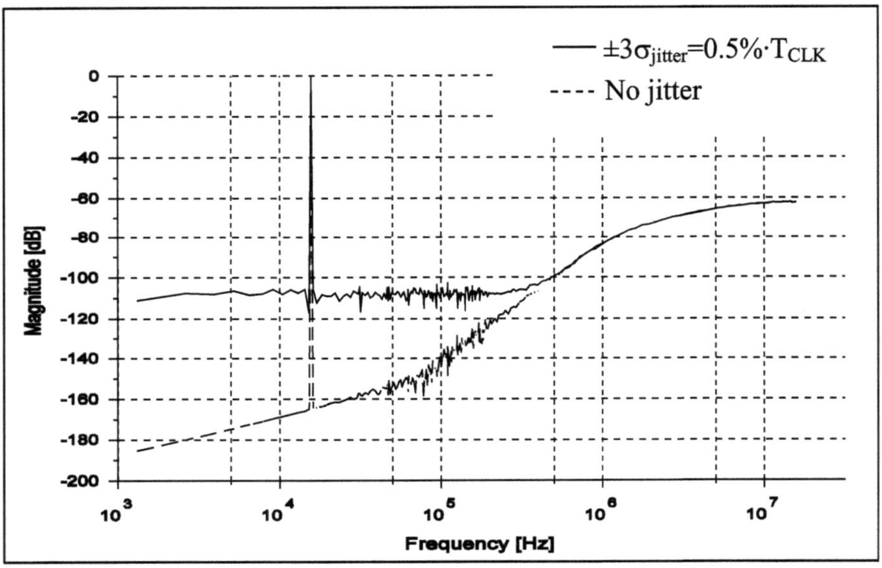

Fig. 5.4 Effect of the clock jitter on the spectrum of a delta-sigma modulator

($n = 12$) should be achieved, then the jitter has to be <6 ps. The effect of the jitter on the spectrum of the 32 MHz, third order modulator is shown in Fig. 5.4, where a jitter with a ± 3 sigma value of 0.5 % of the clock cycle is added. The one-sigma value for a 32 MHz clock can be calculated to 26 ps.

Figure 5.5 shows the degradation of the SNR and the THD versus the clock jitter. Note that THD is basically degrading due to the increase noise floor, as no harmonics are spiking out in Fig. 5.4.

Note that the jitter has two components including the clock jitter itself but also the distortion caused by the switching of the DAC in the feedback loop. If the reference requires some settling time, then this settling error is also added on top of the jitter. If this settling error occurs non-linear to the input voltage, than a

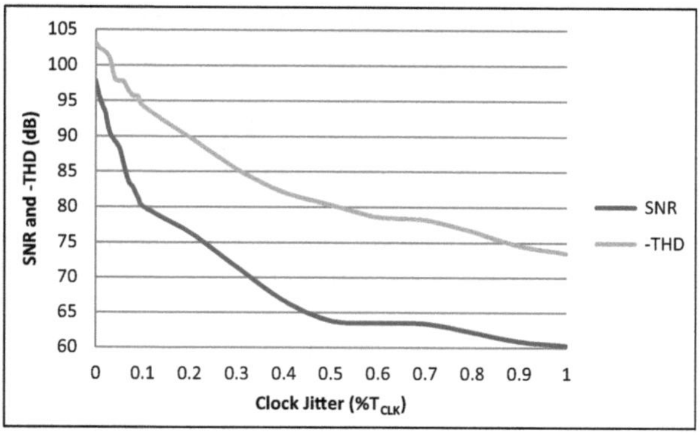

Fig. 5.5 Simulated SNR and THD versus clock jitter of a delta-sigma modulator

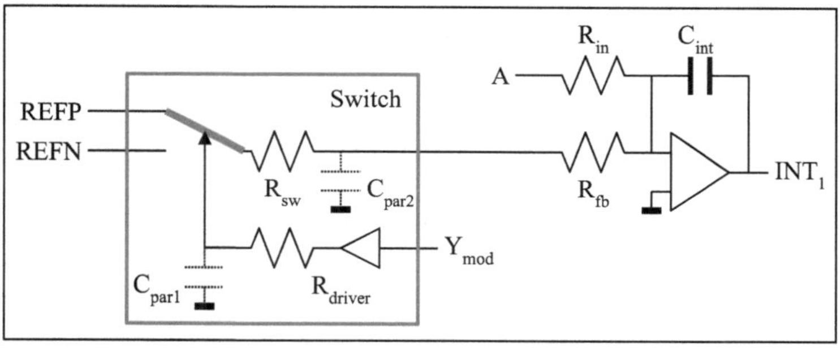

Fig. 5.6 Integrator of a delta-sigma modulator with resistive feedback

non-linear transfer function is the consequence. Papers are therefore discussing different forms of feedback.

5.1.1 Resistive Feedback

The standard resistive feedback (1 bit DAC) is presented in Fig. 5.6 [6]. The input resistor R_{in} and the feedback resistor R_{fb} can be matched very well, which would lead to good gain performance. However, the switching process will cause some settling error of the reference voltage. First of all, the switch transistors between the integrator and the positive and the negative reference potential require a size, so that its on-resistance R_{sw} can be neglected. Otherwise the on-resistance will affect the gain performance and particularly the gain drift as on-resistance will significantly drift over temperature and supply voltage. The linearity is not affected as the operating point of the switch is always the reference and therefore code independent.

The required switch resistance can be estimated the following way. The resistors R_{in} and R_{fb} are both generating noise, which together is $V_n = \sqrt{2 \cdot 4kTR \cdot f_{eff}}$.

Consequently, the required resistance must be low enough to achieve the anticipated noise performance, which is expressed by the SNR.

$$SNR = 20 \cdot \log\left(\frac{A}{V_n}\right)$$

The equation for the SNR can be solved to Eq. (5.3) for the rms-noise V_n. Note that A expresses the rms-value of the input voltage.

$$V_n = \frac{A}{10^{\frac{SNR}{20}}} \tag{5.3}$$

Delta-sigma modulators should be used for high resolution. For resolution <16 bit, the SAR ADC would be more power efficient, faster and has the advantage of taking snapshots. The SNR should therefore be at least 95 dB at a 5 V input range ($A = 1.77$ V), which leads to an rms-noise of 31 μV.

$$V_n = \frac{A}{10^{\frac{SNR}{20}}} = \frac{5}{2\sqrt{2} \cdot 10^{\frac{95}{20}}} = 31 \;\; \mu\text{V}$$

If all other noise sources are neglected beside the resistor noise then

$$V_n = \frac{A}{10^{\frac{SNR}{20}}} = \sqrt{8kTR \cdot f_{eff}}$$

The input resistor can be calculated to

$$R_{in} = R_{fb} = \frac{A^2}{8kT \cdot f_{eff} \cdot 10^{\frac{SNR}{10}}} \tag{5.4}$$

The bandwidth is first limited by the transfer function of the continuous-time delta-sigma modulator, but mostly by the digital filter following the modulator. Consequently, the effective bandwidth can be estimated to the Nyquist frequency at the filter output, which is

$$f_{Nyq} = \frac{f_{clk}}{2 \cdot OSR} \tag{5.5}$$

The 32 MHz modulator with an OSR of 32 will therefore provide a Nyquist frequency of 500 kHz, so that

$$R_{in} = R_{fb} < \frac{(1.77 \;\; \text{V})^2}{8 \cdot 1.38 \cdot 10^{-23} \frac{VAs}{K} \cdot 300 \;\; \text{K} \cdot 500 \;\; \text{kHz} \cdot 10^{\frac{95}{10}}} = 59.8 \;\; \text{k}\Omega$$

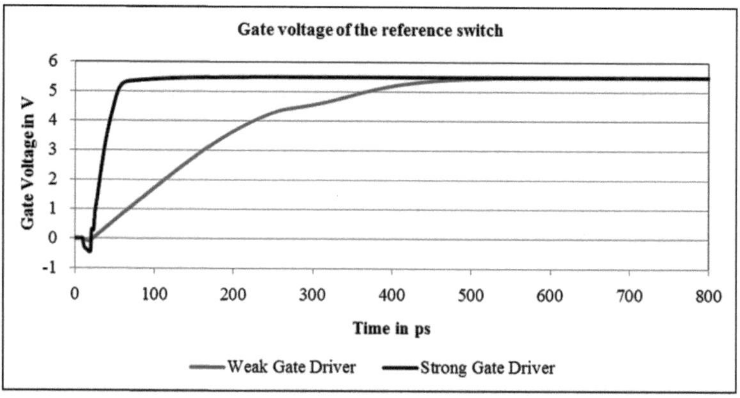

Fig. 5.7 Rising slope of the gate potential of a switch transistor for 5 V and 200 Ω on-resistance

Now that the resistors are determined based on their noise, the integrator capacitors will be chosen to fulfill the required transfer function of the modulator (see Sect. 4.1).

The switch resistance R_{sw} can be estimated by its drift of ±30 % over temperature and supply voltage. This can cause a gain drift Δg expressed with Eq. (5.6) below, which should be significantly smaller than 0.1 %.

$$\Delta g = \frac{\Delta R_{sw}}{R_{fb}} = \frac{\pm 0.3 \cdot R_{sw}}{R_{fb}} < \pm 0.1\ \% \tag{5.6}$$

Based on Eq. (5.6) the switch resistance should finally be

$$R_{sw} = \frac{\Delta g \cdot R_{fb}}{\pm 0.3} < \frac{\pm 0.1 \cdot R_{fb}}{\pm 0.3 \cdot 100} = 200\ \Omega \tag{5.7}$$

The switch transistors will also come with parasitic switch capacitors, which increase with decreasing R_{sw}. The gate capacitance C_{par1} of the transistor in Fig. 5.6 will soften digital edges on the gate (see Fig. 5.7). Appropriate driver circuits are required for the switches. Soft slopes will increase the clock jitter. Additionally, it makes the turn-on process of the transistor sensitive to noise and distortion of the power supply, which will add to the jitter. There is likelihood that the supply distortion is furthermore input voltage dependent, so that nonlinearity might also be added.

The drain capacitance C_{par2} in Fig. 5.6 needs to be reloaded during every switching process by the reference voltage. Some of the capacitance is the drain-gate capacitance, which will further cause charge injection. The reference source will require some settling time, which for an example can be seen in Fig. 5.8. An integration error will occur, during this settling effect causing gain and linearity errors. The curves in Fig. 5.8 were simulated with the assumption that the reference source is externally of the IC, that the wiring on the PCB, package, bondwire and on the IC act as a series inductance of 6 nH and that the wiring resistance on chip is 20 Ω.

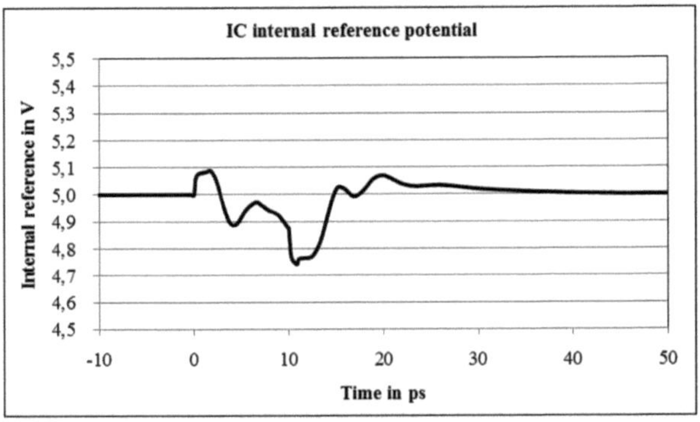

Fig. 5.8 Effect of a switching process inside an integrator on the reference potential

5.1.2 Return-to-Zero Architecture

The error described above will only occur, if the reference is actually switching. It does not occur, if it remains connected. Consequently, if the modulator output Y_{mod} changes its value ($0 \rightarrow 1$ or $1 \rightarrow 0$), an error is generated. This is not the case, if Y_{mod} remains constant during two decisions ($0 \rightarrow 0$ or $1 \rightarrow 1$). The modulator output Y_{mod} will remain constant at 0 or 1 at the negative or positive full-scale inputs. The full-scales will basically not be affected by the switching error. On the other side, the modulator output Y_{mod} switches almost continuously at mid-scale. The error is therefore input voltage dependent and will cause a non-linearity. The error is furthermore temperature and supply voltage dependent and will cause drift.

If the error would occur with every clock, independently on the previous value, then the error would linearly affect transfer function. Such a behavior can be achieved, if the feedback V_{fb} is switching back from *REFP* or *REFN* to 0 V after a certain period of time [7, 8]. In the example of Fig. 5.9, the integrator is connected to *REFP* or *REFN* for 0.75 clock cycles if $Y_{mod} = 0$ and then returns to 0 V for the remaining 0.25 clock cycles.

The full-scale *FS* of the modulator is set by the amount of charge per time, which is provided by the feedback, if the modulator output is 1. The quantization error can be regulated, as long as the charge coming from the input remains less. This is expressed with Eq. (5.8), where the time period is set to one clock cycle T_{CLK}.

$$Q_{FS} = \frac{FS \cdot T_{CLK}}{R_{in}} = Q_{fb} \tag{5.8}$$

The feedback charge for the return-to-zero architecture is decreased by the factor x, if the feedback resistor is connected to *REFP* for only $x \cdot T_{CLK}$. For simplicity, the negative reference potential is estimated to zero.

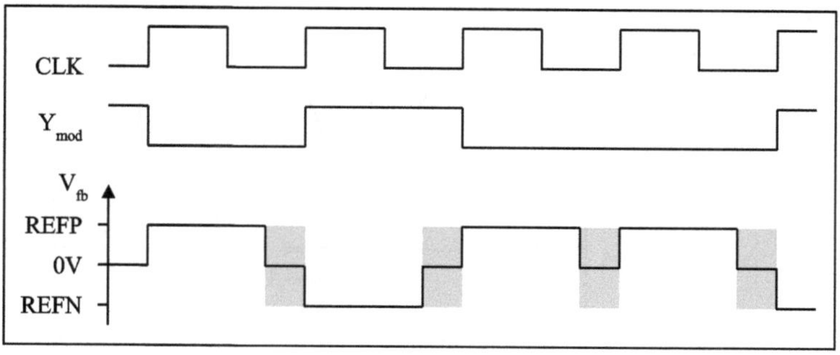

Fig. 5.9 Integrator feedback inside delta-sigma modulators with return-to-zero architecture

$$Q_{fb}(Y_{\text{mod}} = 0) = \frac{REFP \cdot x \cdot T_{CLK}}{R_{fb}} \tag{5.9}$$

This leads to

$$\frac{x \cdot REFP}{R_{fb}} = \frac{FS}{R_{in}} \tag{5.10}$$

Based on Eq. (5.10), the reference needs to be increased by the factor 1/x to remain the same full-scale range. Alternatively, the feedback resistor R_{fb} can be reduced by the factor of x. Both actions would increase the error in charge based on a clock jitter Δt, which is illustrated with Eq. (5.11).

$$\Delta Q_{fb} = \frac{REFP \cdot \Delta t}{x \cdot R_{fb}} \tag{5.11}$$

Return-to-zero architectures will therefore trade a better linearity with a worse noise performance. The return-to-zero architecture is not used for multi-bit architectures (see Sect. 5.1.5), which have multiple feedback switches. Normally, only one of these switches changes its state per clock cycle. The return-to-zero architecture would force all feedback switches to toggle, which would cause a significantly higher distortion.

5.1.3 Capacitive Feedback

The noise based on jitter can be improved, if the charge from the feedback loop is independent of the clock cycle. This could be done by replacing the resistive feedback with a switched-capacitor feedback as shown in Fig. 5.10 [9, 10]. The capacitor is discharged at the end of a clock cycle, so that jitter does not affect the

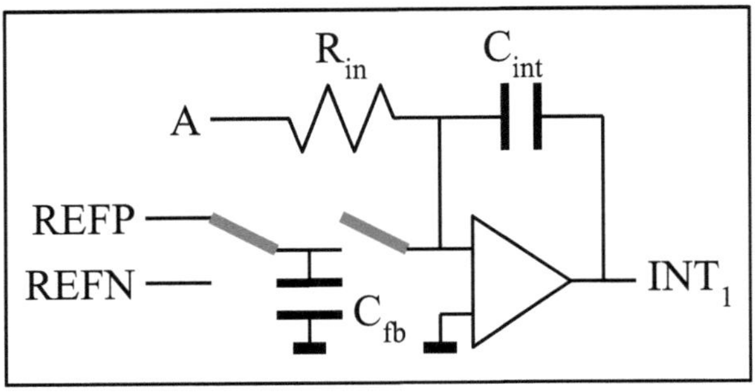

Fig. 5.10 Continuous-time integrator with capacitive feedback

transferred charge. This solution would also prevent the errors that are based on reference settling effect after switching the feedback.

The question that will arise is why not using a complete switch capacitor solution directly. Once the feedback is switched, the amplifier requires a fast settling time and therefore a high bandwidth. Additionally, the gain g of the modulator is set by Eq. (5.12).

$$g = \frac{Y_{\text{mod}}}{A} = \frac{1}{f_{CLK} \cdot C_{fb} \cdot R_{in}} \tag{5.12}$$

The feedback capacitor C_{fb} typically shows low drift, so that the gain drift is basically dependent on the clock frequency f_{CLK} of the modulator and the input resistor R_{in}. An RC-oscillator might be chosen, so that the frequency f_{CLK} decreases with the resistance. This could remain the gain drift limited. Unfortunately, RC-oscillators typically have a poor power supply rejection, so that noise and distortion on the power supply would cause clock jitter, which again would cause noise and distortion on the modulator output Y_{mod}. A clock frequency, which varies with temperature and supply voltage, is mostly undesirable in an application, which processes the output data from the converter in its own clock domain.

Note that the capacitive feedback will have an exponential settling, which has a high current peak at the time the capacitor is connected to the integrator. On the other side, the amplifier of the integrator has a limited gain and bandwidth, which will be discussed in Sect. 5.2. The amplifier input will therefore not be at virtual ground, but a differential input voltage is build up, which causes noise and distortion again.

The continuous-time architecture with capacitive feedback has two advantages compared to the switched-capacitor approach, which might make it easier to use in an application. First, the input is not switched and has a constant input resistance. This is easy to drive in an application, as the high bandwidths for recharging the sample capacitor are not required.

Secondly, the sample capacitor from a switched-capacitor approach will sample the input signal directly with f_{CLK}. The Nyquist theorem says that signals and noise with higher frequency than half the sample rate will be folded to the lower frequencies. An anti-aliasing filter is required. The continuous-time modulator

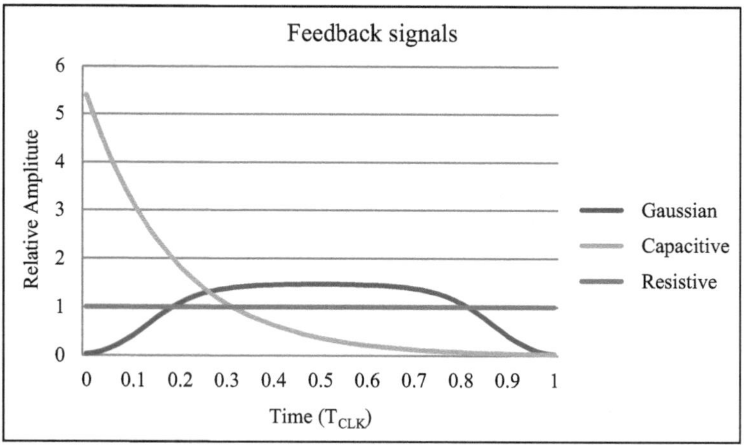

Fig. 5.11 Capacitive, resistive and Gaussian feedback signals in delta-sigma modulators

will sample the input signal not prior to the digitizer. Additionally, the integrator bandwidth, which is roughly 10 times the allowed input bandwidth, is normally significantly lower than the Nyquist frequency of the modulator that is the maximum input bandwidth times the oversampling ratio OSR. The usage of an OSR of 64 and a third order modulator will already provide an attenuation of 48 dB at modulators Nyquist frequency. The continuous-time modulator therefore already acts as the anti-aliasing filter.

Experiments have been done with other feedback signals. A Gaussian signal form might be preferable for example as shown in Fig. 5.11, but the signal generation would be more complicated.

Figure 5.11 shows the relative feedback current for a Gaussian, a capacitive and the resistive feedback during one clock cycle, where $Y_{mod} = 1$. It can be seen that the capacitive feedback has a significantly higher initial current compared to the other signal forms.

5.1.4 Current Feedback

The problem with the resistive feedback is mainly the switch size, which is required to be large for a low on-resistance to avoid gain errors and gain drift. The feedback resistor is basically converting the reference voltage into a reference current. As an alternative approach, a constant reference current can be generated, which is either connected to the integrator if $Y_{mod} = 0$ or to ground if $Y_{mod} = 1$ as shown in Fig. 5.12 [11].

This is also advantageous for a differential approach, where the current will flow onto a positive integrator capacitor if $Y_{mod} = 0$ or onto a negative integrator capacitor if $Y_{mod} = 1$. Care should be taken that the current is not discontinued

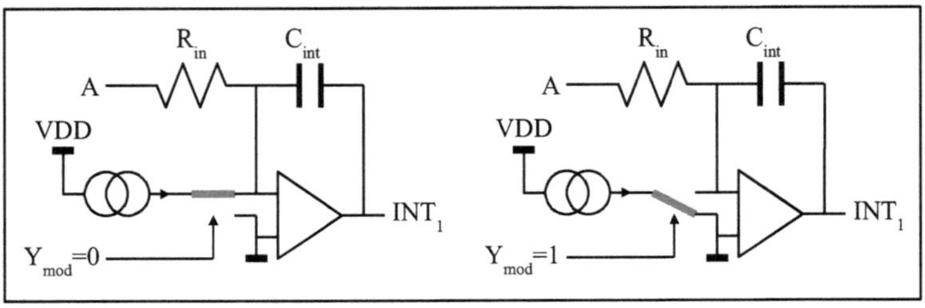

Fig. 5.12 Integrator with current feedback

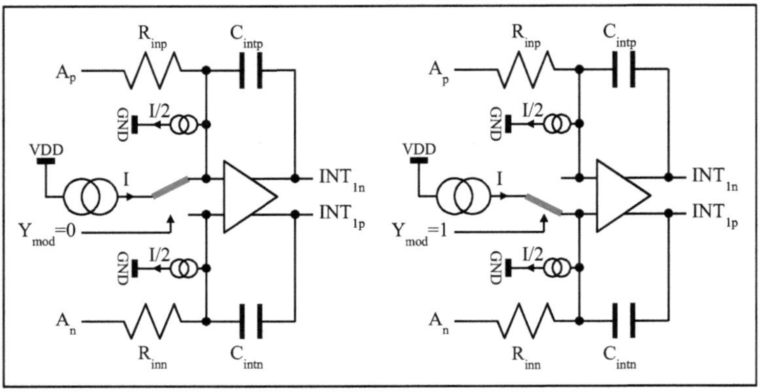

Fig. 5.13 Differential integrator with current feedback

during the switching process (see Fig. 5.13). Also the common-mode current needs to be subtracted on each input node of the amplifier.

The resistance of the switch is uncritical as long as the output resistance of the current source is high. The switch can therefore be designed in a significantly smaller size. Errors caused by the switching process are therefore significantly lower.

Care must be taken that the currents match the input path. It is therefore advantageous, if the feedback current is generated from a reference voltage and a resistor as shown in Fig. 5.14. The amplifier is regulating the voltage across the resistor to the reference voltage *REF*. The currents are then mirrored to the integrator by the PCH transistors. R_{fb} should match well with the input resistor R_{in}. The amplifier offset needs to be small. An auto-zero architecture should be considered. The current source is also simplified. A regulated cascaded current source might be used.

Note that this architecture can easily be modified to a current input delta-sigma modulator as illustrated in Fig. 5.15. The input resistor R_{in} in Fig. 5.12 is basically converting the input voltage A to an input current to the modulator. Removing the input resistor R_{in} will change the modulator from a voltage input to a current input.

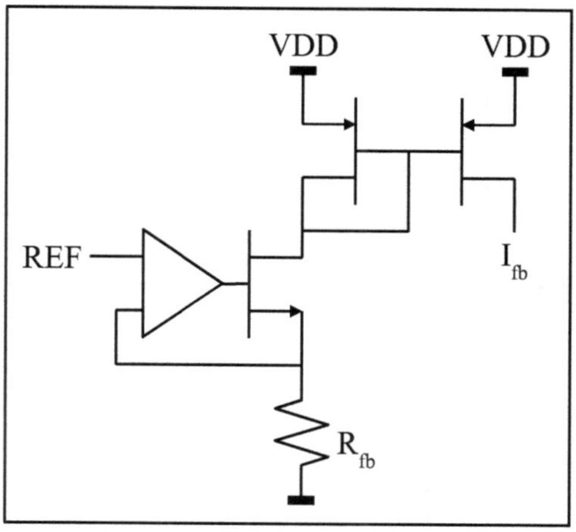

Fig. 5.14 Possible and simplified current source for the feedback in delta-sigma modulators

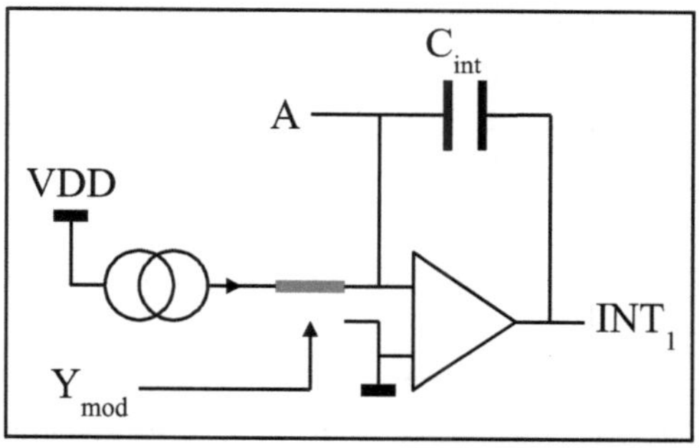

Fig. 5.15 Block diagram of a continuous-time integrator with current input

5.1.5 Multi-bit Feedback

The multi-bit feedback was already discussed in Chap. 4. However, it is gaining importance here, as it reduces the jitter requirements [12]. Reason is that the feedback is not switched between *REFP* and *REFN*, but only jumps the fraction of $2^N - 1$ of the full-scale *FS* as can be seen in Fig. 5.16, where *N* is representing the resolution of the DAC. *k* is representing a particular code in the ADCs transfer function.

This reduces the error caused by the jitter Δt to

$$\varepsilon_{\Delta t} = \frac{\Delta V_{\text{int1}}}{FSR} = \frac{2 \cdot \frac{REF}{2^N - 1} \cdot \Delta t}{2 \cdot REF \cdot T_{CLK}} = \frac{\Delta t}{\left(2^N - 1\right) \cdot T_{CLK}} \tag{5.13}$$

The jitter requirements can therefore be represented by

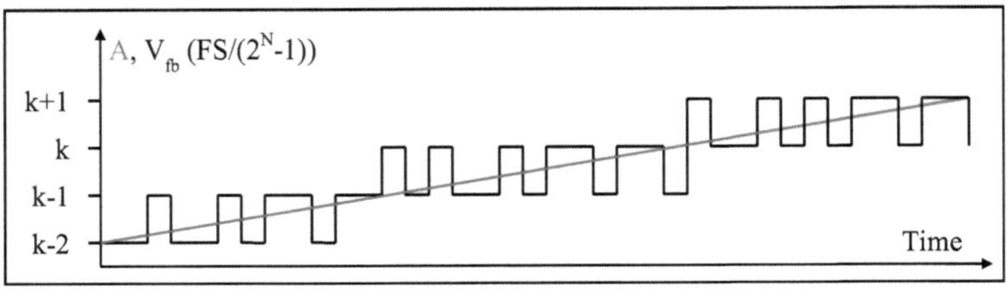

Fig. 5.16 Exemplary bit stream output of a multi-bit delta-sigma converter over time

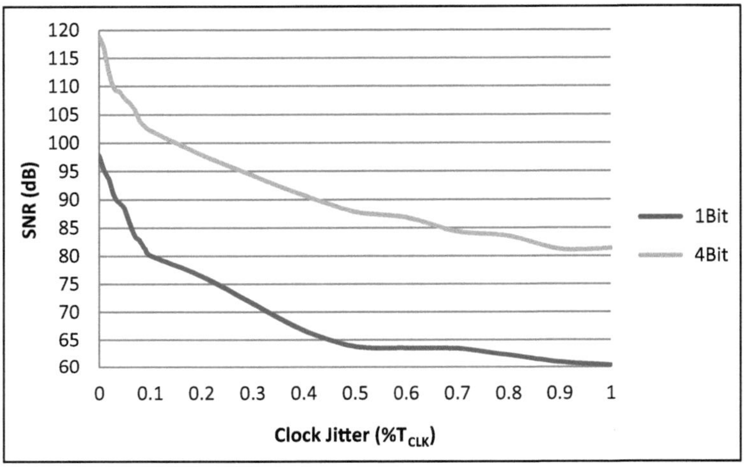

Fig. 5.17 Comparison of the SNR versus clock jitter for a modulator with 1 bit and a 4 bit quantizer

$$\Delta t < \frac{(2^N - 1)T_{CLK}}{2^{n+1}} \tag{5.14}$$

Note that N is representing the resolution of the quantizer and the DAC in the feedback loop and n the anticipated resolution of the delta-sigma ADC. The feedback requires the accuracy of n, which was discussed previously. Trim, calibration or dynamic element matching is required.

For illustration, the effect of the jitter on the SNR was modeled in Scilab for a single bit and a 4 bit quantizer and is shown in Fig. 5.17. The noise is dominated by the jitter once the noise drops below an anticipated SNR. It can be seen that the multi-bit quantizer allows more jitter. In simulation, the SNR with 1 bit quantizer drops below the 90 dB limit at a ± 3 sigma clock jitter of 0.04 % of a clock cycle, while the 4 bit quantizer does the same at 0.4 %. A factor of 8 was expected, but some additional resolution is gained by the reduced quantization noise.

Again, a modulator of third order was used for the modeling, which operates at a clock frequency of 32 MHz. The OSR for the results in Fig. 5.17 was chosen to 128. Figure 5.18 is illustrating the spectrums of modulators with 2 bit and 4 bit ADC.

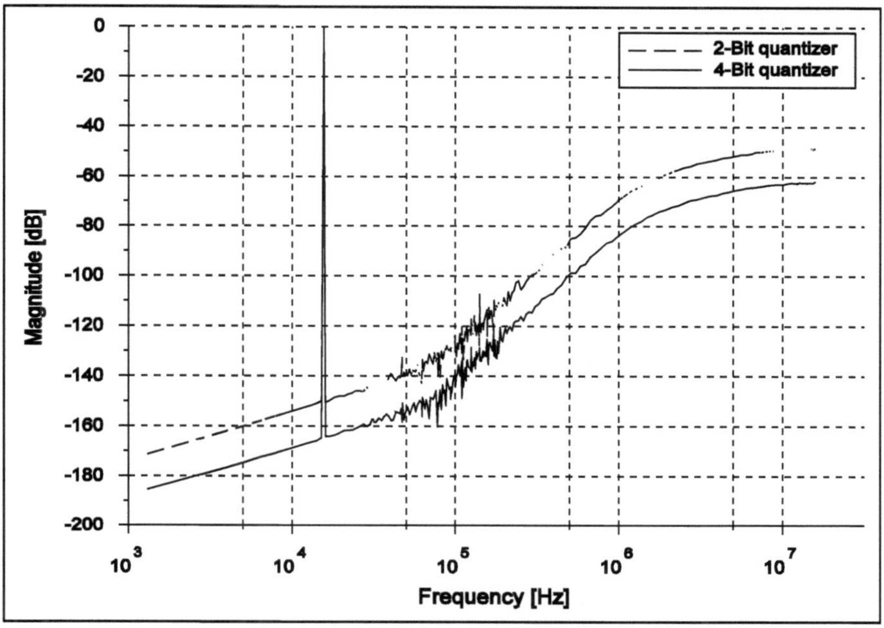

Fig. 5.18 Comparisons of the spectrum of delta-sigma modulators with 2 bit and 4 bit quantizer

The quantization noise of the 4 bit quantizer is generally 12 dB lower compared to the 2 bit quantizer.

Note that the multi-bit feedback is also reducing the gain and bandwidth requirements of the amplifier of the first integrator. This will be discussed further below.

5.2 Excess-Loop Delay

The last section discusses several feedback architectures to minimize the clock jitter. A much higher time delay is caused by the comparator, which will be in the order of nanoseconds instead of picoseconds. Furthermore, a comparator delay will increase if its differential input voltage is small. This is true for comparators, which have differential pairs as input, but also for pure cross coupled latches. Note that the quantizer does not really need accuracy, as it is within the regulation loop, so that a simple cross coupled latch is absolutely sufficient.

The comparator delay will act similar to the jitter, if the feedback switching is directly performed after the comparator decision and will destroy the SNR. It will further add significant THD due to the large code dependency of the comparator delay.

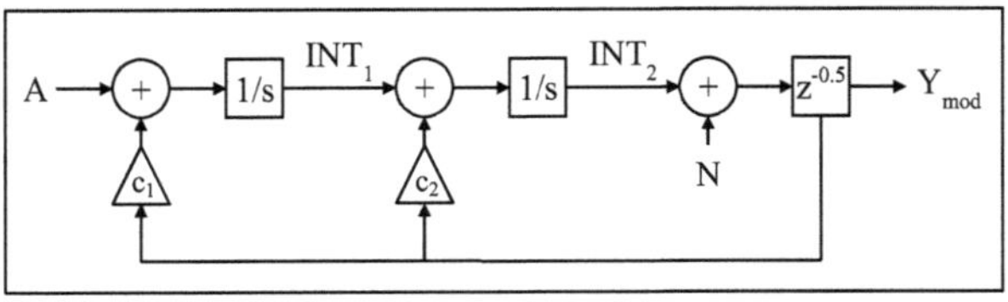

Fig. 5.19 Block diagram of a continuous-time delta-sigma modulator with excess loop delay

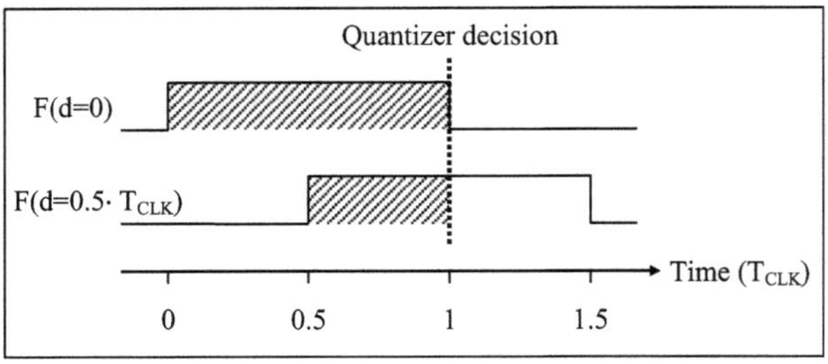

Fig. 5.20 Comparison of feedback signals inside modulators with and without excess loop delay

As a consequence, the feedback cannot directly be connected to the quantizer output, but must be clocked in some way. An asynchronous clock delay can be used, but the delay circuitry will add clock jitter again. Naturally, the feedback will switch with the negative clock edge, if the comparator decision is triggered on the positive clock edge. The delay however will affect the transfer function of the modulator as shown in the block diagram of Fig. 5.19 [13, 14].

c_1 and c_2 are the feedback coefficients. c_1 is directly setting the input voltage range of the modulator. c_2 is added to improve the stability of the modulator's transfer function. The quantizer is adding noise *N* and its decision is delayed by half a clock cycle, which is expressed with the term $z^{-0.5}$.

The error that is caused by $z^{-0.5}$ at the next comparator decision can be estimated by setting the analog input *A* to zero and comparing the voltages INT_1 and INT_2 at the integrator outputs without and with time delay *d*. For simplicity, the integrator outputs should initially be zero. The feedback of the modulator switches then to REF for one clock cycle. The two scenarios are displayed in Fig. 5.20.

$$INT_1(d=0) = \int_0^{T_{CLK}} c_1 dt = [c_1 \cdot t]_0^{T_{CLK}} = c_1 \cdot T_{CLK} \tag{5.15}$$

$$INT_2(d=0) = \int_0^{T_{CLK}} (c_1 \cdot t + c_2)dt = \left[\frac{c_1 \cdot t^2}{2} + c_2 t\right]_0^{T_{CLK}}$$
$$= \frac{c_1 \cdot T_{CLK}^2}{2} + c_2 \cdot T_{CLK} \tag{5.16}$$

$$INT_1(d=0.5) = \int_{0.5\cdot T_{CLK}}^{T_{CLK}} c_1 dt = [c_1 \cdot t]_{0.5\cdot T_{CLK}}^{T_{CLK}} = \frac{c_1 \cdot T_{CLK}}{2} \tag{5.17}$$

$$\begin{aligned} INT_2(d=0.5) &= \int_{0.5\cdot T_{CLK}}^{T_{CLK}} (c_1 \cdot t + c_2) dt = \left[\frac{c_1 \cdot t^2}{2} + c_2 t\right]_{0.5\cdot T_{CLK}}^{T_{CLK}} \\ &= \frac{3}{4} \cdot \frac{c_1 \cdot T_{CLK}^2}{2} + \frac{c_2 \cdot T_{CLK}}{2} \end{aligned} \tag{5.18}$$

The integration errors ε_{INT1} and ε_{INT2} at the next comparator decision can be calculated to

$$\varepsilon_{INT1}(d=0.5) = INT_1(d=0.5) - INT_1(d=0) = -\frac{c_1 \cdot T_{CLK}}{2} \tag{5.19}$$

$$\begin{aligned} \varepsilon_{INT2}(d=0.5) &= INT_2(d=0.5) - INT_2(d=0) \\ &= -\frac{c_1 \cdot T_{CLK}^2}{8} - \frac{c_2 \cdot T_{CLK}}{2} \end{aligned} \tag{5.20}$$

The effect of excess loop delay on the SNR is shown in Fig. 5.21. A single bit architecture was chosen as it is more sensitive for instability compared to multi-bit architectures. The OSR was chosen to 32.

The excess loop delay can be compensated by a modified transfer function that takes the error expressed in Eqs. (5.17) and (5.18) into account. The equations can therefore be written more general by substituting $T_{CLK}/2$ by $c_d \cdot T_{CLK}$, where c_d is the delay divided by one clock cycle.

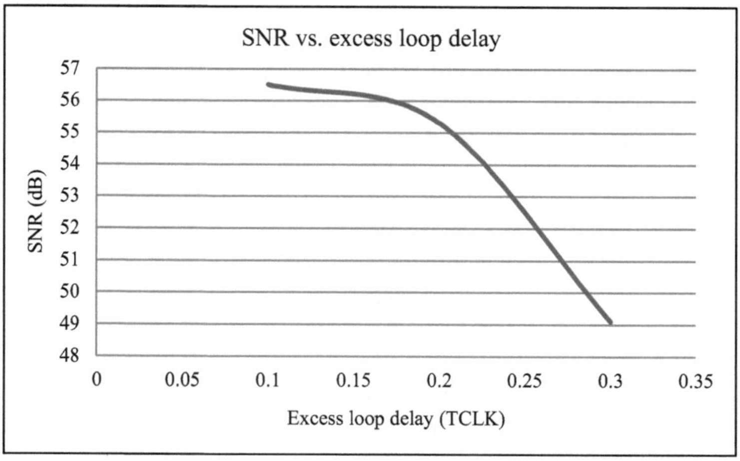

Fig. 5.21 Effect of the excess loop delay on the SNR for a second order delta-sigma modulator

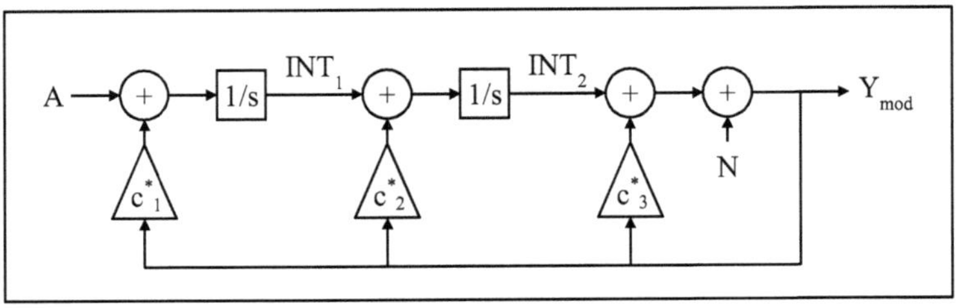

Fig. 5.22 Block diagram of a modulator with compensation of the excess loop delay

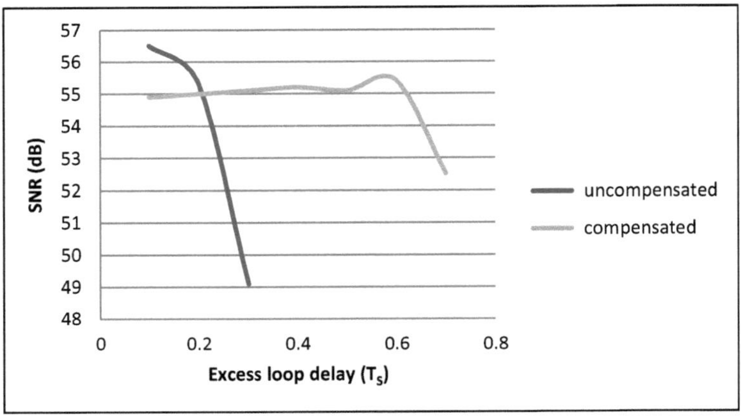

Fig. 5.23 SNR performance of a modulator with and without excess loop compensation

$$\varepsilon_{INT1}(c_d) = -c_1 \cdot c_d \cdot T_{CLK} \tag{5.21}$$

$$\varepsilon_{INT2}(c_d) = -\frac{c_1 \cdot c_d^2 \cdot T_{CLK}^2}{2} - c_2 \cdot c_d \cdot T_{CLK} \tag{5.22}$$

This difference in voltage could now be added directly to the integrator outputs by modifying the feedback coefficient c_2 and by an additional feedback with the coefficient $c_3{}^*$ to the output of the second integrator as shown in Fig. 5.22.

The new coefficients follow the Eqs. (5.23) through (5.25).

$$c_1^* = c_1 \tag{5.23}$$

$$c_2^* = c_2 + c_1 \cdot c_d \tag{5.24}$$

$$c_3^* = \frac{c_1 \cdot c_d^2}{2} + c_2 \cdot c_d \tag{5.25}$$

A comparison in SNR of a modulator with and without excess loop delay compensation is shown in Fig. 5.23. The spectrums with excess loop delays of 30 % of a clock cycle are compared in Fig. 5.24.

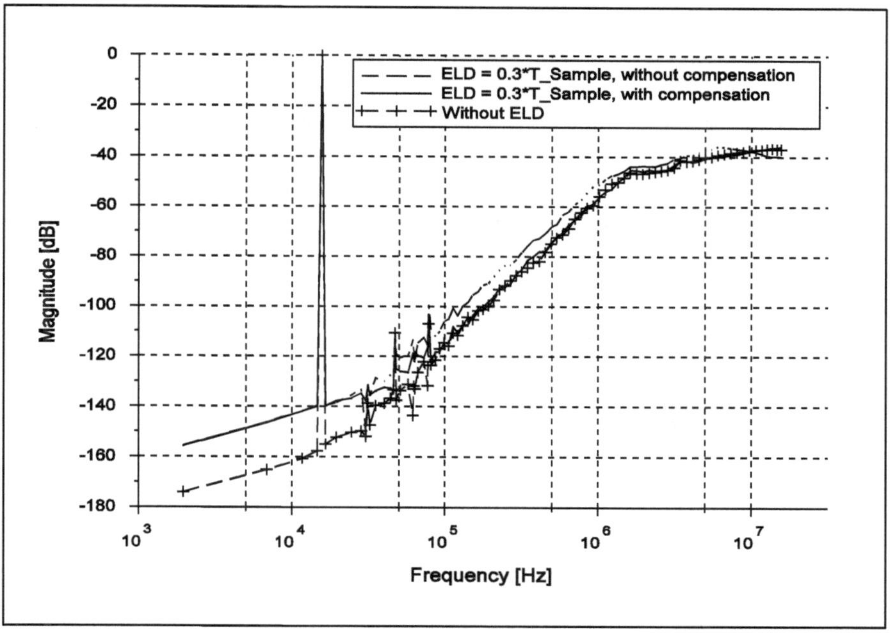

Fig. 5.24 Spectrums of modulators with excess loop delay, without delay and compensated delay

The spectrum of the compensated excess loop delay follows the ideal spectrum without delay at higher frequencies and the uncompensated spectrum at low frequencies. The increased noise floor at lower frequencies however is so low that it does not severely affect the SNR, if the other known noise sources are added.

Until now, architectural issues such as the jitter and ELD were discussed. In the following the parameter requirements of the integrators will be evaluated.

5.3 Integrator Requirements

The advantage of the continuous-time delta-sigma modulator should be the amplifier requirements in terms of gain and bandwidth. This should be investigated more closely and compared to the switched-capacitor architecture in this section.

Therefore, the requirements of the switched-capacitor architecture are evaluated first. It requires two settling times, one for charging the sample capacitors and one for the integration phase. The bandwidth can be estimated with the equation below, if exponential settling is assumed.

$$f_{-3\text{dB}} > \frac{\ln(2) \cdot (n+1)}{2\pi \cdot \frac{T_{CLK}}{2}}$$

If for example, a signal with a maximum bandwidth of 500 kHz should be sampled to a 16 bit accuracy n, then the output data-rate f_{data} must be higher than 1MSPS. The output data-rate is the clock frequency divided by the oversampling ratio OSR. A typical value for the OSR is 32, which should further be used in this example. The clock period T_{CLK} can therefore be calculated to

$$T_{CLK} = \frac{1}{OSR \cdot f_{data}} = \frac{1}{32 \cdot 1 \text{ MHz}} = 31.25 \text{ ns} \tag{5.26}$$

The required unity gain bandwidth is therefore

$$f_{-3\text{dB}} > \frac{\ln(2) \cdot (16+1)}{2\pi \cdot \frac{31.25 \text{ ns}}{2}} = 120 \text{ MHz} \tag{5.27}$$

Equation (5.27) is neglecting the time, which is lost for providing non-overlapping clocks (~1 ns per clock edge) and for the comparator decision (~2 ns), which would reduce the remaining time from 31.25 ns to 27.25 ns and drive the bandwidth requirements to 137 MHz. This bandwidth is actually required under worst case conditions, which are often found at slow transistor models and hot temperatures. The bandwidth at a temperature of 25 °C and typical models might be 50 % higher than at the worst case conditions, so that a typical bandwidth of 206 MHz is required for the analog input source and for the amplifier of the integrator. Nevertheless, the calculation should be continued with the calculated value of 120 MHz.

In a next step, the thermal noise density of the amplifier in the switched-capacitor approach should be calculated assuming a 1 V peak-to-peak input range, which equals 350 mV_{rms}. It should further be estimated that 15 effective bits are anticipated, which would reflect a signal-to-noise ratio of 92 dB. The rms-noise V_{rms} can then be calculated to

$$V_{rms} < \frac{350 \text{ mV}_{rms}}{10^{\frac{SNR}{20}}} = 8.8 \text{ µV}_{rms} \tag{5.28}$$

The thermal noise will dominate due to the high bandwidth. The noise will first be integrated to the full effective noise bandwidth of the first amplifier, which is $\frac{\pi}{2} \cdot 120$ MHz. The noise of the amplifier will be sampled on the capacitors of the integrator, which can be estimated to a first order system.

The sampling will reduce the bandwidth to $\frac{f_{CLK}}{2}$, but will also fold the noise equally into the band from 0 Hz to the Nyquist frequency $\frac{f_{CLK}}{2}$, so that the sampling will remain the total noise. Only the decimation filter will reduce the effective bandwidth by the oversampling ratio OSR. The effective bandwidth f_{eff} for this example is therefore

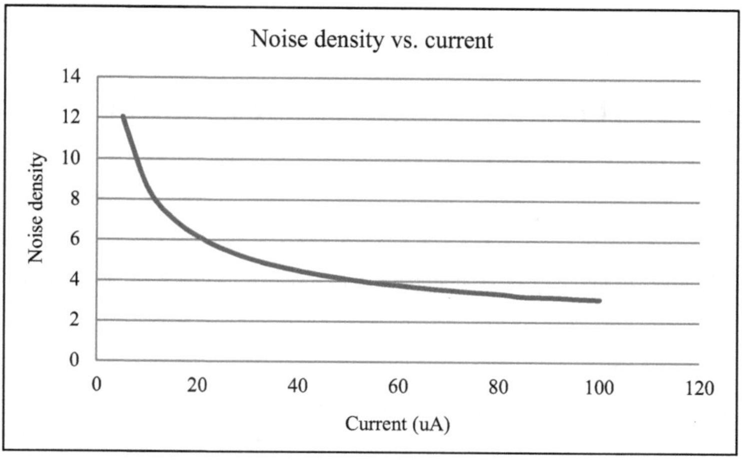

Fig. 5.25 Input related noise density of an integrator versus current consumption

$$f_{eff} = \frac{\frac{\pi}{2} \cdot f_{-3\text{dB}}}{OSR} = \frac{\frac{\pi}{2} \cdot 120\ \text{MHz}}{32} = 5.89\ \text{MHz} \tag{5.29}$$

The integrators required noise density n_{int} can therefore be calculated to

$$n_{\text{int}} < \frac{V_{rms}}{\sqrt{f_{eff}}} = 3.6 \frac{\text{nV}}{\sqrt{\text{Hz}}} \tag{5.30}$$

If the input range is chosen to 1 V_{pp} and the modulator is utilizing a 1.25 V reference voltage, then a 1.5 V or a 1.8 V semiconductor technology might be chosen, which typically has a minimum gate length of 150–180 nm. The required bandwidth of typically 120 MHz is doable. Noise densities, which are better than the $3.6 \frac{\text{nV}}{\sqrt{\text{Hz}}}$ from Eq. (5.30), require a significant current through the amplifiers input stage. This is illustrated in Fig. 5.25 for a 1.8 V process.

The integrator noise is not the only noise source. The sampling noise kT/C was mentioned previously, where the thermal noise of the resistance of the input switch is getting filtered by the sampling capacitor. As the switch resistance is low, the initial bandwidth is wide. The noise is then folded into a bandwidth of half the clock frequency by the sampling process. The high frequency noise is then suppressed by the digital low-pass filter, which reduces the bandwidth by the oversampling ratio again. Thus the kT/C noise is also reduced by $\sqrt{OSR}$. This will allow an increased noise level. The sample capacitor must be larger than

$$C_{\text{int}} = \frac{kT}{OSR \cdot V_n^2} = \frac{1.38 \cdot 10^{-23} \frac{VAs}{K} \cdot 300\ \text{K}}{32 \cdot (8.8\ \mu\text{V})^2} = 1.67\ \text{pF} \tag{5.31}$$

A limited gain of the amplifier will cause a non-ideal virtual ground. The amplifier will have a small remaining differential input voltage, which is a function of the

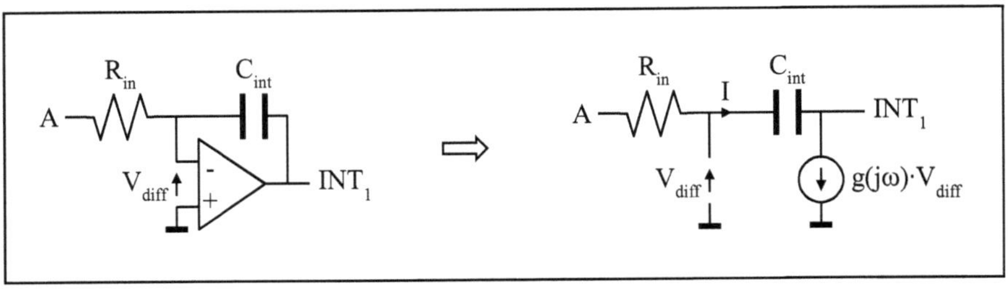

Fig. 5.26 Equivalent circuitry of an integrator with resistive feedback to calculate the transfer function

output voltage. Thus the resulting error is dependent on the integrator output voltage and therefore also on the derivation of the input voltage. A non-sufficient gain will therefore affect the total harmonic distortion THD. It is recommended to model the anticipated delta-sigma architecture together with non-ideal amplifiers and the particular signal conditions and to set the gain based on simulation results.

The values above should now be compared to results from continuous-time modulators. Limited gain and bandwidth will influence the transfer function of an integrator, which will be shown in the calculation below. These calculations are based on the equivalent amplifier circuitry of Fig. 5.26.

If the amplifier is ideal, then its differential input voltage V_{diff} is 0 V leading to the transfer function

$$F_{ideal}(j\omega) = -\frac{1}{j\omega R_{in}C_{\text{int}}} = -\frac{f_i}{jf} \tag{5.32}$$

with f_i being the 0 dB-frequency of the integrator ($f_i = 1/(2\pi R_{in}C_{int})$).

A real amplifier has limited gain and bandwidth, which in Fig. 5.26 is expressed with $g(j\omega)$. The amplifier output INT_1 equals $g(j\omega)\cdot V_{diff}$. Therefore, the following equations apply.

$$INT_1 = g(j\omega)\cdot V_{diff} \Leftrightarrow V_{diff} = \frac{INT_1}{g(j\omega)} \tag{5.33}$$

$$INT_1 = -V_{diff} - \frac{I}{j\omega C_{\text{int}}} \tag{5.34}$$

$$I = \frac{A + V_{diff}}{R_{in}} \tag{5.35}$$

Equations (5.33) through (5.35) lead to

$$INT_1 = -\frac{INT_1}{g(j\omega)} - \frac{A + \frac{INT_1}{g(j\omega)}}{j\omega R_{in}C_{\text{int}}}$$

The transfer function is

$$F(j\omega) = \frac{INT_1}{A} = -\frac{g(j\omega)}{1 + j\omega R_{in} C_{\text{int}}[1 + g(j\omega)]} \tag{5.36}$$

Note that Eq. (5.36) will lead for large $g(j\omega)$ to Eq. (5.32). Now the amplifier has limited DC-gain G and limited -3 dB-bandwidth f_a, so that

$$g(j\omega) = \frac{G}{1 + j\frac{\omega}{2\pi f_a}} \tag{5.37}$$

The effect of the non-ideal amplifier on the integrators transfer function is illustrated in Fig. 5.27. It can be seen that sufficient DC gain is required on one side and that the gain-bandwidth-product GBW of the amplifier needs to be significantly higher (at least an order) than the integrators 0 dB-frequency. Otherwise, the transfer function of the integrator will be significantly influenced, which then will affect the transfer function of the modulator.

If the reduced bandwidth of the modulator gets in the order or lower than the signal frequency, then the signal amplitude will be attenuated inside the modulators transfer function. The consequence will be a lower SNR. The internally attenuated signal amplitude will show up as attenuated feedback, which is illustrated in Fig. 5.28 in comparison to Fig. 5.29. The attenuated feedback will increase the difference between the input signal A and the feedback signal FB, so that the current onto the integrator capacitor increases together with the virtual ground (V_{diff}) of the amplifier, which can be seen in Figs. 5.28 and 5.29 as well.

The effect of integrators with limited gain-bandwidth-product GBW on the SNR is illustrated below in Fig. 5.30 in an example of a third order modulator. The

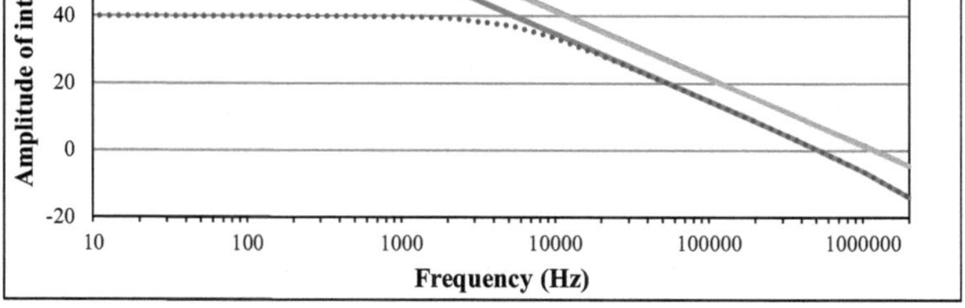

Fig. 5.27 Transfer function of an integrator with ideal and non-ideal amplifier

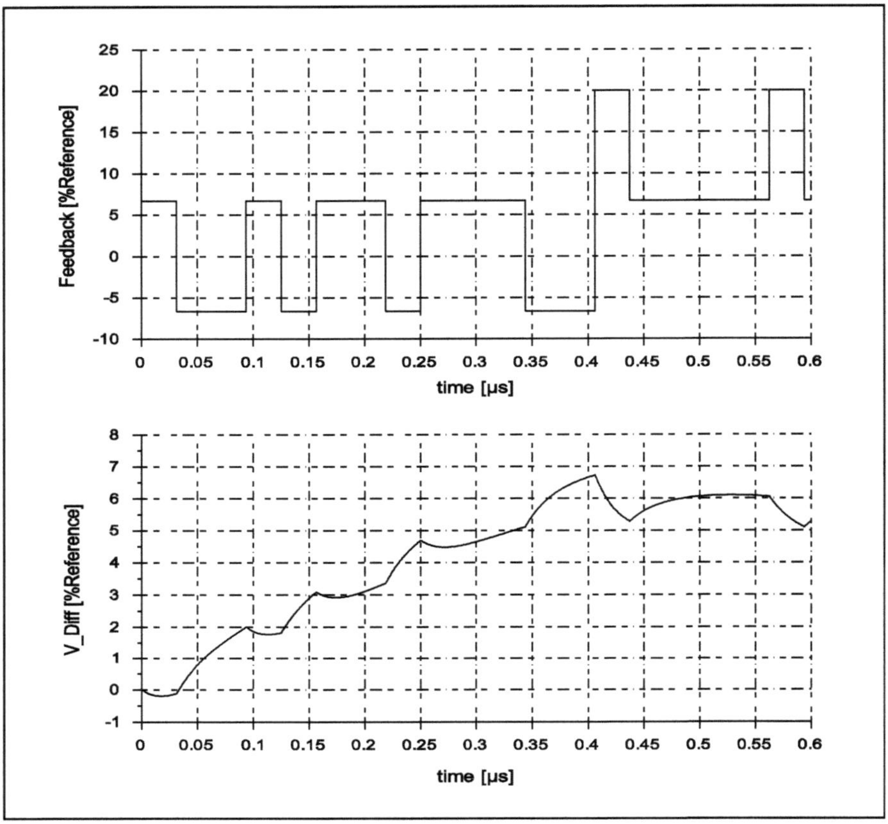

Fig. 5.28 Modulator feedback and amplifier input for an amplifier with limited GBW of 1 MHz

modulator operates at 32 MHz clock frequency and an OSR of 32. The input signal is a 500 kHz sine wave. Curves for the SNR are shown, where

(a) The GBW of the amplifier of the first integrator is limited and the other two integrators are ideal
(b) The GBW of the amplifier of the second integrator is limited and the first and third integrators are ideal
(c) The GBW of the amplifier of the third integrator is limited and the first and second integrators are ideal

For this particular modulator of third order and a clock frequency of 32 MHz, the gain should be higher than 80 dB and the GBW should exceed 100 MHz, which is roughly two orders above the integrators 0 dB-frequency of 1.2 MHz.

The effect of a limited GBW on the spectrum is shown in Fig. 5.31, which compares modulator outputs with an ideal first integrator, with an integrator that's amplifier is limited to 100 MHz and 1 MHz.

As an interesting result, also the continuous-time modulator needs amplifiers with high gain and bandwidth. However, the bandwidth requirement is not

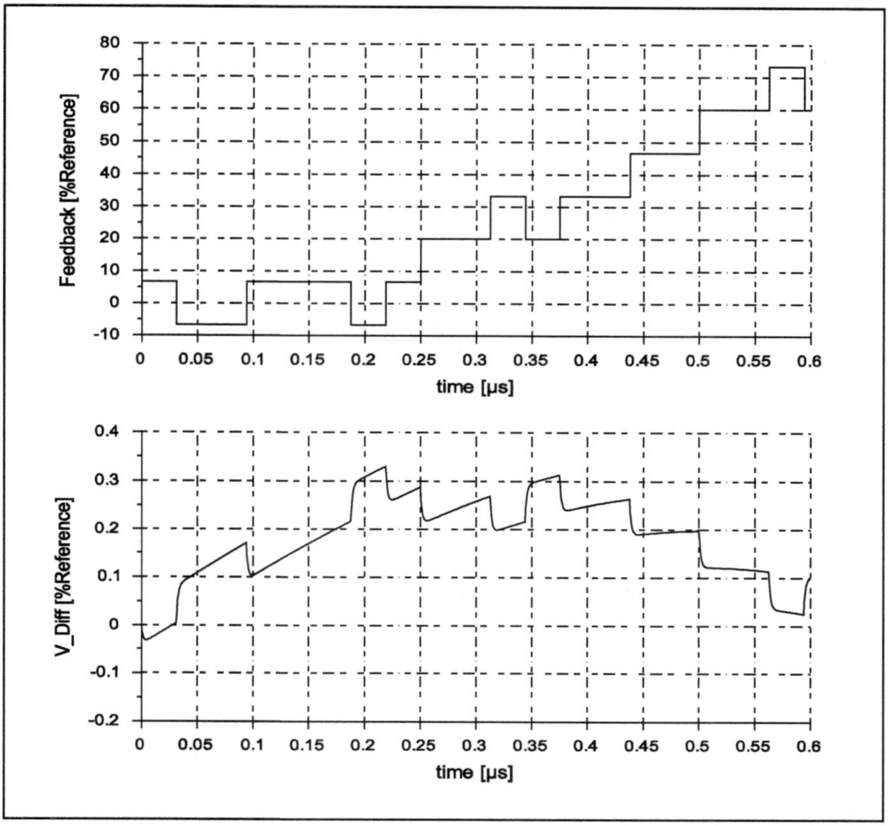

Fig. 5.29 Modulator feedback and amplifier input for an amplifier with limited GBW of 100 MHz

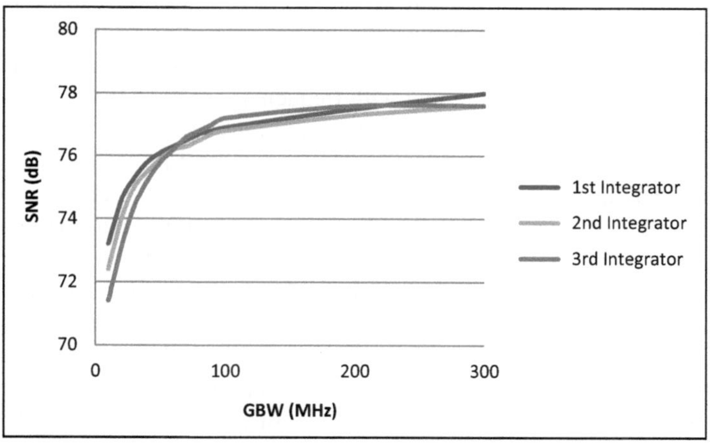

Fig. 5.30 Effect of an amplifier with GBW = 1 MHz inside specific integrators to the SNR of a ΔΣ ADC

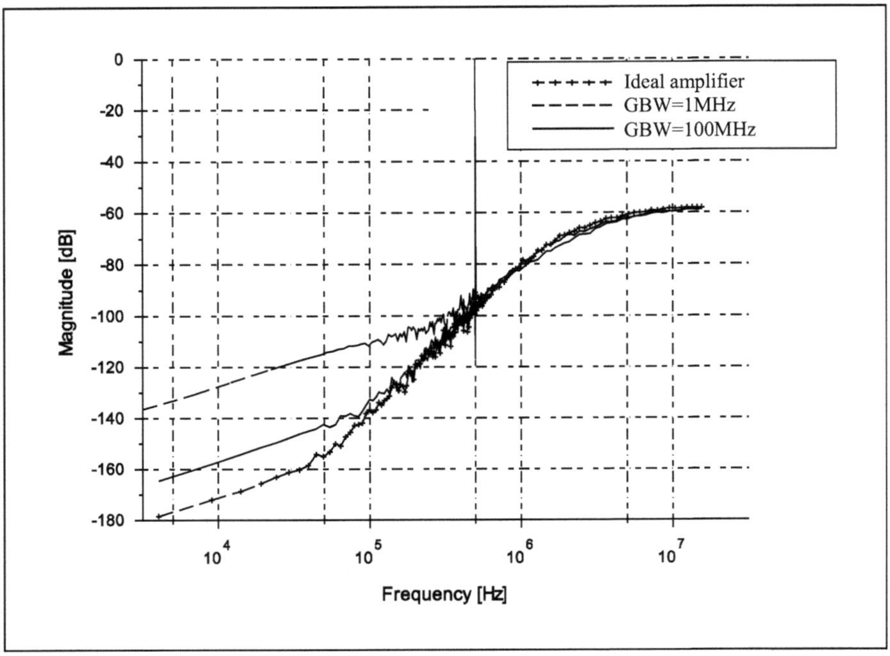

Fig. 5.31 Spectrums of delta-sigma modulators using amplifiers of different GBWs

dependent on the clock frequency, but on the maximum signal frequency. Distortion at high frequency such as switching effects will be suppressed by the bandwidth of the modulator. The requirements on gain and bandwidth should in any case be modeled sufficiently for the architecture in use.

Note that there are many applications, where a signal must be converted with high resolution, if the bandwidth is low. The same signal can have lower resolution if the signal bandwidth is increasing. The current measurement in motor drives or power converters is such an example. The normal control loop, which requires high resolution such as 12 bits, is running with frequencies around 16 kHz. In case of a failure, the current can jump to high values and must be turned off within a few microseconds to prevent severe damage. This overcurrent protection requires high bandwidth and therefore high clock frequency around 20 MHz, but only low resolution such as 6 bits. The bandwidth of the switched-capacitor approach must be designed for the 20 MHz operation. The bandwidth of the continuous-time modulator is only dependent on the signal bandwidth in the high resolution mode.

The previous example for the anticipated 500 kHz input signal is designed for position sensing with encoders in electrical motor drives. The encoder is a disk of glass with radial stripes. When the motor is rotating, the stripes are detected with photo diodes. The angular position or speed is measured by counting the stripes. The output current of the diode has a waveform, which is similar to a sine wave, when the stripe passes the diode. Therefore, additional resolution can be obtained by measuring the actual current through the diode. The additional resolution is important for the speed calculation, if the motor rotates slowly and the number of counts is low. The higher the number of counts per time, the higher is the signal frequency, but the less important is the additional resolution. So again, high resolution is required for input signals with low frequencies and low resolution is sufficient for signals with high frequency. The bandwidth of the amplifier in the continuous-time modulator only needs to support high performance at low signal frequencies.

In the following, the requirements for the noise are evaluated. Due to the sampling process, the noise of the amplifier of the switched capacitor approach was folded into a frequency band of half the clock frequency. If the bandwidth of the amplifier is for example 128 MHz and the clock frequency is 32 MHz, then the noise density in the band of 0–16 MHz increases by the factor of $\sqrt{8}$.

This is not the case for the continuous-time modulator. The noise of the integrator is filtered by the bandwidth of the modulator, which typically is around 10-times the maximum signal frequency and therefore lower than half the clock frequency. The digital low-pass filter will then reduce the bandwidth to half the output data rate f_{data}, which was anticipated to be 500 kHz, so that the noise density of the integrator n_{int} must be less than

$$n_{\text{int}} < \frac{V_{rms}}{\sqrt{\frac{f_{data}}{2}}} = \frac{8.8\ \mu\text{V}}{\sqrt{500\ \text{kSPS}}} = 12.4\frac{\text{nV}}{\sqrt{\text{Hz}}} \tag{5.38}$$

The requirements on the noise density are significantly lower compared to the switched-capacitor approach, which is important to achieve a low power consumption.

5.4 CM Feedback

The continuous-time modulator has significantly less switching involved than time-discrete modulators. Charge injection might be less of an issue if disregarding the errors that are adding to the jitter. A single ended approach might be considered. However, the differential approach brings advantages.

First of all, it can increase the internal signal amplitude by a factor of two without increasing the supply range. This will increase the SNR or reduce the sensitivity. As a second advantage, which was also mentioned before, the differential approach allows choppering of at least the amplifier of the first integrator. This way, offset and offset drift might be significantly reduced. However, care must be taken that the distortion caused by the switching effect does not cause distortion to the analog input or feedback path. If the chopper frequency is a multiple of the data rate, then the distortion will fall into frequencies, where the transfer function has notches. The distortion would be suppressed.

The fully differential amplifier will regulate the differential input voltage to virtual ground. It does not set the common-mode voltage. Therefore, an additional

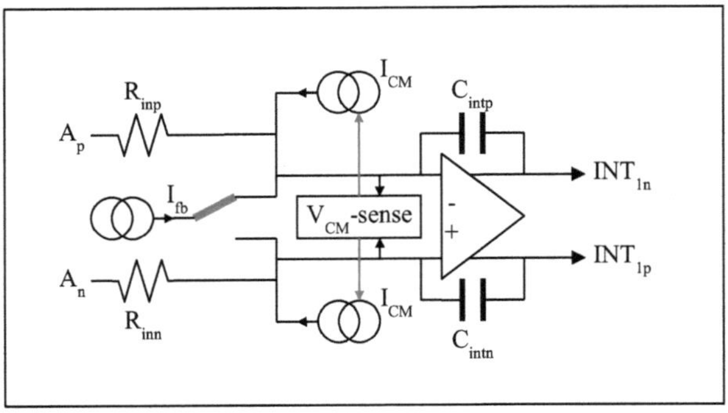

Fig. 5.32 Differential integrator with current sources I_{CM} for common-mode feedback

common-mode feedback is required [15]. This common-mode feedback could for example be implemented with regulated current sources as shown in Fig. 5.32.

A mismatch ΔI_{CM} of the currents I_{CM} will add to the offset of the amplifier. A differential input voltage of

$$\Delta A = A_p - A_n = R_{in} \cdot \Delta I_{CM} \tag{5.39}$$

would be required to compensate the erroneous current. However, the feedback current could be choppered together with the amplifier to suppress the offset.

This feedback current opens interesting possibilities. The analog input nodes only connect to resistive inputs, which are insensitive to high voltages. The common-mode input voltages can therefore be significantly higher than the supply voltage without destroying chip internal transistors. Signals with large common-mode range can be converted on modulators with low supply voltage.

This is helpful in industrial applications, where machinery and control circuitry can be located in different areas and where the ground potentials differ. Such different ground potentials act as common-mode voltage for the receiving side. A wide common-mode range would be helpful.

References

1. M. Ortmanns, F. Gerfes, Continuous-Time Sigma-Delta A/D Conversion, Springer, New York/Dordrecht/Heidelberg/London, 2005.
2. Sebastian Raschbacher, Modellierung, Simulation und Fehleranalyse eines zeitkontinuierlichen ΔΣ Modulators, bachelor thesis at the Friedrich-Alexander University Erlangen, 2012.
3. J.M. de la Rosa, Sigma-delta modulators: tutorial overview, design guide, and state-of-the-art survey, in the journal of Transactions on Circuits and Systems, Volume 58, Issue 1, pp. 1-21, 2011.
4. J.A. Cherry, W.M. Snelgrove, Continuous-time delta-sigma modulators for high-speed A/D conversion - theory, practice and fundamental performance limits, Kluwer Academic Publishers, New York, 2002.

5. Lucien Breems, Johan H. Huijsing, Continuous-Time Sigma-Delta Modulation for A/D Conversion in Radio Receivers, Kluwer Academic Publishers, Bosten/Dordrecht/London, 2010.
6. Yung-Chou Lin, Wen-Hung Hsieh, Chung-Chih Hung, A continuous-time delta-sigma modulator using feedback resistors, in the proceedings of the International Symposium on VLSI Design, Automation and Test (VLSI-DAT '09), pp. 243-246, 2009.
7. O. Oliaei, H. Aboushady, Jitter effects in continuous-time ΣΔ modulators with delayed return-to-zero feedback, in the proceedings of the IEEE International Conference on Electronics, Circuits and Systems, pp. 351-354, 1998.
8. Kong-Pang Pun, S. Chatterjee, P.R. Kinget, A 0.5-V 74-dB SNDR 25-kHz continuous-time delta-sigma modulator with a return-to-open DAC, in the IEEE journal of Solid-State Circuits, Volume 42, Issue 3, pp. 496-507, 2007.
9. Yan Chen, Kong-Pang Pun, A 0.5 V 2-1 cascaded continuous-time delta-sigma modulator synthesized with a new method, in the proceedings of the 53rd IEEE International Midwest Symposium on Circuits and Systems (MWSCAS), pp. 1165-1168, 2010.
10. Jun-Gi Jo, Jinho Noh, Changsik Yoo, A 20-MHz bandwidth continuous-time sigma-delta modulator with jitter immunity improved full clock period SCR (FSCR) DAC and high-speed DWA, in the IEEE journal of Solid-State Circuits, Volume 46, Issue 11, pp. 2469-2477, 2011.
11. K. Matsukawa, Y. Mitani, M. Takayama, K. Obata, A 69.8 dB SNDR 3rd-order continuous time delta-sigma modulator with an ultimate low power tuning system for a worldwide digital TV-receiver, in the proceedings of the IEEE Custom Integrated Circuits Conference (CICC), pp. 1-4, 2010.
12. V. Dhanasekaran, M. Gambhir, M.M. Elsayed, E. Sánchez-Sinencio, Continuous time multi-bit ΔΣ ADC using time domain quantizer and feedback element, in the IEEE journal of Solid-State Circuits, Volume 46, Issue 3, pp. 639-650, 2011.
13. M. Keller, A. Buhmann, J. Sauerbrey, M. Ortmanns, Y. Manoli, A comparative study on excess-loop-delay compensation techniques for continuous-time sigma-delta modulators, in the IEEE journal on Circuits and Systems, Volume 55, Issue 11, 2008.
14. W. Gao,O. Shoaei, W.M. Snelgrove, Excess loop delay in continuous-time delta-sigma modulators and the compensation solution, in the proceedings of the International Symposium on Circuits and Systems (ISCAS 1997), pp. 65-68, 1997.
15. L. Samid, Y. Manoli, An ultra-low power continuous-time sigma delta modulator, in the proceedings on the IEEE International Conference on Electronics, Circuits and Systems (ICECS 2003), Volume 2, pp. 782-785, 2003.

Chapter 6
External Driver Circuitry and Test of ADCs

High performance SAR (Chaps. 2 and 3) and delta-sigma ADCs (Chaps. 4 and 5) are presented previously. However, an ADC can only perform to its specification (see Chap. 1), if the external connections are adequate [1]. Whenever the ADCs performance is pushed into new dimensions, then the external driver circuitry needs to be reconsidered. The analog inputs and the reference inputs are most critical, because a fast and accurate settling is required at a good noise performance. The ground and supply voltages need to be solid and stable. Particular requirements are discussed in the next Sect. 6.1.

The external circuitry is also required for a proper test solution. The principles of the different test methods are explained in Sect. 6.2. The performance needs to be optimized for noise and repeatability. The performance of the production test has to be guaranteed and monitored. A quality flow is therefore described in Sect. 6.3.

6.1 External Connections

In former years, the test solution in the laboratory showed a varying DNL when modifying the sampling time. This pointed to a problem with the settling behavior of the external input operational amplifier (OPA). In a first attempt, faster and faster amplifiers were tested, but the behavior did not improve. The higher bandwidth furthermore added noise to the input signal and degraded the SNR.

Therefore, an extensive investigation on the external connections was done starting from the driving input amplifier [2–4].

6.1.1 Driving Amplifier

The analog input voltage of the SAR ADC is connected to a sample capacitor. Before the conversion can start, the capacitor has to be charged to the input voltage

F. Ohnhäuser, *Analog-Digital Converters for Industrial Applications Including an Introduction to Digital-Analog Converters*, DOI 10.1007/978-3-662-47020-6_6

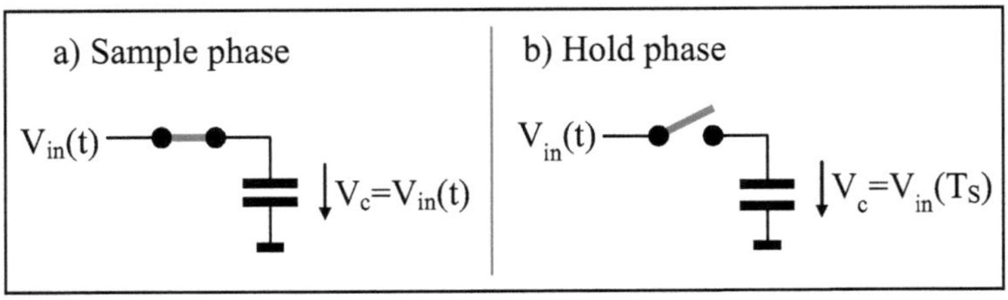

Fig. 6.1 Simplified sample and hold circuit during (**a**) sample and (**b**) hold mode

V_{in} (see Fig. 6.1a). The input switch is closed. The signal is held after the sampling time T_s by opening the input switch (Fig. 6.1b). Now, the converter can process the signal. After the conversion is finished, the sample capacitor is switched back to the analog input voltage.

The capacitor needs to be fully recharged during the sampling time of the converter. If a first order settling is estimated, then Eq. (2.13) is valid again, where n was the resolution of the ADC and T_s the available time to recharge the capacitor. This time is called the sampling or acquisition time.

$$f_{-3\text{dB}} = \frac{\ln(2)\cdot(n+1)}{2\pi\cdot T_s}$$

The acquisition time, which is specified in the datasheet, often expresses the time of the complete conversion cycle (sampling plus conversion time) T_{tot} minus the amount of time T_{conv}, which was used for the pure conversion.

The time T_{tot} is expressing the inverted value of the conversion rate. For a 500 kSPS converter, T_{tot} is 2 μs. If the conversion of a 16 bit converter takes 16 clock cycles, then the specified number for the acquisition time is 2 μs minus 16 clock cycles with exemplary 100 ns each, which results in $T_s = 400$ ns. This number does not consider the switching process between the last bit decisions to the sample mode. It also does not consider that the sampling might be finished asynchronously to the clock signal *CLK* with a conversion start signal *CONVST*. Unfortunate timing between *CLK* and *CONVST* might reduce the sampling time by another clock cycle, so that only 3 clock cycles or 300 ns remain for the sampling. The bandwidth of the driver circuitry needs to be at least

$$f_{-3\text{dB}} = \frac{\ln(2)\cdot(n+1)}{2\pi\cdot T_s} = 6.25\ \text{MHz}$$

If the output current of the driving operational amplifier is limited, then the settling does not occur exponentially but linear to the time, the operational amplifier is slewing (Fig. 6.2).

The slewing phase increases the settling time, which then has to be estimated out of the slewing time and the exponential settling.

Some converters provide an initial discharge, where the sampling capacitor is pre-charged to either ground for unipolar ADCs or to the reference voltage for

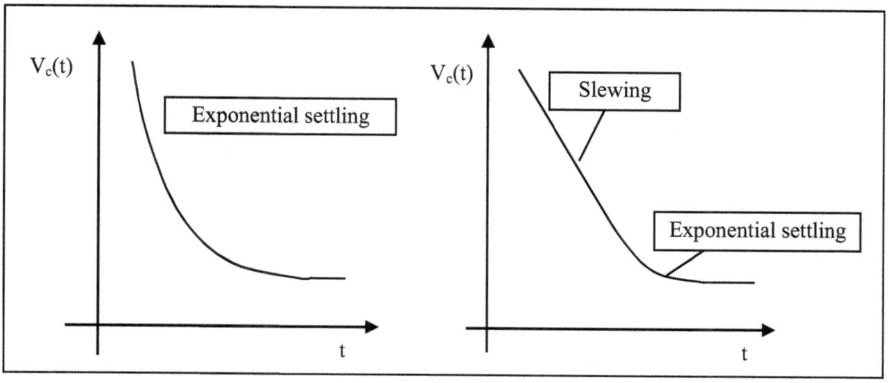

Fig. 6.2 Settling behavior of an operational amplifier that is recharging a capacitive load

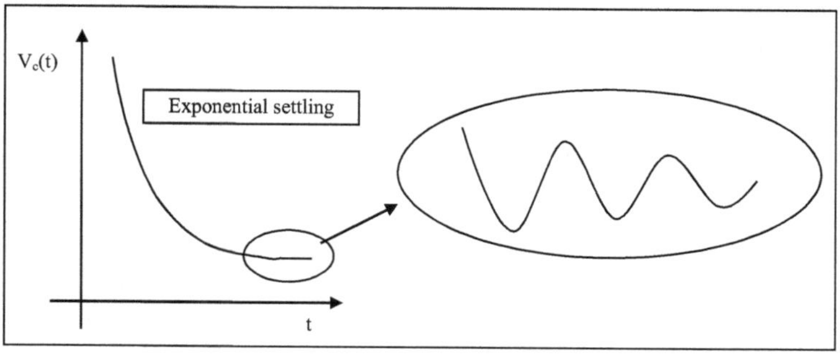

Fig. 6.3 Settling behavior of an amplifier with low phase margin recharging a capacitor

bipolar ADCs [5]. The exponential settling follows the Eq. (6.1). If the settling time T_s is constant for each conversion, then the exponential term is becoming a constant value c.

$$V_c(T_s) = V_{in} \cdot \left(1 - e^{-\frac{T_s}{\tau}}\right) = c \cdot V_{in} \tag{6.1}$$

The ADCs performance with respect to linearity and noise is not influenced, if the input capacitor is not fully charged to the input voltage, but a gain error is caused. As a disadvantage, the initial discharge will consume time, which is not available for the amplifiers settling process. Also, Eq. (6.1) is only true for a pure exponential settling and not if a slewing phase is involved.

Unfortunately, most amplifiers have problems when driving capacitive loads. Capacitors tend to reduce the phase margin of the driving amplifier and make them unstable. The output of the amplifier can show some oscillation (Fig. 6.3).

The oscillation is normally so small that it cannot be seen with an oscilloscope, but it can be measured with the analog-to-digital converter in the following way. The ADC input voltage needs to be applied to the driving operational amplifier.

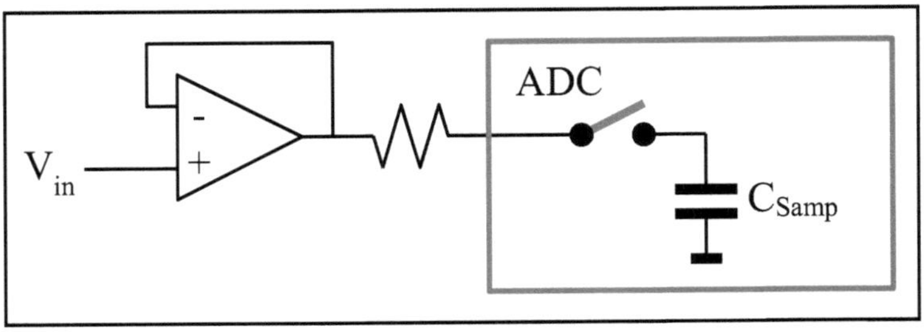

Fig. 6.4 Analog driver circuit to an ADC with resistor between OPA and ADC

When the output code of the converter is oscillating with increasing acquisition time, then the amplifier is oscillating.

To prevent the amplifier from oscillation, it can be decoupled from the capacitor using a resistor between the amplifier and the capacitor (Fig. 6.4).

The maximum value of the resistor R_{max} can be calculated with the time constant τ from Eq. (2.12)

$$\tau = R_{\max} \cdot C_{samp} = \frac{T_s}{\ln(2) \cdot (n+1)}$$

If the sampling capacitor of the 16 bit converter with 300 ns sampling time is 20 pF, then $R_{max} = 1.2\ \text{k}\Omega$. Sometimes, a resistor is already included in the input path of the converter.

This resistor has the additional advantage that the maximum output current from the amplifier is limited to I_{max}, which might prevent the driving amplifier from slewing and makes the settling behavior more predictable

$$I_{\max} = \frac{FSR}{R_{\max}} = \frac{5\ \text{V}}{1.2\ \text{k}\Omega} = 4.2\ \text{mA} \tag{6.2}$$

The resistor R_{max} forms a low-pass filter together with the sampling capacitor C_{samp}, which is also reducing the system noise.

An RC combination can be used instead of the resistor. For this case, the external capacitor C_{ext} has to be added to the internal capacitor C_{samp} in the above formula. Also the input resistance of the ADC needs to be added, which in most cases can be neglected, as the input resistance of the converter is only a few Ohms.

The driving amplifier is adding noise to the application. To ensure that the amplifier is not dominating the system noise, its rms-noise should be smaller than the noise of the converter. Let's assume that the rms-noise of the converter is around 30 μV. With the calculated 6.25 MHz bandwidth, the noise density n_{amp} of the amplifier should be smaller than

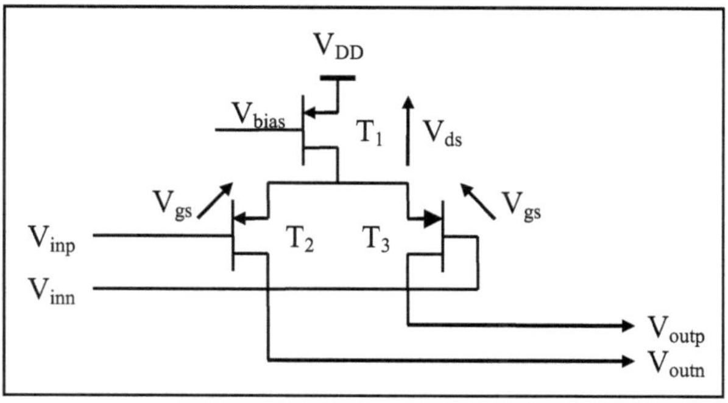

Fig. 6.5 Input differential pair of an operational amplifier

$$n_{amp} = \frac{30\ \mu\text{V}}{\sqrt{\frac{\pi}{2}} \cdot 6.25\ \text{MHz}} = 9.5 \frac{\text{nV}}{\sqrt{\text{Hz}}}$$

This equation takes into consideration that also noise above $f_{-3\text{dB}}$ is adding to the total noise. If the ADC has differential inputs, where the positive and the negative input needs to be driven by amplifiers, then noise is injected twice, which adds a factor of $\frac{1}{\sqrt{2}}$ to the calculation of the maximum noise density.

Another consideration in the input path is the type of the driving amplifier. The input stage of an operational amplifier (Fig. 6.5) is a differential pair, which is very similar to the one of a comparator. The common-mode input voltage of such a differential pair is limited, because the input transistors require a proper gate-source voltage $V_{GS} \approx -1.0$ V and the current source a proper drain-source voltage $V_{DS} \approx -0.2$ V. The maximum input voltage for an input stage with PMOS transistors is limited to the supply voltage V_{DD} minus 1.2 V.

Many applications run from a single supply voltage, which typically is 5 V or 3.3 V. This limits the input voltage range of the ADC. To keep it as big as possible, the full voltage span is used for the signal path and rail-to-rail amplifiers are required. The rail-to-rail input of amplifiers is realized by having two input stages in parallel, a differential pair with PMOS transistors for the lower voltage range and a differential pair with NMOS transistors for the upper voltage range.

Unfortunately, both differential pairs will show different gain and offset, so that the transfer function of the amplifier shows distortion at the input voltage, where the cross-over occurs. This is shown in Fig. 6.6. This switching point is typically between 3 V and 4 V for a 5 V supply. It will limit the linearity and the THD performance. Furthermore, it will cause power supply rejection issues. Fortunately, there are some workarounds. One is using special rail-to-rail amplifier with excellent THD performance; the other solution avoids the switching point with the choice of the amplifier configuration.

There are a few low THD rail-to-rail operational amplifiers (OPA) on the market [6]. One implementation is shown in Fig. 6.7, where two differential pairs are avoided by adding a charge pump to the input stage. This charge pump will raise the

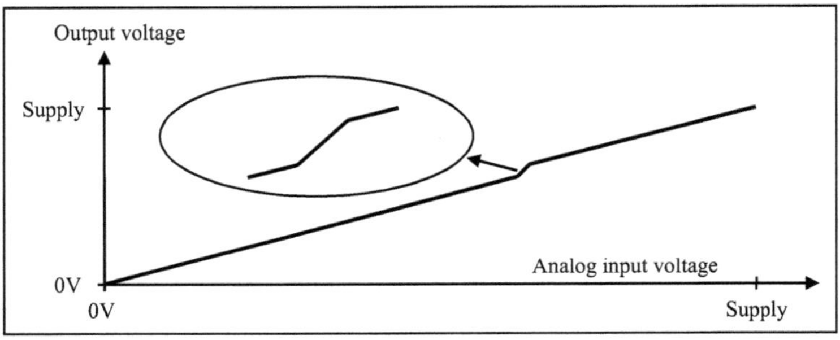

Fig. 6.6 Transfer function of a typical CMOS rail-to-rail amplifier in buffer configuration

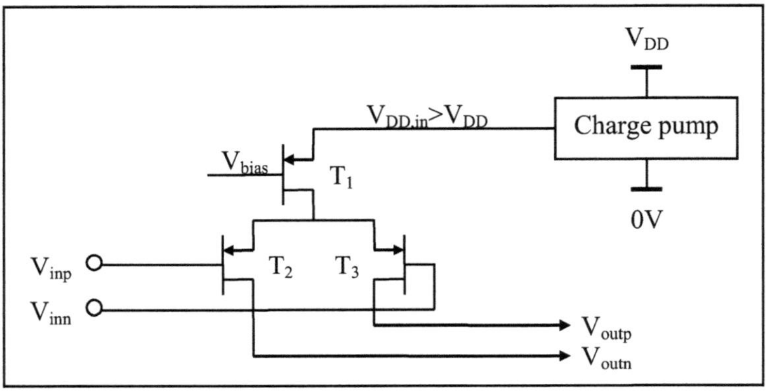

Fig. 6.7 Input differential pair of a low THD amplifier using a charge pump

supply of the input stage $V_{DD,in}$ above the system supply voltage V_{DD}, so that the PMOS input stage can support the full rail-to-rail input range.

Alternatively, the input range of the amplifier can be limited by operating it in inverting configuration or in a gain configuration. The alternatives are shown in Fig. 6.8. Both examples assume a bipolar ADC, where the input range is ±REF around the reference voltage.

The input of the inverting solution is inverted twice, once by the amplifier and once by the ADC. The positive input to the amplifier is connected to the reference voltage, so that the operating point of the amplifier is automatically different from the amplifiers cross-over voltage.

In the second solution, the input signal V_{in} is divided by two, so that the positive input of the amplifier is between 0 V and *REF* and therefore less than the cross-over voltage. The signal is then gained up by two again to use the full input range of the ADC.

The matching of the resistors and of their temperature coefficient are limited, which will cause offset and gain errors that will drift over temperature. Also the common-mode rejection is limited by the resistor matching and the resistors

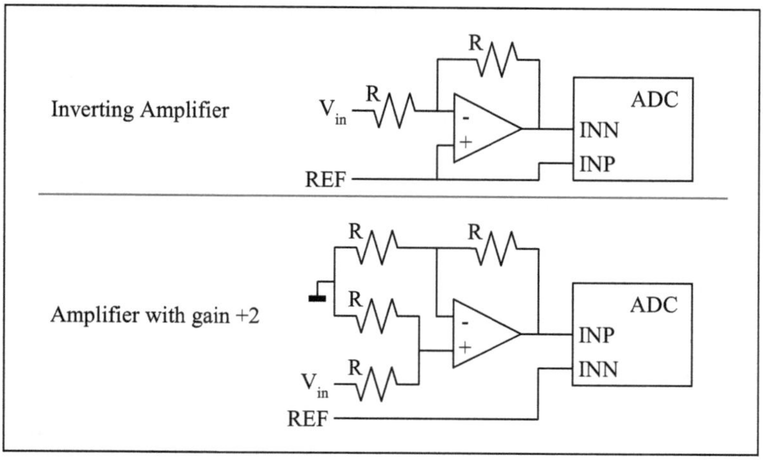

Fig. 6.8 Driver configurations for ADCs limiting the amplifiers input range

decrease the noise performance [7]. Note that both configurations in Fig. 6.8 gain up the amplifiers noise by a factor of 2.

Table 6.1 below shows an example of the most relevant specifications of an amplifier, which is used to drive ADCs.

Another common issue in industrial applications is that signals with the industrial ± 10 V input range should be connected to an ADC with 5 V input range due to the limited amount of ADCs with larger input range. Typically, the ± 10 V signal is then divided resistively as shown in Fig. 6.9 for a differential signal.

The amplitude must be reduced by a factor of 4, so that the resistor ratio is chosen $R_{1x} = 3 \cdot R_{2x}$. V_{shift} would add a common-mode voltage, so that the common-mode output voltage is at 2.5 V. Often, a common-mode voltage is associated to the input voltage, so that INP and INN could be expressed as

$$INP = V_{CM} + \frac{Vdiff}{2}$$
$$INN = V_{CM} - \frac{Vdiff}{2}$$

The resistive divider has a very limited common-mode rejection, which could be calculated in the following. For the calculation, V_{shift} is assumed to 0 V. This would lead to the following equation:

$$V_{out} = \left(\frac{R_{2p}}{R_{1p} + R_{2p}} + \frac{R_{2n}}{R_{1n} + R_{2n}}\right) \cdot \frac{V_{diff}}{2} + \left(\frac{R_{2p}}{R_{1p} + R_{2p}} - \frac{R_{2n}}{R_{1n} + R_{2n}}\right) \cdot V_{CM} \tag{6.3}$$

The common-mode rejection is defined as the output voltage change as a derivative of the common-mode input voltage:

Table 6.1 Exemplary section of the OPA365 datasheet from Texas Instruments [6]

Parameter	Min	Typ	Max	Unit
Input voltage range				
Common-mode voltage range	VSS-0.1		VDD + 0.1	V
Common-mode rejection ratio	100	120		dB
Offset voltage				
Input offset voltage		100	200	μV
Drift		1		μV/°C
vs. power supply		10	100	μV/V
Noise				
Input voltage noise, f = 0.1–10 Hz		5		μV_{pp}
Input voltage noise density		4.5		nV/sqrt(Hz)
Frequency responds				
Gain-bandwidth product		50		MHz
Total harmonic distortion + noise		0.0004		%
Output				
Voltage output swing from rail		10	20	mV
Power supplies				
VDD	2.2		5.5	V
Quiescent current		4.6	5	mA
Temperature range				
Specified performance	−40		+125	°C

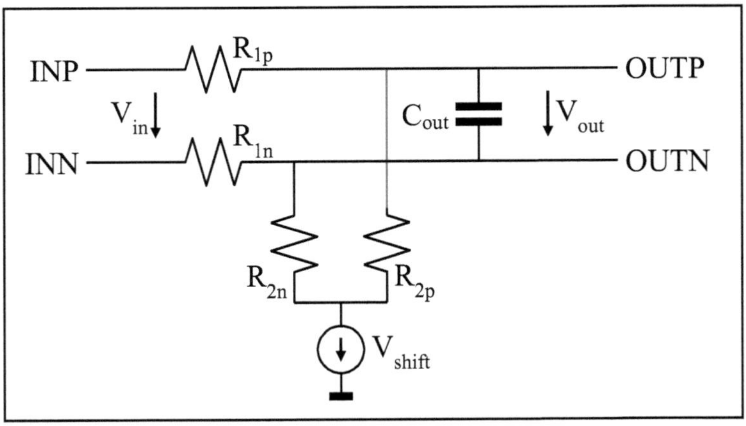

Fig. 6.9 Differential resistive divider to reduce the amplitude of a differential ±10 V signal

$$\frac{dV_{out}}{dV_{CM}} = \frac{R_{2p}}{R_{1p} + R_{2p}} - \frac{R_{2n}}{R_{1n} + R_{2n}} \tag{6.4}$$

If resistors with 1 % accuracy are used, then the maximum of Eq. (6.4) would be:

$$\left.\frac{dV_{out}}{dV_{CM}}\right|_{\max} \approx \frac{1.01 \cdot R_2}{0.99 \cdot R_1 + 1.01 \cdot R_2} - \frac{0.99 \cdot R_2}{1.01 \cdot R_1 + 0.99 \cdot R_2} \approx 0.0075$$

The common-mode rejection can therefore be limited to 42.5 dB. If such a resistive divider is used, then 0.1 % accurate resistors should be used. Note that instrumentation amplifiers are available that can perform this task with high common-mode rejection.

The time constant and the bandwidth are described with Eqs. (6.5) and (6.6). They are helpful to decide, if the sample and hold capacitors can be recharged without active buffer.

$$\tau = 2 \cdot \frac{R_1 \cdot R_2}{R_1 + R_2} \cdot C_{out} \tag{6.5}$$

$$f_{-3\mathrm{dB}} = \frac{1}{4\pi \cdot \frac{R_1 \cdot R_2}{R_1 + R_2} \cdot C_{out}} \tag{6.6}$$

6.1.2 Reference Input

Design requirements of the reference voltage are already discussed in Sect. 2.5 [8]. The requirements to the external components are dependent on the internal structure. If an internal reference buffer is provided, then the reference voltage applied to the converter has to be stable and low noise.

Very important, the reference is the differential voltage between the reference voltage *REFP* and ground *REFN* (see Fig. 6.10). This is causing a problem on single supply applications, because the ground is not free of supply currents. These currents might cause noise and distortion. Care must be taken on the printed circuit board (PCB) layout that these return currents do not cross sensitive inputs.

The best strategy inside the ADCs is to keep the supply ground and the reference ground completely separated all the way to the analog ground plane of the PCB. This includes separated ESD structures, bond pads, bond wires and pins inside of the ADC. If a 16bit ADC has a supply current of only 5 mA and the resistance of the ESD cell is 0.2 Ω, then the internal ground is already at 1 mV or 16 LSB. If reference ground *REFN* and supply ground *GND* share the same pad, then a pure DC current will cause a gain error. Current spikes however might be code dependent and add noise and linearity errors.

On external components, the ground pins for the reference capacitor C_{REF} and the supply capacitor should have separated vias. Compensation capacitors should be placed differentially, before the pins are connected to a plane.

This is especially true for the reference capacitor of Fig. 6.10. This capacitor has to stabilize the differential voltage at the reference input *REFP* and its ground *REFN*. If spikes on the supply ground are coupled into the reference ground, then the capacitor will also give this spike to the reference input and the differential reference remains stable.

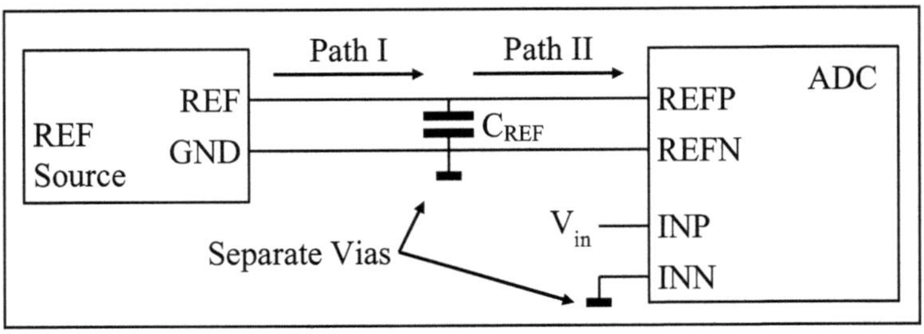

Fig. 6.10 Reference source with compensation capacitor C_{REF} driving an ADC

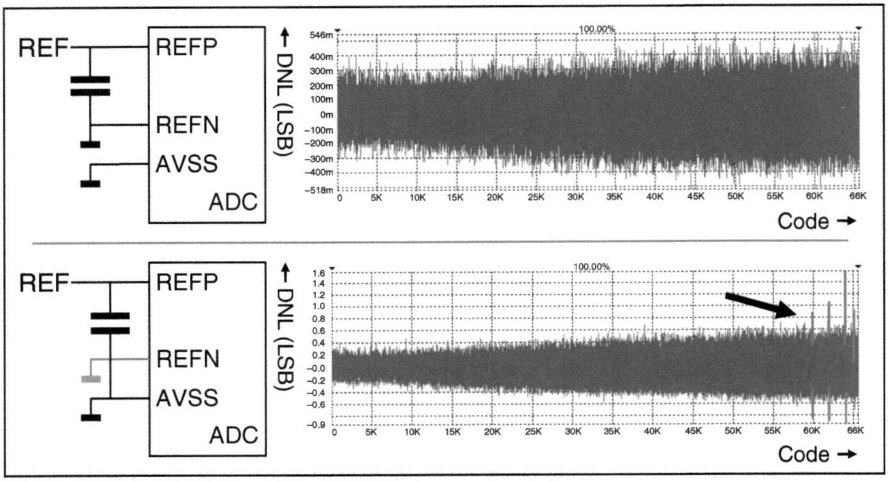

Fig. 6.11 DNL of SAR ADC with (**a**) correct capacitor connection and (**b**) incorrect connection

Figure 6.11 compares the differential linearity, where (a) the reference capacitor is connected as shown in Fig. 6.10 and where (b) the capacitor is connected to the analog ground plain instead of the negative reference input pin.

The structure of Fig. 6.10 has certain requirements to the reference source. The load rejection has to be low enough and the reference source has to remain stable, even if it has to drive a large capacitor. The capacitor is acting as a voltage source for the spikes coming from the ADC. The resistance in path II of Fig. 6.10 between the capacitor and the ADC will not cause linearity errors as long as the internal capacitors of the CDAC are fully recharged during the required settling time, which typically is half a clock cycle.

Path I of Fig. 6.10, which is located between the reference source and the capacitor, is very critical, because the DC current, which the external capacitor forces out of the reference, will cause a voltage drop across any resistance in path I, identical to the load rejection of the reference itself.

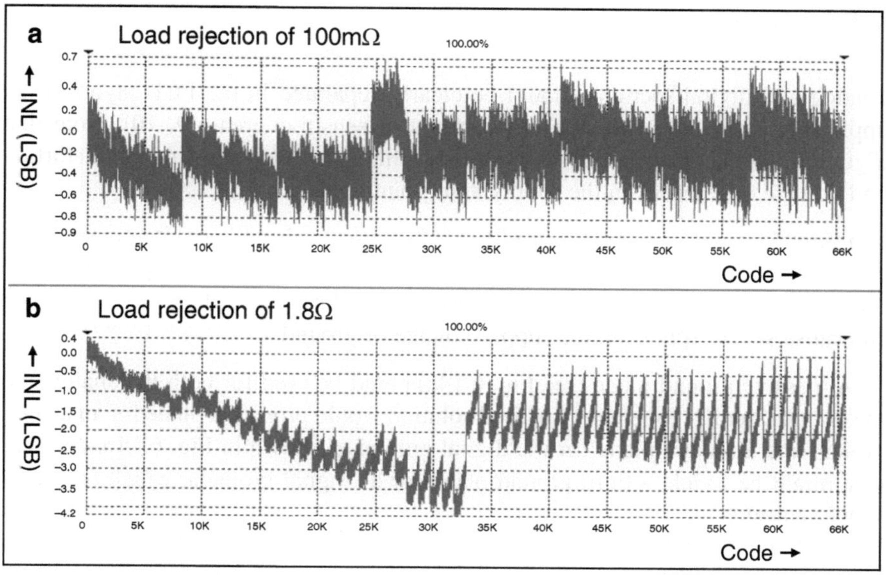

Fig. 6.12 Comparison of SAR ADC's INL with a resistor in path I of (**a**) 0.1 Ω and (**b**) 1.8 Ω

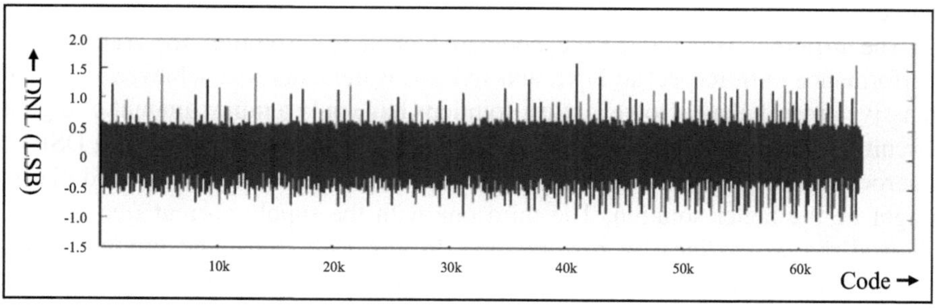

Fig. 6.13 DNL of a 1MSPS, 16bit SAR ADC with 100 Ω resistor in path II

Figure 6.12 shows a comparison of (a) a correct connection with minimized resistance between the reference source and the external capacitor and (b) a connection that has 1.8 Ω added in series. The INL reaches −4.2 LSB.

The resistance in the reference path mainly affects the INL. The reference input currents to the ADC do not change significantly from code to code, so that the DNL remains good. The resistance from the reference switches is in the order of 40–100 Ω anyhow. If an external reference resistance in path II from Fig. 6.10 increases too much, then the settling suffers and starts to affect the DNL as well. An example with 100 Ω is illustrated in Fig. 6.13.

This chapter is showing that the ground is very critical especially on single supply applications. The next chapter will therefore suggest a grounding and a supply scheme.

6.1.3 Grounding and Supply

Analog and digital sections should be clearly separated in the PCB layout of an application to avoid critical return currents in sensitive areas [9, 10]. Note that digital ground currents and spikes from switching supplies can generate distortion and voltage differences in the supply and ground planes.

Caution Low power analog amplifier often use push-pull architectures. Also, dynamic biasing could be used in the reference buffer or the comparator of the ADC. These schemes increase the supply current during a settling process, which can cause severe spikes on the supply voltage or ground.

A massive ground plane (see Fig. 6.14a) could be used for both the analog and the digital circuitry together. A second solution would provide a plane for an analog ground and a second plane for the digital ground as shown in Fig. 6.14b. Another idea might be to add a third ground plane for all signal grounds, which is free of supply currents.

Separated ground planes definitely need to be shorted at the converter, because currents through the converters digital outputs will otherwise have long return paths. In addition, spikes can cause the local digital ground to be above or below a diode drop of the analog ground. This way, ESD diodes can open inside the ADC and cause essential distortion.

The different structures were compared on a test solution for ADCs. Best performance with respect to DNL and INL measurements was achieved with one massive ground plane. However, test solutions have only a minor amount of digital circuitry, which in addition is all running synchronously to the ADC. If a DSP or microcontroller is added onto a PCB, the second structure with two ground planes might be the better solution. The third one with the supply current free ground, showed the worst linearity performance. In any case, separated ground planes should not overlap to minimize capacitive coupling between them.

A fully differential structure can provide a good, but not an ideal power supply rejection. Typical values for ADCs are around 70 dB like shown in Fig. 6.15, where the power supply rejection ratio is -75 dB for low frequencies. At first, the power supply rejection drops with increasing frequency, but it improves again to higher frequencies due to the low-pass behavior of the internal blocks. Therefore, the supply needs to be clean. Ideal would be the usage of a linear regulator. Switching supplies are not recommended.

If the analog and the digital circuitry are supplied out of the same voltage source, then some filtering for the analog section is recommended. For low power products, a resistor can be placed between the 5 V supply and the converter. The resistor will form a low-pass structure together with the compensation capacitor of the supply.

If a voltage drop of 100 mV is allowed and a converter consumes 5 mA, then a resistor with 20 Ω can be placed between the pin and the general supply voltage. The resistor of 20 Ω together with a 4.7 μF capacitor will generate a corner frequency of 1.7 kHz, so that high frequency noise in the supply will be filtered

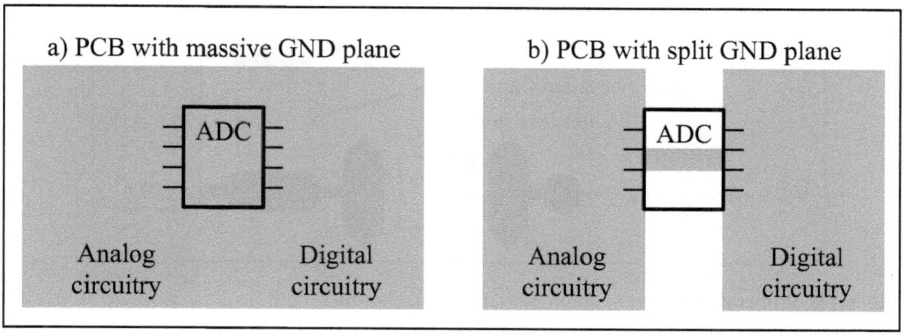

Fig. 6.14 PCB concept with (**a**) massive ground plane and (**b**) separated ground planes

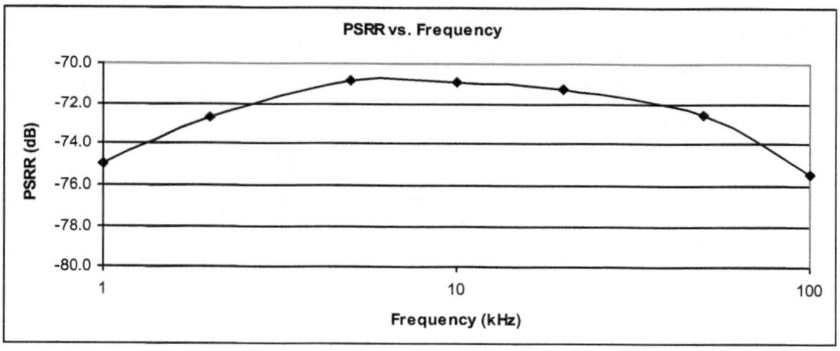

Fig. 6.15 Typical PSRR over frequency of a SAR ADC [11]

effectively. An excellent alternative to the resistor are ferrites, which have a low DC resistance and an increasing impedance with frequency. They can provide a lower voltage drop compared to the resistor with at least similar filter functionality.

The combination of the resistor and the decoupling capacitors are not only protecting a certain device from supply noise, but also the electric environment from distortion coming from this particular device. Again, low power circuits often only reduce the quiescent current of the device, but the peak currents remain unchanged. An internal reference buffer for example has to recharge the CDAC capacitors within one clock cycle and therefore needs to provide the required peak current. These peak currents also create distortion in the ground and supply plane.

Therefore, a good practice for the PCB layout of the decoupling capacitor is to connect the supply and ground pins to the capacitor first and then to the dedicated layer like shown in Fig. 6.16.

6.1.4 *Digital Interface*

The interface can influence the performance of ADCs even beside the clock jitter [12]. If the digital output driver is switching, then any parasitic external load capacitance of the digital line has to be recharged through the ADC. If a serial

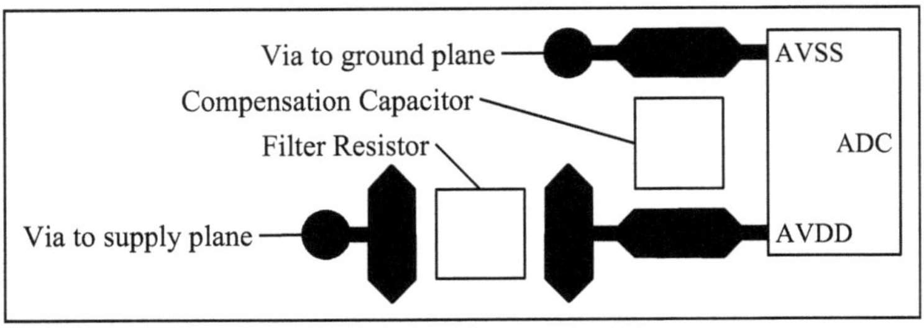

Fig. 6.16 Proper connection of a compensation capacitor on a PCB

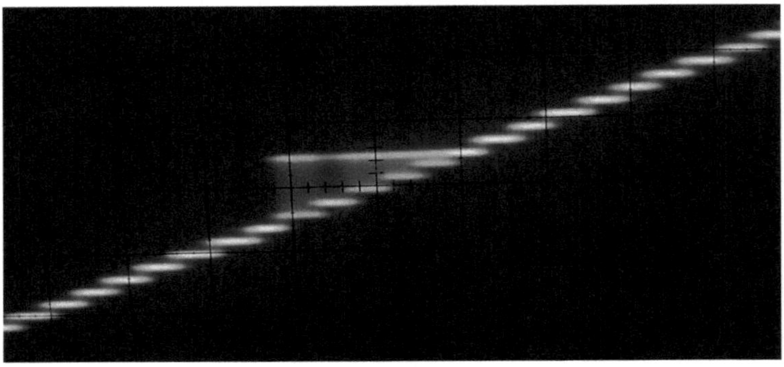

Fig. 6.17 Staircase with distortion caused by a parallel bus access during the conversion

data output is switching constantly between a 5 V digital supply and ground with a frequency of 10 MHz and has to recharge an external parasitic capacitance of 30 pF, then the average current can be calculated to

$$C = \frac{\Delta Q}{\Delta V} = \frac{I \cdot \Delta t}{\Delta V} = \frac{I}{\Delta V \cdot f} \Rightarrow I = C \cdot \Delta V \cdot f = 30\ \text{pF} \cdot 5\ \text{V} \cdot 10\ \text{MHz} = 1.5\ \text{mA}$$

The switching process is causing problems in several ways. The recharge currents are causing spikes on the supply with an amplitude of several 100 mV per 50 Ω output driver, which can couple into analog signals. Analog signals should not be wired next to digital signals or supplies, because bondwires and the general wiring also generate a transformer with low coupling. If analog signals are bonded next to digital supplies, then the transformer circuit adds dynamic currents into the analog signal. Digital signals can also couple into analog signals via the bondpad and the ESD structures, which have an essential parasitic capacitance to the substrate of the integrated circuit.

Problems can be avoided, if the ADC has a separated analog and digital supply. The analog and the digital pins of the ADC should be located separately. The analog pins can be guarded by the analog ground and supply pins. If this is not possible, then slow moving digital signals could be placed next to analog signals, like a chip-select signal *CS*, a power-down signal *PD* or a read pin *RD*.

Figure 6.17 shows an example, where data is read from a SAR ADC via a parallel interface during a particular bit decision. The most critical point in time is the hold time of the sample and hold stage followed by the comparator decisions, where the comparator input voltage is less than an LSB. It was discussed previously that the digital signals can either couple noise into signals and substrate through parasitic capacitors. Also, they can generate supply oscillation, which then couples into the analog signals.

6.2 Test Methods and Production Test

The production test does not only ensure the functionality of a product, but also its specifications [13]. To achieve high-end performance, essential test time is required, so that the production test is the biggest cost adder. This is especially true, when high accuracy test machines (see Fig. 6.18) must be chosen due to the linearity and noise performance of the analog signal generators.

The test machine consists a main frame, which provides resources such as the general power supply, computing or cooling, and the test head that includes all test resources such as power supplies, reference voltages, analog signal sources or digital resources. A device-under-test (DUT) is applied to the test head by the handler. A handler interface board (HIB) is connected between the test head and the handler. The HIB (Fig. 6.19) has the function to route the resource of test head to the pins of the DUT. The handler is injecting a packaged IC into a contactor, which is mounted on the HIB. The contactor with integrated springs ensures good contact and a long life time of the test hardware. A manual actuator as shown in Fig. 6.19 is used if tests are performed without handler.

ICs can also be tested on wafer level with the help of a prober. The wafer is placed inside the prober, which is doing the positioning. The test head is contacting to an IC on the wafer via a probe interface board (PIB) and probe-needles, which are shown in Fig. 6.20.

The specifications are mainly separated in three sections: the DC specifications of the ADC's transfer function, the AC specifications and the power dissipation. Test methods for these specifications and the improvements for the latest ADC generation are discussed in this chapter.

6.2.1 DC Specifications

The DC specifications are the offset, the gain, the differential non-linearity and the integral non-linearity, which are all discussed in Sect. 1.1. All of them express the accuracy of the transfer function.

The easiest way to measure the performance of an ADC is to apply a precise low noise voltage to the ADC input. The transfer function can be monitored by stepping the precision voltage through the input range of the ADC and simultaneously recording the output of the ADC (see Fig. 6.21).

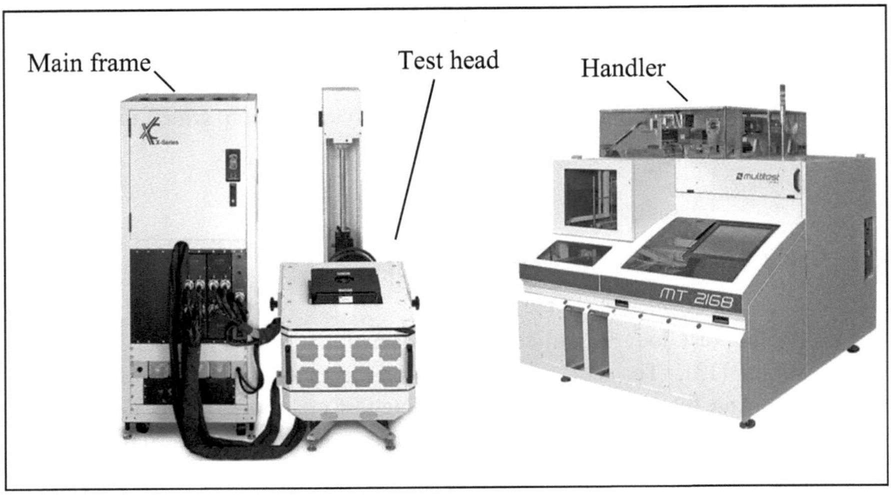

Fig. 6.18 Automated test machine and handler (with friendly support of Xcerra)

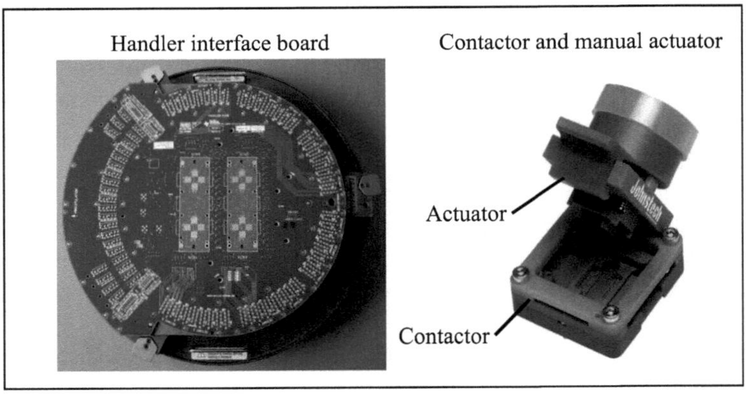

Fig. 6.19 Handler interface board and contactor with manual actuator

Different noise sources (input signal, thermal and flicker noise of the ADC, kT/C-noise of the sample and hold) cause the transition voltage to be noisy. Therefore, the data of the circuitry of Fig. 6.21 can be analyzed with the histogram method, where each code should occur with a certain amount of hits. The actual code length and the DNL are evaluated by counting the hits of the code of interest and comparing the number to the average hits per code, if the input signal is a straight ramp. Other input signal forms are sometimes utilized such as sine waves. In this case, the number of hits of the code of interest must be compared with the expected value.

Usually, the ramp is applied to the input of the ADC [14–16]. The steepness of the ramp might be chosen, so that there is an average of at least 8 hits per code. The ramp should be repeated often enough so that the influence of the noise is

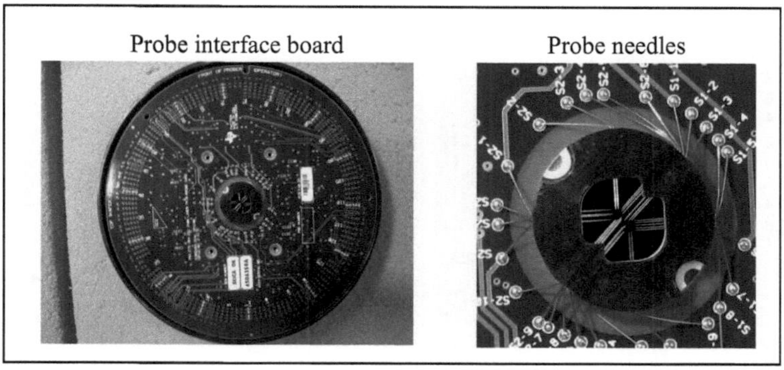

Fig. 6.20 Probe interface board with probe needles

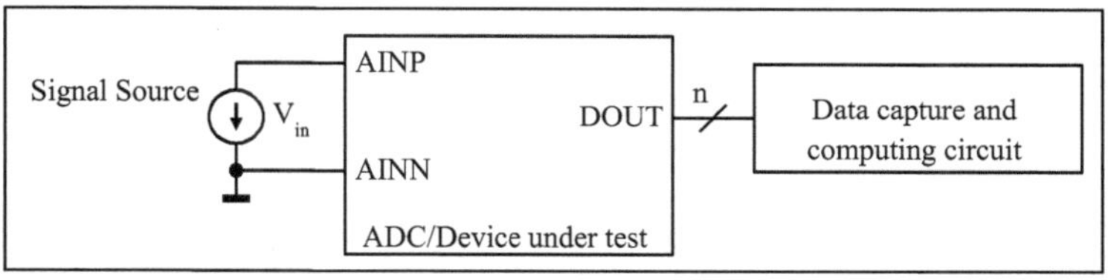

Fig. 6.21 Simple circuit to compute the transfer function of an ADC

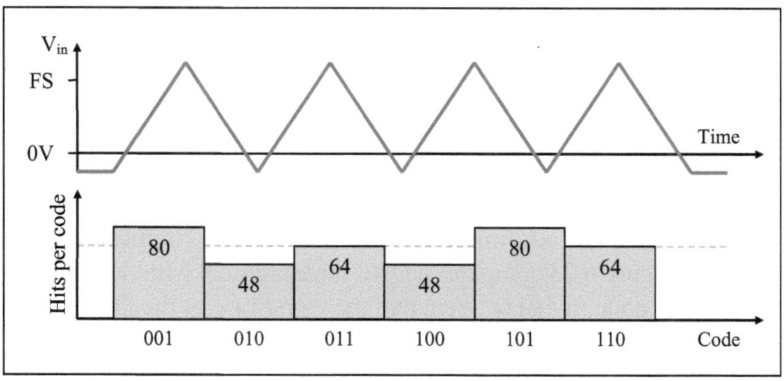

Fig. 6.22 Histogram test counting the occurrence of ADC output codes to calculate linearity

suppressed. The histogram test of Fig. 6.22 uses 4 up and 4 down ramps, because of limited memory of the digital resources in the particular case. Every single ramp comes with an average of 8 hits per code, so that a total of 64 hits per code are expected.

Code 011 and code 110 have exactly 64 hits, so that their DNL is 0 LSB. Code 010 and 100 show 48 hits per code, so that the code length is 0.75 LSB, which is expressed in a DNL of −0.25 LSB. In the same way, code 001 and 101, which have 80 hits per code, have a DNL of +0.25 LSB.

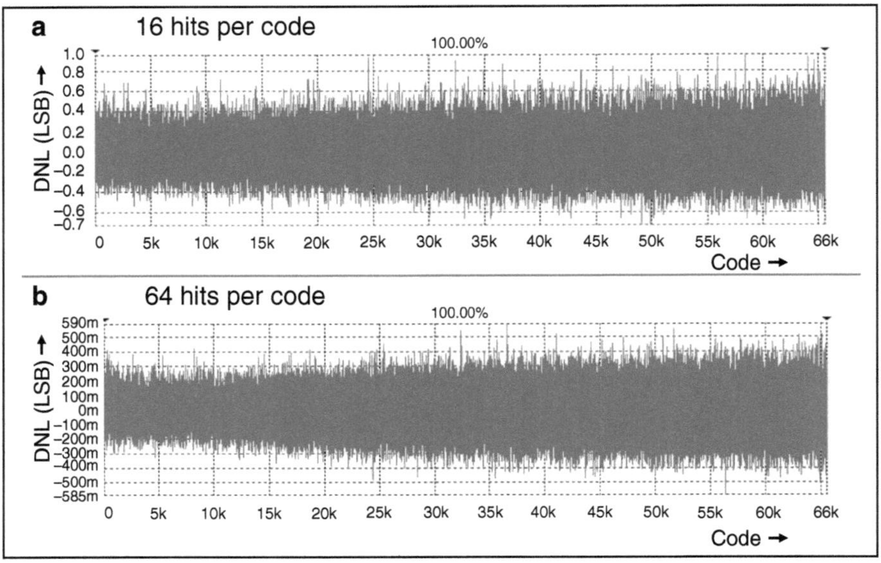

Fig. 6.23 DNL of a SAR ADC measured with histogram test with (**a**) 16 hits per code and (**b**) 64 hits per code

The required amount of hits per code can be calculated out of the ADCs peak-to-peak noise, which is here estimated to 6 sigma. If the ADCs input voltage would be exactly at a transition, then the monitored output codes would be in the range of ±3 sigma of the noise around the code transition. Consequently, some codes might occur more often than others even if they have an ideal DNL. Doubling the average hits per code would increase the measurement accuracy by 3 dB.

If for example the peak-to-peak noise is 4.8 LSB, then the standard deviation of the test is 0.8 LSB. The standard deviation would be lowered to 0.4 LSB, if the measurement is repeated four times, 0.2 LSB if measured 16 times and 0.1 LSB if tested 64 times. Figure 6.23 compares a DNL measurement with (a) 16 hits per code and (b) 64 hits per code. It can be seen, that the worst case of the DNL is 1LSB with 16 hits per code and 0.6 LSB with 64 hits per code.

A high performance linearity test can only be achieved with a low noise signal source and with a significant amount of average hits per code h_{av}, which has a huge impact on the test time T_{ramp} and therefore on the test costs. For a 16 bit converter ($n = 16$), 64 hits per code and a conversion rate of 500 kSPS, the pure data acquisition time is 8.4 s.

$$T_{ramp} = \frac{2^n \cdot h_{av}}{f_{conv}} = 8.4 \text{ s} \tag{6.7}$$

On top of the noise issue, signal generators typically have a limited power supply rejection, so that the 50/60 Hz can be measured in the data output of the ADC.

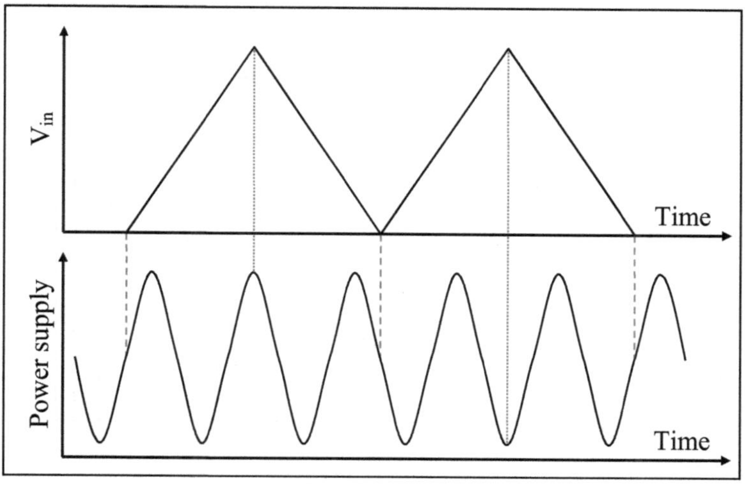

Fig. 6.24 ADC input V_{in} for a histogram test timed to optimize for power supply rejection

Steps were taken to cancel the influence of the power supply. This can be done by synchronizing the ramp to the power supply cycle.

Figure 6.23 illustrates that the influence of the power supply can be suppressed, when at least two up and two down ramps are performed, while each ramp takes k plus a quarter power line cycles.

In the example of Fig. 6.24 the power supply is high, when the ramp reaches its full-scale range for the first time and it is low the second time. The average influence is neglected.

In average, each code in Fig. 6.22 came out to have 64 hits, so that the gain is ideal. The gain would have been too small, if for example 60 hits per code would have been detected. This would express a gain error of $-6.25\ \%$.

The offset error can be detected by forcing 0 V to the input of the converter. The average output code is expressing the offset error. Unfortunately, this is only possible for bipolar converters. If a unipolar ADC shows a positive offset voltage, then the converter will continuously output zeros disregarding the value of the offset. Therefore, a small DC input voltage V_{DC} has to be forced to the ADC. V_{DC} has to be larger than the maximum offset error. The offset is then calculated with the average output code $code_{av}$ minus V_{DC} expressed in LSB.

$$Offset\ [LSB] = code_{av} - \frac{V_{DC}}{1\ \text{LSB}} \tag{6.8}$$

A digital-to-analog converter is used to generate the ramp. The accuracy of the DAC has to be higher than the one of the ADC, because every linearity error, offset error or gain error of the DAC adds an error to the measurement of the transfer function of the ADC. Therefore, the voltage source is typically calibrated with a

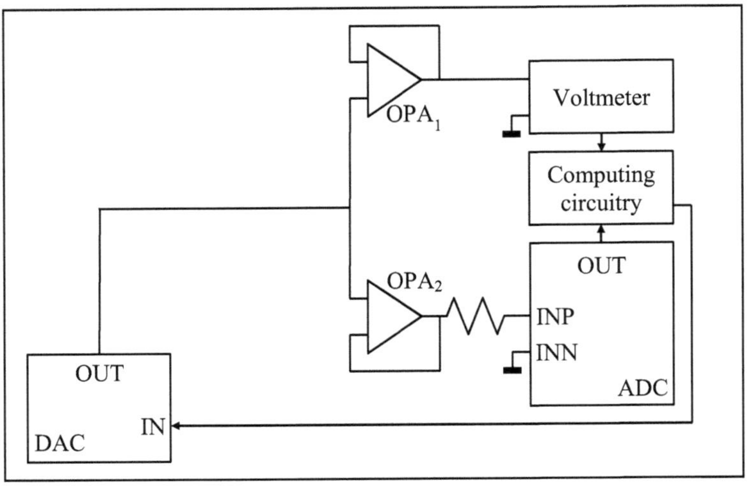

Fig. 6.25 Measurement circuit for the linearity of ADCs including a voltmeter for calibration

precision voltmeter before starting the production test. A test solution then relies on the stability of the DAC.

A precision voltmeter usually captures the input voltage with an integral measurement over a power line cycle (50 or 60 Hz) to eliminate the influence of the power supply.

The sample and hold capacitor of the ADC will generate spikes to the input circuitry. Therefore, an adequate input circuitry (OPA_2) needs to be added between the DAC and the ADC. The voltmeter and its wiring to the DAC might cause distortion. The voltmeter should be decoupled with a low offset operational amplifier (see Fig. 6.25, OPA_1) or should alternatively be switched off with a multiplexer.

Each amplifier will add some offset, which also has to be calibrated by measuring a DC voltage directly at the inputs of the ADC first and then over the decoupled path. During the calibration of the offset, the ADC needs to be turned off; otherwise the spikes mentioned above will add another error.

If the resolution or performance of the ADC is equal or close to that of the DAC, then the test setup is not accurate enough. This is particularly true for the differential and integral non-linearity. The performance might be increased by changing the ramp to a sine wave, which is filtered and runs synchronously to the power supply with 50/60 Hz [17, 18].

Stacking two DACs on top of each other as shown in Fig. 6.27 can increase the resolution of the test circuit as well. The two DAC output voltages are added, using a gain of one for DAC_1, but dividing the output voltage of DAC_2 by a factor of x. If $x = 128$, another 7bit of resolution will be added. DAC_1, which sets the coarse voltage, is called the pedestal DAC.

The gain adjustment (factor 1 and $1/x$) and the adder of the two DAC output voltages can be implemented with an operational amplifier that is connected in inverting configuration with two input voltages as shown in Fig. 6.26. The negative input of the amplifier is at virtual ground, so that the input currents are $I_{in1} = \frac{V_{in1}}{R_{in1}}$ and $I_{in2} = \frac{V_{in2}}{R_{in2}}$. As the input of the amplifier is high impedance, the currents are added ($I_{fb} = I_{in1} + I_{in2}$) and flow through the feedback resistor R_{fb}. If $R_{in1} = R_{fb}$ and $R_{in2} = x \cdot R_{fb}$, then the output voltage can be calculated as

$$V_{out} = -\left(V_{in1} + \frac{V_{in2}}{x}\right)$$

A capacitor C_{lp} should be placed in parallel to R_{fb}. C_{lp} and R_{fb} form a low pass filter together and are reducing the signal bandwidth to minimize noise. The bandwidth, however, should exceed the conversion rate for sufficient settling.

$$f_{-3\text{dB}} = \frac{1}{2\pi R_{fb} C_{lp}} > f_{conv}$$

This method does increase the resolution, but not the linearity of the input signal. However, it can be very useful, if used together with a feedback loop. This method called servo loop is explained in the following [19, 20].

The feedback in the servo loop circuitry of Fig. 6.27 is closed at the output of the converter. A computing circuitry compares the output of the ADC with the code of interest and sets DAC_2 accordingly, so that the input voltage at the ADC converges to the transition voltage V_{tr} of the code of interest. This voltage is then measured directly with the voltmeter. The pedestal DAC DAC_1 is set to the code of interest, so that the input voltage of the ADC is initially already close to the transition voltage. This ensures that the DAC_2 is not running out of range.

In case that the output code of the ADC is less or equal to the code of interest, the value of DAC_2, now called the servo DAC, is increased; otherwise it is decreased by one step. When settled, the input voltage of the ADC will toggle with very high

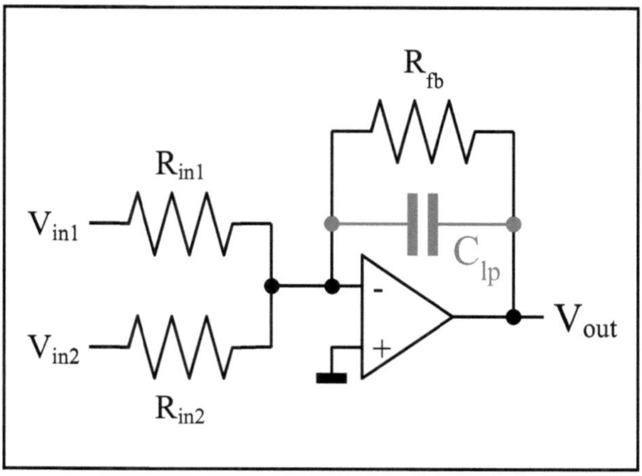

Fig. 6.26 Amplifier configured as analog adder

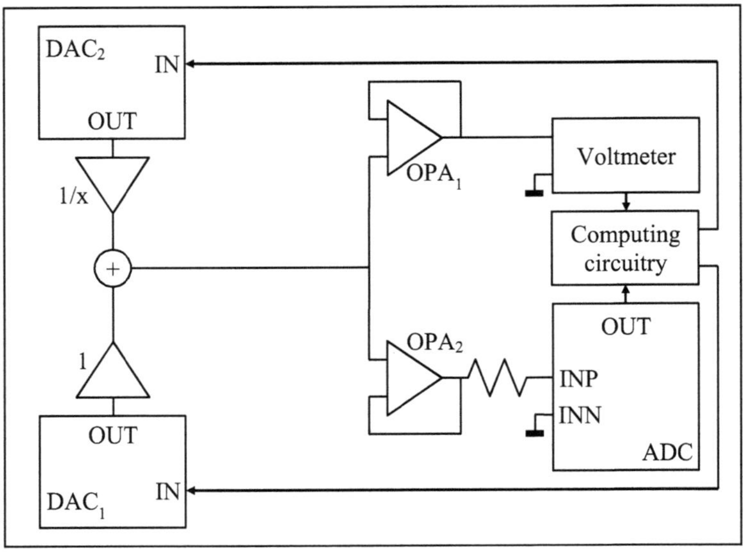

Fig. 6.27 Servo loop circuit to measure the linearity of ADCs

accuracy around the transition voltage. The integral measurement of the voltmeter will further average the small steps over one power line cycle.

The servo loop shows the additional advantage that high precision DACs are not necessary anymore and that the performance of ADCs can be measured with an accuracy that is equal or higher than that of the DACs.

Figure 6.28 shows the settling of the input voltage of the ADC within a servo loop system to the transition voltage of interest. At time 0, the pedestal DAC is set to apply a voltage that is close to the anticipated transition voltage V_{tr}. As the ADC output is still below or equal to the code of interest during the times 1–2, the servo DAC increases the ADC input voltage step by step. At times 3 and 5, the input voltage equals with the transition voltage. The noise of the system will decide, if the output of the ADC is below, equal or above the code of interest and at time 6, the system noise even causes a wrong decision. The integral measurement of the voltmeter can start to measure the transition voltage as soon as the code of interest and the code above both occurred at least once. The integral voltage measurement over one or more power line cycles will average the system noise, so that a high repeatability is achieved.

The performance of a test solution is typically expressed in the repeatability, where the same unit is tested over and over again. The standard deviation of the test results should be as narrow as possible. While the standard deviation of the ramp method with 64 hits per code for a 16 bit SAR ADC is in the range of 0.1 LSB for the DNL of a particular code, the servo loop increases the performance to 0.05 LSB for a 16 bit ADC.

The wiring of the servo loop circuit can be simplified by replacing the servo DAC with an integrator that generates the fine adjustment to the ADC input. Identical to the servo DAC, the integrator increases the input voltage, if the output

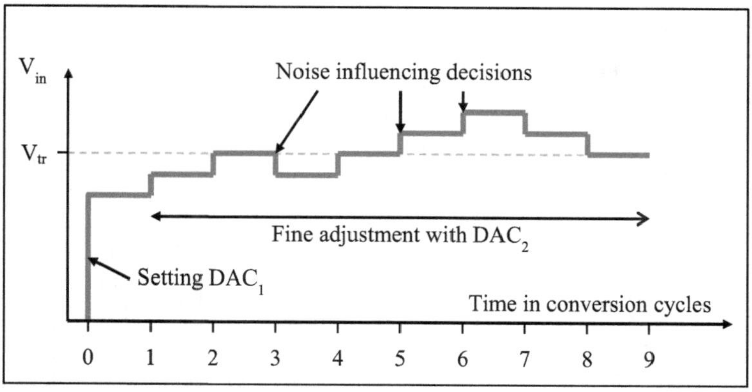

Fig. 6.28 Typical settling of the ADC input voltage V_{in} in the servo loop system

code is less or equal to the code of interest. It reduces the input voltage, if the output code is above the code of interest. For this purpose, the input of the integrator is either switched to the reference voltage *REF* or to its inverted voltage *–REF*. The switch is controlled by the computing circuit. An implementation of a system with an integrator is shown in Fig. 6.29.

A PC might provide the code of interest. The control circuit just compares the code of interest from the PC with the output code of the ADC and sets the switch position at the input of the integrator. This can be done with a few discrete gates. The PC waits for a while, so that the input voltage of the ADC can settle, and then starts the integral measurement of the voltmeter. After reading the result from the voltmeter, the PC sets the next code of interest.

The software and control circuitry are very simple. This method is therefore also ideal for development engineers of applications to compare the performance of different ADCs.

The change of the output voltage of the integrator during a conversion time Δt can be calculated as following. A constant positive or negative reference voltage is applied to the resistor of the integrator R_i. As the node at the operational amplifier is at virtual ground, the input current can be calculated to $I_i = \pm\frac{REF}{R_i}$. This current is charging or discharging the capacitor C_i

$$C_i = \frac{\Delta Q}{\Delta V_i} = \frac{I_i \cdot \Delta t}{\Delta V_i} \Leftrightarrow \Delta V_i = \frac{I_i \cdot \Delta t}{C_i} = \frac{\pm REF}{f_{conv} \cdot R_i \cdot C_i}$$

If the reference is $REF = 10$ V, the conversion rate $f_{conv} = 500$ kSPS, $R_i = 100$ kΩ and $C_i = 1$ µF, then the integrator output voltage V_i will change by $\Delta V_i = 200$ µV. In addition, this integrator output voltage will be divided by the factor x. If x is 100, then the input voltage of the ADC changes by only 2 µV per conversion cycle. In this way, high accuracy measurements are possible.

Unfortunately, the servo loop consumes a lot of test time. The voltmeter has to measure at least over one power line cycle, which equals 20 ms at a 50 Hz supply.

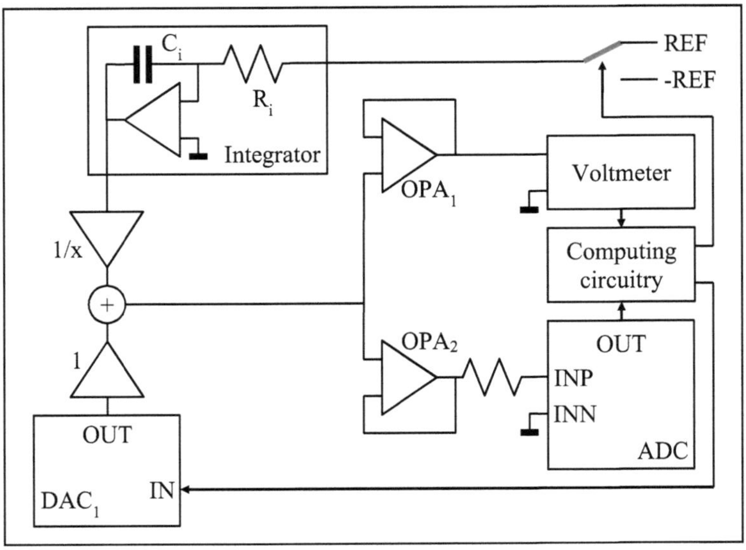

Fig. 6.29 Servo loop circuit to measure the linearity of ADCs using an integrator

Roughly the same amount of time needs to be added to set the pedestal DAC and to let it and the servo loop settle. The measurement of all 65535 code transitions of a 16 bit converter would roughly consume 45 min (40 ms · 65535), which is unacceptable for a production test.

In some cases however, not all transitions need to be measured. The CDAC of SAR ADCs relies on the matching of the capacitors, which can be as good as 0.1 %. In this way, the lower 10 bits are typical uncritical and do not have to be measured. A set of critical codes can be put together. This is particularly true for ADCs, which are trimmed on wafer level and which might be subject of the packaging shift.

The MSBs of the CDAC for example are trimmed. The trim solution might for some reasons miss-trim an ADC or a packaging shift causes these particular codes to be long or short. The DNLs of these particular codes have to be measured. For a 16 bit converter only the MSB transition (code 32767), the MSB-1 transition (code 16383), MSB-2 (code 8191), MSB-3 (code 4095), MSB-4 (code 2047) and the MSB-5 (code 1023) have to be measured together with their combination codes.

In addition, dynamic effects might generate long and short codes, which again occur, when big capacitors are switching (see Sect. 2.3.4). This is especially true, if the MSB capacitor is already set to the reference and the MSB-1 or MSB-2 capacitor switch from ground to the reference (codes 40959 and 40960, 49151 and 49152, 57343 and 57344).

The method of generating a code set is only allowed as long as particular codes are especially critical (see Fig. 6.30a) and the specification is wide. The DNL of Fig. 6.30b does not show critical codes. If the specification would be set tight to ±0.5 LSB or ±1 LSB, then a reduced code set is not allowed and the servo loop is not usable.

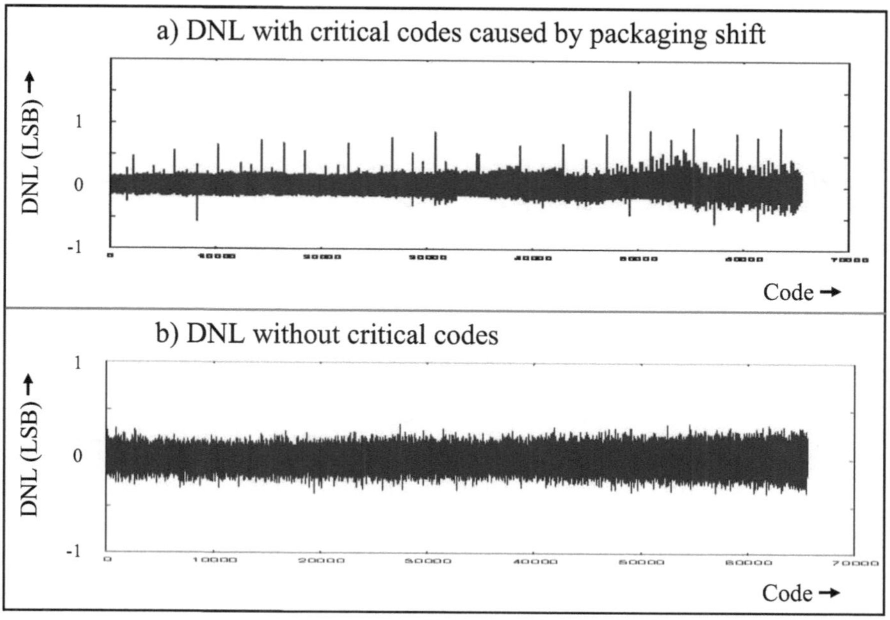

Fig. 6.30 Exemplary DNL of SAR ADCs with and without critical codes. (**a**) DNL with critical codes caused by packaging shift. (**b**) DNL without critical codes

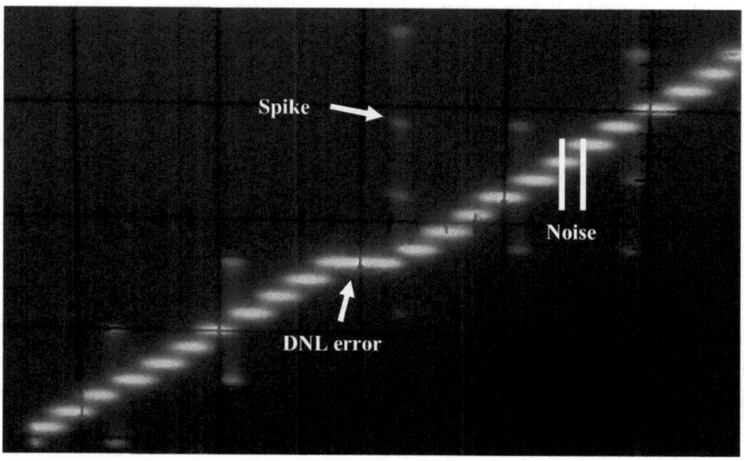

Fig. 6.31 Plot of a staircase showing a transfer function with spikes, DNL error and noise

Taking histograms or making integral measurements with the servo loop circuit all use the advantage of averaging. Some effects like spikes in the middle of a transfer function (Fig. 6.31) might not be seen. Therefore, it is always necessary to make the transfer function visible during the debugging process of the ADC development.

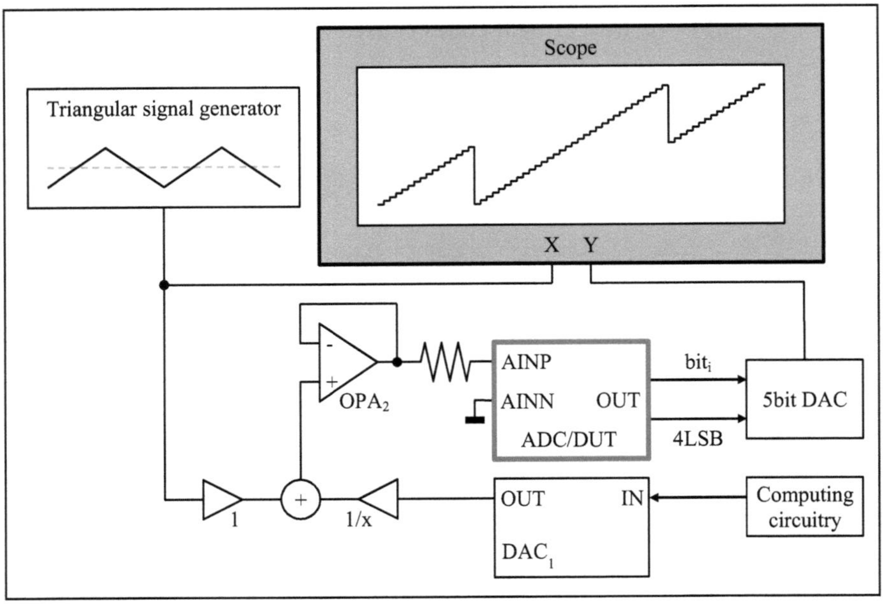

Fig. 6.32 Block diagram for a measurement circuitry to illustrate a staircase

This can be achieved by adding a slow ramp to the pedestal voltage and therefore to the input of the ADC. The ramp, which can also be realized with the integrator, is also directly connected to the *X* input of a scope. The least significant bits of the ADC output are applied to a DAC with low resolution and high speed. The DAC output voltage is then connected to the *Y* input of the oscilloscope.

To find a certain bit of interest bit_i on the scope, it can be applied as the most significant bit to the DAC. The DAC can either be a discrete IC (see Fig. 6.32) or it can be implemented with binary weighted resistors.

A plot of an erroneous stair case is shown in Fig. 6.31. The staircase above shows spikes in the transfer function at every fourth code. In this case, the spikes were generated by an incorrect timing. The particular device was also untrimmed, so that the bit of interest in the middle of Fig. 6.31 has two or three times the width if compared with the other codes. Also the codes are overlapping each other. This transition noise is generated by the sum of noise sources, which are discussed in Chap. 2.

6.2.2 AC Tests

The AC measurement is the biggest challenge for the production test of high-end SAR ADCs, due to the lack of adequate sine generators. The noise of the signal generators in test machines is typically specified with 82 dB for frequencies above

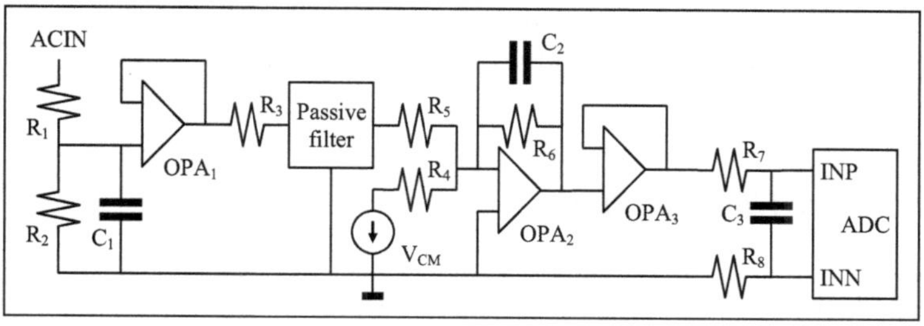

Fig. 6.33 Possible AC path on a handler interface board for ADCs

2 kHz, while the converters achieve up to 93 dB with single-ended inputs or 98 dB with differential inputs. The AC parameters of many high-speed and high-performance SAR converters are therefore either only specified typically or they are tested with an input frequency of less than 2 kHz. Additional filtering of the signal on the board is required.

Special passive high-order band-pass and low-pass filters are available on the market, but these come with a limited THD performance. The best performance, which was achieved with a 100 kHz 5 pole band-pass filter, reached 92 dB in SNR and −91 dB in THD. The THD of high performance ADCs is below −95 dB. Options need to be placed into the schematic of the final test solution to optimize the operating point of the signal source and the filter function. The implemented solution is described in Fig. 6.33.

The signal is generated in the test head of the test machine and has some wiring associated before it gets to the test board. In this path, noise might be coupled into the signal. The SNR of the input signal is then directly proportional to the amplitude of the signal. It might therefore be good to use high signal amplitudes all the way to the handler interface board, where it is then adjusted to the required amplitude with a resistive divider. This divider can also be used to perform some first filtering, where the parallel resistance of R_1 and R_2 together with C_1 form a first pole.

The capacitive load C_1 might again lower the THD performance of the signal source, so that its value needs to be optimized. For the same reason, a first amplifier in buffer configuration OPA_1 is added, because the inductances and capacitors inside the passive filter might influence the performance of the signal source.

If the passive filter is in use, then the amplitude of the signal source has to be adjusted with R_1 and R_2 to the filters ideal operating point. The resistor R_3 is adjusting the input impedance to the filter. In the same way, R_5 is adjusting the output impedance. In the next step, the output voltage of the filter has to be adjusted to the input range of the converter.

This is performed by the second operational amplifier OPA_2, which is connected in inverting configuration with two inputs, where the gain can be set with the ratio of the feedback resistor R_6 to the input resistors R_5 and R_4. A negative common-mode voltage V_{CM} is applied to R_4. This is required, if a passive band-pass filter is used, which removes the DC component of the input signal. The bandwidth of OPA_2 is optimized with the capacitor C_2.

The third operational amplifier together with R_7, R_8 and C_3 forms the standard driving input circuitry to the converter, which was already described previously.

6.2.3 Power Consumption

The third block of specifications describes the power consumption of a product in different modes like normal operation, nap mode or power-down mode. For these tests, only the supply current is measured as the voltage to the ADC is forced from the test machine and therefore is known.

The current measurement in the normal operation is typically uncritical. The voltage sources inside the test machine have a current measure unit included, which has sufficient accuracy.

The currents in power-down operation, however, are typically in the nano-Ampere range, which causes two problems. First, the leakage current of the supply capacitors can exceed the leakage current of the ADC and second, the accuracy of the current measure unit is not sufficient anymore.

The capacitor issue can partly be solved by connecting the capacitor via a relay to the converter. This has the disadvantage that the capacitor needs to be placed further away from the device-under-test (DUT), so that inductance is added between the ADC and the decoupling capacitor. Ringing on the supply voltage is therefore more likely.

Typically, two decoupling capacitors are in use. One is a ceramic capacitor C_{Cer} (typically 100 nF), which has a lower impedance than a tantalum capacitor C_{Tan}. The ceramic capacitor is placed directly to the DUT, while it is less critical to have the tantalum capacitor further away (see Fig. 6.34). The remaining leakage current can be calibrated, when the ADC is not inserted.

A test machine typically includes higher accuracy instruments than the current measure unit of the voltage source. However, these instruments are located in the test machine, so that the current has to be routed from the voltage source to the HIB, from there to the higher accuracy current meter, which is inside the test machine again, and back again to the HIB. The chance of leakage current in this path is pretty high. An alternative solution is shown in Fig. 6.34.

During the power-down measurement, the current is not routed through the decoupling capacitor C_{Tan}, but through a 1 MΩ resistor R_1. A current of 1 nA is therefore transformed into 1 mV, which can be measured with sufficient accuracy using a precision voltmeter with a 1 GΩ input resistance.

6.2.4 Functionality and Continuity Testing

The chapters above describe the measurements of all the performance specifications. They do not necessarily test all the digital functionality or all pins, so that it is

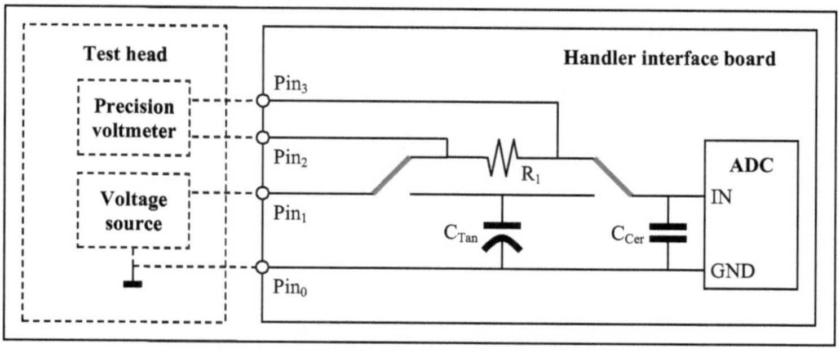

Fig. 6.34 Possible schematic for supply current measurement on a HIB for ADCs

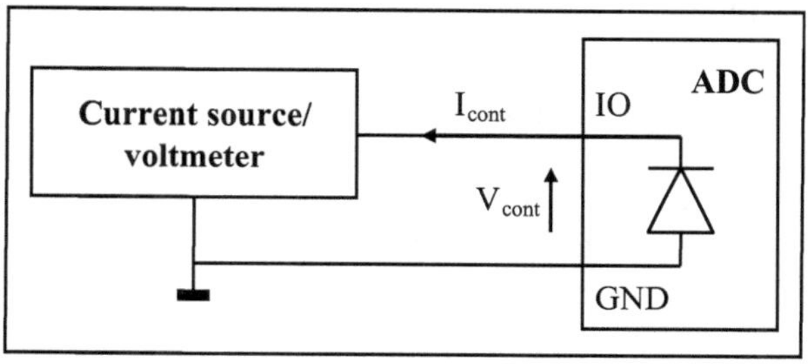

Fig. 6.35 Typical measurement setup for continuity

still possible that digital gates are defected or bondwires are missing. Furthermore, bad contacts might exist between the DUT and the test board. This is controlled with the functionality and the continuity test.

The functionality test is very dependent on the interface design of the ADC. If it is complex, then an automated scan test should be used. However, most interfaces are straight forward, so that a simple digital input pattern can do the job.

Every pin is protected by an ESD-cell (ESD = Electro Static Discharge) to prevent damage on the chip during an ESD event. Typical input or output (I/O) cells include a diode to ground, which clamps the on-chip voltage, if the input of a pin is driven below ground. The continuity test is forcing a small current into the pin (see Fig. 6.35). This current will cause the typical voltage drop across the diode V_{cont}, which can then be measured.

The continuity testing should be exercised before and after the performance test. If V_{cont} is out of specification, then either the DUT has a bad contact or a bondwire is missing or the ESD cell is damaged.

The production test is proving the performance of each device that is tested. Furthermore, it can be used to ensure the quality of a production setup over life and the quality of the test machine. A production flow for quality control is shown in the next chapter.

6.3 Quality Control in the Mass Production [21]

The production of analog and digital ICs differs essentially. Digital ICs provide a functionality, which is tested with scan patterns. Analog ICs specify the performance of their functionality. This performance will furthermore drift over temperatures. The test equipment has a limited accuracy and repeatability, which has to be controlled regularly. Several tools need to be defined to ensure a long-term and stable performance. These tools are called 'Temperature Characterization', 'Lab-to-Final Correlation', 'Capability Study', 'Correlation Lockout' and the 'Quality Control Run'.

Analog ICs do not only specify the digital logic levels, functionality and speed, but also an analog accuracy. For operational amplifiers, these specifications could be a maximum offset error of 100 μV, an open loop gain above 120 dB, a noise density as low as 1.1 $\frac{nV}{\sqrt{Hz}}$, a Total Harmonic Distortion of −100 dB or the power consumption. Typical specifications of ADCs are the offset error, the gain error, differential and integral non-linearity, the AC performance and the power dissipation. The test accuracy for ADCs often has to be in the order of microvolts.

If these critical parameters are assured with minimum and maximum values in the datasheet, then these parameters need to be measured for each single device. A pure functional test or statistical test is not sufficient. In addition, guardbands have to be implemented between the datasheet limits and the final test limits. These guardbands will cover for parameter shifts like temperature drifts or limited test accuracies. A correlation test is implemented to ensure a long-term accuracy of the test system and a quality control run is added at the end of the production test to ensure that the test system remained stable during the production.

6.3.1 *Characterization*

Parameters of analog ICs are drifting over temperature. The open-loop-gain of an operational amplifier for example would shift, which affects the regulation of the output voltage, so that the amplifiers offset will also drift. The specifications inside a datasheet are ensured over a defined temperature range, which is typically −40 °C to 85 °C for industrial products. Ideally, an IC needs to be tested over the complete temperature range in the production test. Unfortunately, this is technically and economically unrealistic. For high end products, the costs for the test at room-temperature are already exceeding the die costs. Furthermore, at negative temperatures, the humidity of the air is condensing and freezing on the test boards. The condensation causes leakage currents and shorts, which destroy the repeatability of the test.

An alternative is an extensive characterization of the IC behavior over the temperature range during the development. The data of the characterization is used to determine the average drift $average_{drift}$ and the standard deviation $stdev_{drift}$

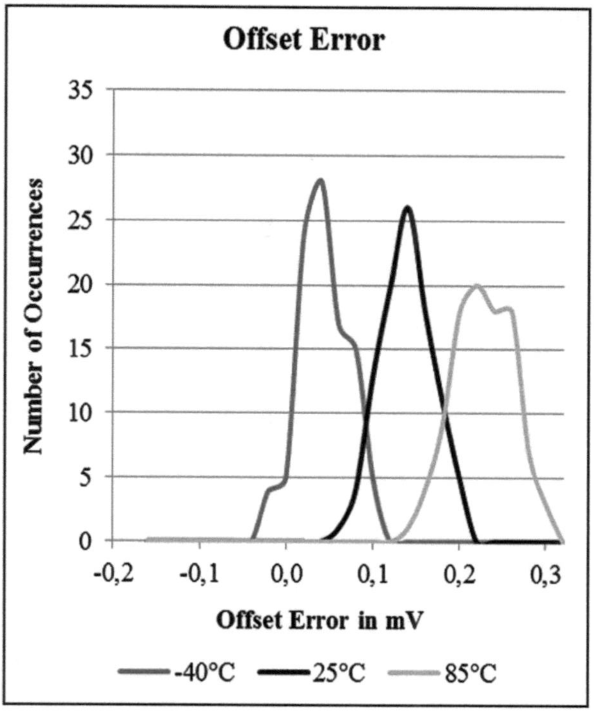

Fig. 6.36 Offset at various temperatures

of the drift for each specification in the datasheet. The average drift plus three times the standard deviation is used to determine a maximum drift. In this way, the final production test can run at room temperature only, if the maximum drift is subtracted from the minimum and maximum specification limits. The production test at room temperature does not compare the results to the specification limits of the datasheet but to the enhanced limits, which are called final test limits. The delta between the datasheet specifications and final test limits are called guardbands.

If for example the offset of an Analog-to-Digital Converter is specified to ± 1 mV, and the maximum positive and negative drift were evaluated to +158 μV and −168 μV, then the final test limits are 832 μV and −842 μV. A part can only be sold, if it passes these final test limits.

This example is further illustrated in Figs. 6.36, 6.37 and Table 6.2. Figure 6.36 shows the distribution of the offset of 100 ADCs at different temperatures. The offset drift is then calculated for each ADC separately. The distribution of the drift is shown in Fig. 6.37.

The accurate calculation of the guardbands for the offset regarding Figs. 6.36 and 6.37 is illustrated in Table 6.2 with a limited amount of exemplary samples. First the offset drift per unit is calculated for the temperature rising from 25 °C, which is used for the production test, to 85 °C. The same is done for temperatures falling from 25 °C to −40 °C. For both temperature ranges, the average drift and the standard deviation are evaluated together with the guardband.

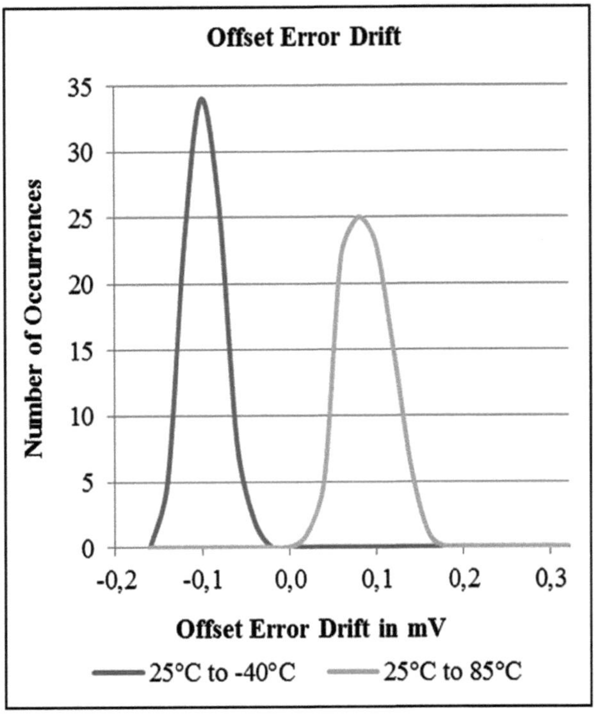

Fig. 6.37 Distribution of offset drift

Table 6.2 Calculation of the guardband with respect to the temperature characterization

	Offset error and offset error drift in mV				
Unit	−40 °C	25 °C	85 °C	−40 °C to 25 °C	85 °C to 25 °C
1	0.02	0.12	0.20	−0.11	0.07
2	0.04	0.14	0.23	−0.10	0.09
3	0.09	0.19	0.28	−0.06	0.13
4	0.02	0.12	0.20	−0.11	0.07
5	0.07	0.16	0.26	−0.08	0.11
6	0.03	0.13	0.21	−0.10	0.08
7	0.05	0.15	0.24	−0.09	0.10
8	−0.01	0.08	0.16	−0.14	0.03
9	0.05	0.14	0.23	−0.09	0.09
10	0.08	0.17	0.27	−0.07	0.12
Average	0.044	0.140	0.228	*−0.096*	*0.088*
Stdev	0.027	0.029	0.034	*0.021*	*0.027*
Guardband				−0.158	0.168
Specification				−1.000	1.000
Test limit				−0.842	0.832

6.3.2 Capability Study

The characterization above studies the drift over different temperatures. In a similar way, the capability study evaluates the shift of parameters for different test systems. A contaminated contactor could cause a high resistive ground connection of the DUT, so that internal settling times are increasing. In addition, the accuracy of the test equipment itself is limited.

Furthermore, the test systems have the capability of testing several devices in parallel. Therefore, a handler interface board as shown in Fig. 6.38 has to provide several sites, which are routed back to different resources in the test machine. Even if these sites are kept symmetrical on the PCB layout, the wiring to the resources has to differ and will cause a shift of parameters, which can be more or less significant.

A device with parameters, which are out of specification, might be tested ‘good’ due to the test tolerances. The goal of the capability study is the generation of guardbands for test inaccuracies. Different test boards, test sites and test machines are taken into account.

For the capability study, 24 units are tested at each site of at least two HIBs. The measurement is then repeated on at least one more test machine. Figure 6.39 is showing typical data of such a capability study for the example of the offset measurement.

$t1$ and $t2$ express the used test systems, $b1$ and $b2$ the two HIBs and $s1$, $s2$, $s3$ and $s4$ the particular test site on each HIB. This data is used to calculate the standard deviation of the test accuracy $stdev_{test}$ for each specification.

Identically to the temperature characterization, three times the standard deviation is added to the guardband. If the standard deviation of the temperature and the capability study show both a Gaussian distribution, then they can be added geometrically. In this way, the total guardband can be calculated to

$$Guardband = average_{drift} \pm 3 \cdot \sqrt{stdev_{drift}^2 + stdev_{test}^2} \tag{6.9}$$

In the example of Fig. 6.39, the test accuracy of the offset has a standard deviation of 37 μV, so that the final guardbands are calculated as shown in Table 6.3.

The example shows that devices are tested ‘bad’, if the offset exceeds the limits of 0.775 mV and −0.777 mV. These limits are nearly one quarter of the guaranteed specification.

6.3.3 Lab-to-Final Correlation

The capability study above is used to cover variation of different test setups. Unfortunately, this will not guarantee that a parameter is measured correctly. If for example the bandwidth of the driving amplifier of the ADC input is too low, then particularly the MSB transition can show distortion. This effect could partly be

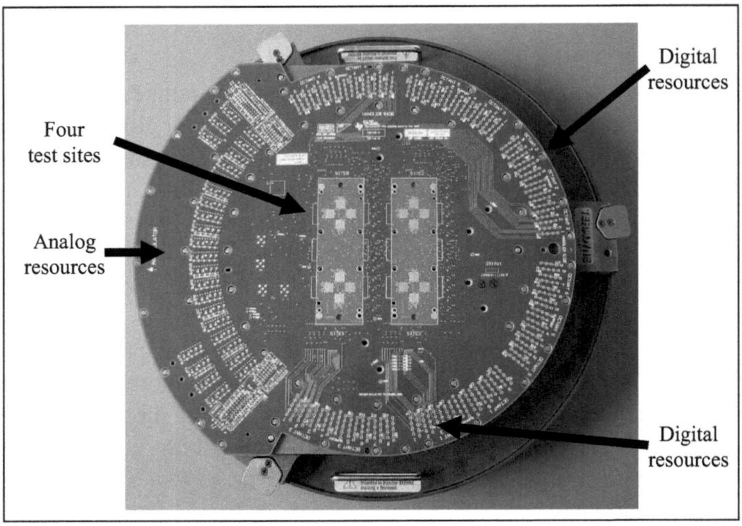

Fig. 6.38 Hardware interface board (HIB) for production test with four test sites

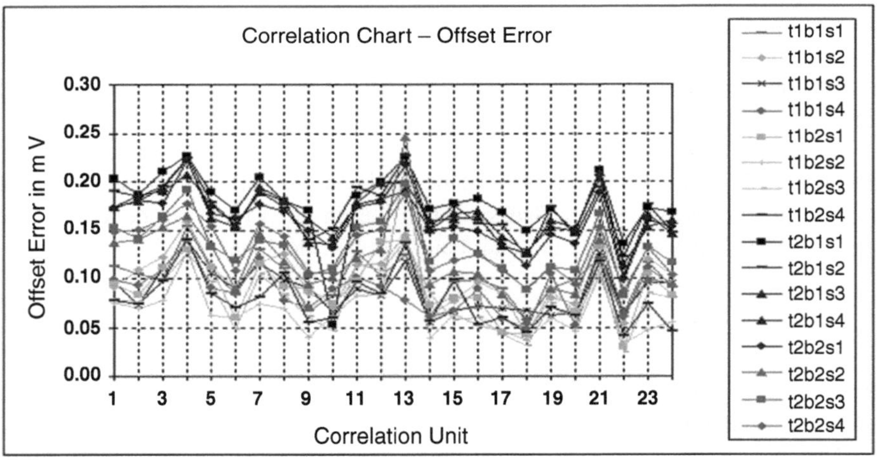

Fig. 6.39 Exemplary capability study of the offset of an ADC

compensated by trimming the length of the MSB differently. If the trim and the production test are done together on the final test system, then the customer might receive miss-trimmed products. The lab-to-final correlation should avoid such problems.

For high-end products, two different test solutions are typically in use. One is used in the laboratory by the development engineers to study and to continually improve the performance of a product. The second test solution uses the test machine for the mass production.

Table 6.3 Test limits calculated with guardband for temperature drift and test accuracy

Average drift	−0.096	0.088	
Stdev drift	0.021	0.027	
Stdev test	0.037	0.037	
Guardband	−0.223	0.225	$= \text{Average}_{\text{drift}} \pm 3 \cdot \text{sqrt}(\text{stdev}_{\text{drift}}{}^2 + \text{stdev}_{\text{test}}{}^2)$
Specification	−1.000	1.000	
Test limit	**−0.777**	**0.775**	= Specification − Guardband

For the lab-to-final correlation, ten units are tested on both test setups. The production test is only approved, if the data between the two correlates within the capability guardband. The lab-to-final correlation therefore ensures the initial accuracy of the production test, but it does not guarantee the performance of the production test setup over time. This is controlled by the correlation lockout.

6.3.4 *Correlation Lockout*

The capability study is used to determine a guardband for a maximum error of a test solution. For a long-term stability of the test solution, the test setup needs to be controlled, so that its inaccuracy does not exceed the maximum allowed limit, which was determined in the capability study. Such a case can for example occur, when a contactor is contaminated or worn out, what easily can happen after testing millions of ICs. The contamination would increase the contact resistance, which can influence the offset or linearity, if the affected contacts include analog input pins or supply pins.

For the correlation lockout, the data of ten known units is stored in the system as part of the development process. Before a production test is started, these ten units are retested and the data is compared with the initial data. If it does not differ by more than the guardband of the capability study, then the production test can be started. If the data differs by more than the guardband, then the test setup is not valid and has to be improved.

6.3.5 *Quality Control Run*

The correlation lockout proves that the test accuracy is valid at the start of the production. Unfortunately, components might break during the production test. For example, a relay might break, so that voltages remain active after turning off the supply voltage of a DUT. This can damage the IC or cause latch-up. As another example, an IC can get clamped inside a contactor. All following ICs are only pressed on top of the stuck IC, so that the stuck IC is tested over and over again instead of testing the following ICs.

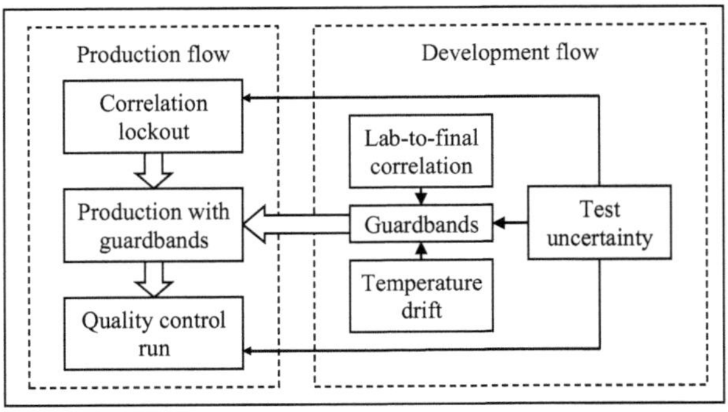

Fig. 6.40 Flowchart for quality control in IC production and development

Such problems can be identified by testing a number of X devices a second time, where none of the retested devices must fail. Here, the devices are not tested to the final test limits, but to the specifications, which are guaranteed in the datasheet.

The hypothesis, which assumes that the production test did not run correctly, should be disproved with a probability of error P that is less than 1 ppm. If the retest is assumed to run with X untested devices, where normally less than 95 % are functional, then the probability of error P can be calculated to

$$P = 0.95^X < 0.000001$$

This equals the probability that none of X untested devices would fail the retest. If no device fails the retest, then the hypothesis is assumed to be wrong or in other words the production test did run correctly. X can then be solved to

$$X > \frac{\log(0.000001)}{\log(0.95)} = 270$$

270 devices need to be retested to support a probability of error below 1 ppm.

6.3.6 *Quality Flow*

The complete quality flow is shown in Fig. 6.40. This flowchart is separated into two areas. The first one covers the tasks, which have to be implemented during the development process. These are the Lab-to-Final Correlation as well as the temperature characterization and the capability study, which are used to calculate guardbands for the production test.

The production flow is shown in the second area. The correlation lockout needs to be passed, before the production test can start. If it passes, the production to the final test limits can be started. After finishing the production, the quality control run needs to be performed. The devices can only be shipped to customers, if all these quality measures passed.

References

1. Guenter Peltz, *Praxis der Analog- und Digitalwandler, Grundlagen, Versuchs- und Anwendungsschaltungen*, Elektor Verlag, Aachen 1990.
2. Frank Ohnhaeuser, *Aussteuern der Eingänge von A/D-Wandlern*, in the journal Design & Verifikation (Germany), Vol. 10, pp. 22-25, October 2002.
3. Frank Ohnhaeuser, *Driving the inputs of Analog to Digital Converters*, in the journal Technologiot (Israel), Vol. 5, pp. 62-64, May 2003.
4. Bonnie C. Baker, *Driving the Analog Inputs of a SAR A/D Converter*, Microchip Technology, 2003, URL: http://ww1.microchip.com/downloads/en/AppNotes/00246a.pdf, status of 3/11/2015.
5. Christopher Peter Hurrell, Bruce Edward Amazeen, *Successive approximation analog-to-digital converter with pre-loaded SAR*, US Patent 6828927, Analog Devices, 2004, URL: http://www.google.com.ar/patents/US6828927, status of 3/11/2015.
6. Product datasheet of the OPA365, Texas Instruments, 2009, URL: http://focus.ti.com/lit/ds/symlink/opa365.pdf, status of 3/11/2015.
7. Frank Ohnhaeuser, *Adjusting signal levels in analogue systems*, in the journal Electronics Weekly (UK), Vol. 10, pp. 30-31, September 2004.
8. Frank Ohnhaeuser, *Piloter l'entrée de reference d'un CAN a approximations successives*, in the journal Electronique (France), Vol 11, pp. 62-64, November 2003.
9. Application Bulletin, *Analog-to-Digital Converter - Grounding Practices Affect System Performance*, Burr-Brown, 1994, URL: http://focus.ti.com/lit/an/sbaa052/sbaa052.pdf, status of 3/11/2015.
10. Don Brockman, Arnold Williams, *Ground Rules for High-Speed Circuits*, Analog Devices, URL: http://www.analog.com/static/imported-files/application_notes/573185595135053265003392$32#595605308453835483630769AN214.pdf, status of 3/11/2015.
11. Frank Ohnhäuser, *Theory and realization of high-end analog-to-digital converters (ADC) based on the principle of successive approximation (SAR)*, PhD Thesis at the University of Erlangen-Nürnberg, Erlangen, 2008.
12. Brad Brannon, Allen Barlow, *Aperture Uncertainty and ADC System Performance*, Analog Devices, 2006 URL: http://www.analog.com/static/imported-files/application_notes/59756494064912342505447175991257024546937062255921511183854180687755AN501_a.pdf, status of 3/11/2015.
13. M. Burns, G.W. Roberts, *An introduction to mixed-signal IC test and measurement*, Oxford University Press, 2001.
14. Solomon Max, *Ramp Testing of ADC Transition Levels using finite Resolution Ramp*, in the proceedings of the International Test Conference (ITC), pp.495-501, 2001.
15. W. Lee, Y. Liao, J. Hsu, Y. Hwang, J. Chen, *A high precision ramp generator for low cost ADC test*, in the proceedings of the 9th IEEE International Conference on Solid-State and Integrated-Circuits Technology, pp. 2103-2106, 2008.
16. F. AlegriaP. Arpaia, A. da Cruz Serra, P. Daponte, *Performance analysis of an ADC histogram test using small triangular waves*, in the IEEE journal of Transactions on Instrumentation and Measurement, Volume 51, Issue 4, pp. 723-729, 2002.
17. F. Stefani, A. Moschitta, D. Macii, P. Carbone, D. Petri, Fast estimation of ADC nonlinearities using the Sinewave Histogram Test, in the journal Elsevier Measurement, Volume 39, No. 3, pp. 232-237, December 2006.
18. J.M. Janik, *Estimation of A/D Converter Nonlinearities from Complex Spectrum*, in the proceedings of the International Workshop for ADC Modeling And Testing (IWADC 2003), pp. 205-208, 2003.
19. F. Ohnhaeuser, *Measuring the DC Accuracy of High Performance Analog-to-Digital Converters (ADCs)*, in the proceedings of the International Workshop for ADC Modeling and Testing (IWADC 2003), pp. 71-74, 2003.

20. C. Wegener, M.P. Kennedy, *Extending the servo-loop for ADC transition level measurements under dynamic input conditions*, in the proceedings of the IEEE Instrumentation and Measurement Technology Conference 2003, Volume 3, pp. 2031-2036, 2001.
21. Frank Ohnhaeuser, *Quality control of high performance analog ICs in mass production*, in the journal Electronics Component World (Great Britain), 01/2008.

Chapter 7
Digital-to-Analog Converters

The last chapters present design details of ADCs in detail with the focus on SAR ADCs in the Chaps. 2 and 3 and delta-sigma ADCs in the Chaps. 4 and 5. Chapter 6 discusses the test and verification of ADCs as well as the external circuitry around the ADC in an application. This chapter will now complement the topic of data converters with an introduction of digital-to-analog converters (DACs). In industrial applications, they are often used as the controlling element. The system performance will then fully rely on the DAC and the absolute values such as INL, offset or gain are important. If they are used inside a control loop, which will increment or decrement the DAC value, then equal step size is required, which is reflected in the DNL.

Similar to ADCs, different architectures are available, which differ in speed and resolution. The most common architectures are the current steering DAC, the string DAC, the R-2R architecture and the delta-sigma approach. Figure 7.1 illustrates their performance in resolution versus conversion rate.

This chapter will first briefly discuss the performance parameters of DACs and will then describe the four architectures in more detail.

7.1 DAC Specifications

All specifications that were discussed for ADCs generally in Chap. 1 also apply for DACs. As the first main difference, the transfer function, which is illustrated in Fig. 7.2, surely has the digital code as input (x-axis) and the analog voltage (y-axis) as the output. Again, analog output voltages of adjacent codes should have equal distance and therefore should be located on an ideal line.

Non-equal voltage steps such as ΔV_{010}, ΔV_{011} or ΔV_{101} in Fig. 7.3 would generate non-linearities. Similarly to the ADC, the difference between the ideal voltage step and the real voltage step forms the differential non-linearity. The differential non-linearities add up to the integral non-linearity, which is often

F. Ohnhäuser, *Analog-Digital Converters for Industrial Applications Including an Introduction to Digital-Analog Converters*, DOI 10.1007/978-3-662-47020-6_7

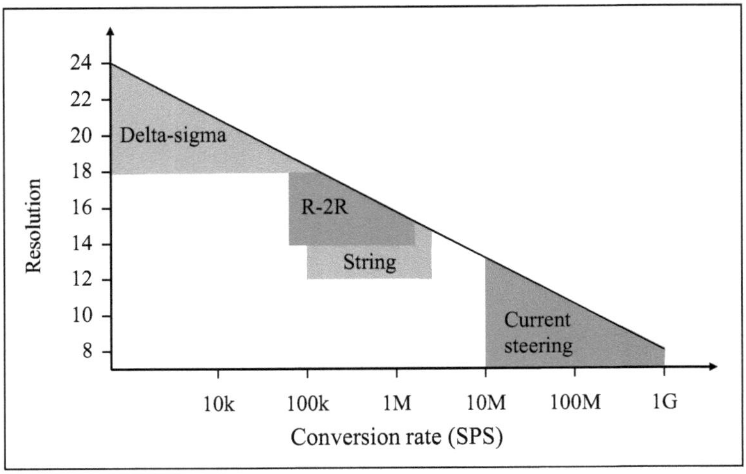

Fig. 7.1 Performance overview for various DAC architectures

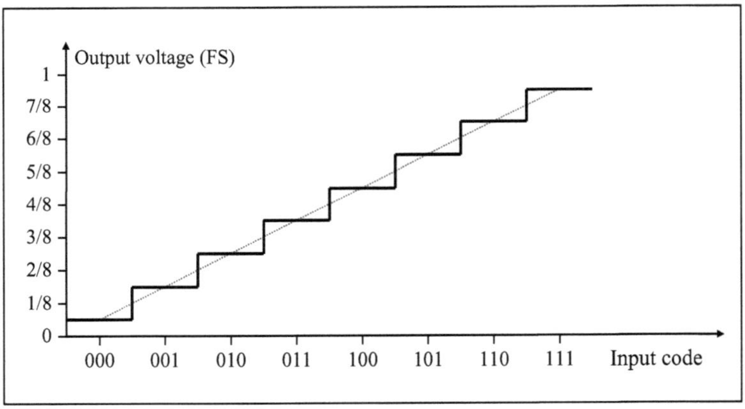

Fig. 7.2 Ideal DAC transfer function

simply called linearity or relative accuracy. The linearity is the difference of the actual voltage to the straight line.

The steps size from code 001 to 010 and 010 to 011 is 50 % larger than the average step size. Consequently, the DNL is 0.5 LSB each. As a specialty, the output of code 101 is less than the output voltage of code 100. Such a negative step size is called non-monotonic. These parameters are shown in Table 7.1 as a typical example. Often, DACs also specify the long-term output voltage drift.

The parameters of DACs differ compared to the ones from ADCs in their noise specification. DACs have an analog output voltage and are therefore mostly specified similarly to operational amplifier with their 1/f-noise and their thermal noise density. Similarly, DACs are mostly specifying the settling time and the slew rate.

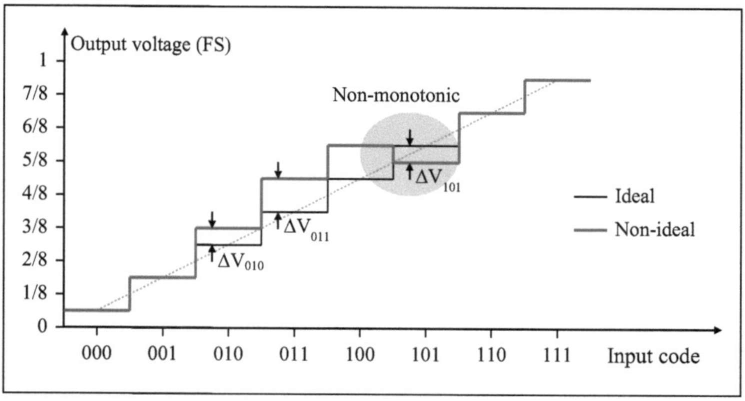

Fig. 7.3 Non-ideal DAC transfer function showing linearity and monotonic errors

Table 7.1 Exemplary DC parameters of DACs extracted from the DAC9881 from Texas Instruments [1]

Parameter	Min	Typ	Max	Unit
Accuracy				
Integral linearity error		±1	±2	LSB
Differential linearity error		±0.5	+1	LSB
Monotonicity	18			Bit
Gain error		±16	±32	LSB
Gain error temperature drift		±0.25	±0.4	ppm/°C
Zero error		±16	±32	LSB
Zero temperature drift		±0.25	±0.8	ppm/°C
PSRR			32	LSB/V

Table 7.2 Exemplary dynamic parameters of DAC9881 from Texas Instruments [1]

Parameter	Min	Typ	Max	Unit
Dynamic performance				
Settling time		5		μs
Slew rate		2.5		V/μs
Code change glitch		37		nV-s
Output noise voltage density f = 1 kHz to 100 kHz		24	30	nV/sqrt(Hz)
Output noise voltage f = 0.1 Hz to 10 Hz		2		μV_{pp}

'Code change glitch' is another interesting parameter. A change of the DAC's input code will trigger the voltage change of the analog DAC circuitry. During the transition time, the output voltage is uncertain and will cause an error of the output voltage, which often has the form of a voltage glitch. The glitch energy is specified by this parameter (Table 7.2).

The further parameters such as the full-scale range, the digital IO voltage levels or the power consumption are very similar to ADC parameters. In the following the most important DAC architectures are discussed starting with the current steering DAC.

7.2 Current Steering DAC

The fastest architecture of the commercially available DACs is the current steering DAC [2, 3] as shown in Fig. 7.4. Current sources are generated that are binary weighted. Depending on the input code, they are switched to a positive V_{op} or a negative V_{on} output node. The output can either be this differential current as done for the DAC5670 from Texas Instruments [4] or the current is connected to resistors, which will convert the current to an output voltage as it is done in the example of Fig. 7.4.

In this example, the total current I_{max} of

$$I_{\max} = I \cdot \left(1 - \frac{1}{2^6}\right) \tag{7.1}$$

can flow through the positive resistance R_{op} and the negative resistance R_{on}, which will lead to a positive full-scale voltage PFS and a negative full-scale voltage NFS of

$$\begin{aligned} PFS &= I_{\max} \cdot R_{op} \quad \text{and} \\ NFS &= -I_{\max} \cdot R_{on} \end{aligned} \tag{7.2}$$

It can easily be seen that the differential non-linearity is strongly dependent on the matching of the binary weighted currents. The full-scale voltages and therefore the gain error are further dependent on the correct absolute value of the currents and the resistors as well. The resistors should certainly be equal. If this is not the case, an offset voltage is generated. Further integral non-linearity can be caused by the voltage coefficient of the resistors. Consequently, the DC accuracy of current steering DACs is limited.

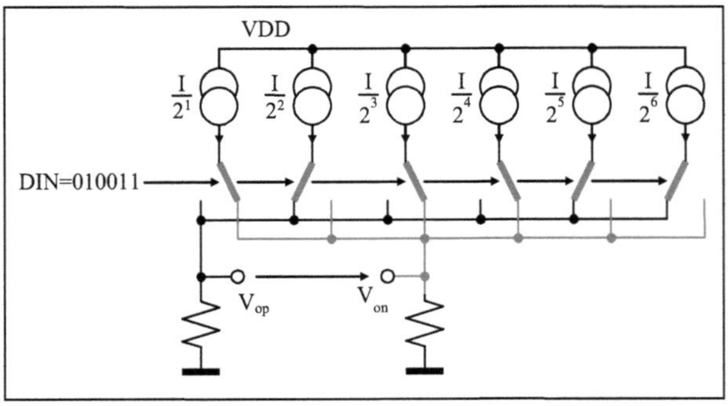

Fig. 7.4 Block diagram of a current steering DAC

The current through a saturated MOS transistor is defined as by Eq. (7.3), where W is the width and L the length of the transistor, V_{gs} describes the gate-source voltage, V_{DS} the drain-source voltage and V_{thres} the threshold voltage. a and b are process parameters.

$$I_D = a \cdot \frac{W}{L} \cdot \left(V_{gs} - V_{thres}\right)^2 (1 + b \cdot V_{DS}) \tag{7.3}$$

Even if choosing W and L large, the matching is limited to around 1 %. It is further important that the current does not change with the output voltage. Therefore, a cascaded current source is required as illustrated in Fig. 7.5.

Note that good matching current sources utilize equal transistors in terms of width and length and only change the number of transistors, which is expressed with the multiplier M of a transistor. This is similar to utilizing unity capacitors for the CDAC in SAR ADCs. The current is set by the transistor T_{cs}. The cascading transistors ensure that the drain-source voltages of the current sources remain constant over the output voltage and are larger than the saturation voltage. The operating point of the cascades is set by the transistor T_{cas}. Calibration routines are typically used to improve the DNL for non-matching current sources.

The large current source transistors would cause long turn-on and turn-off times of the currents. The current sources are therefore operating continuously and the currents are only steered by the switches that are controlled by the digital inputs DIN_x. Note that the MSB is again named bit 1.

Note that similar to the CDAC, the current steering DAC is often separated into a MSB DAC and an LSB DAC. A 12 bit DAC might for example have 6 MSB current sources and 6 LSB current sources. Otherwise, the ratio between the LSB current source and the MSB current source might be 1:2048, which surely would be

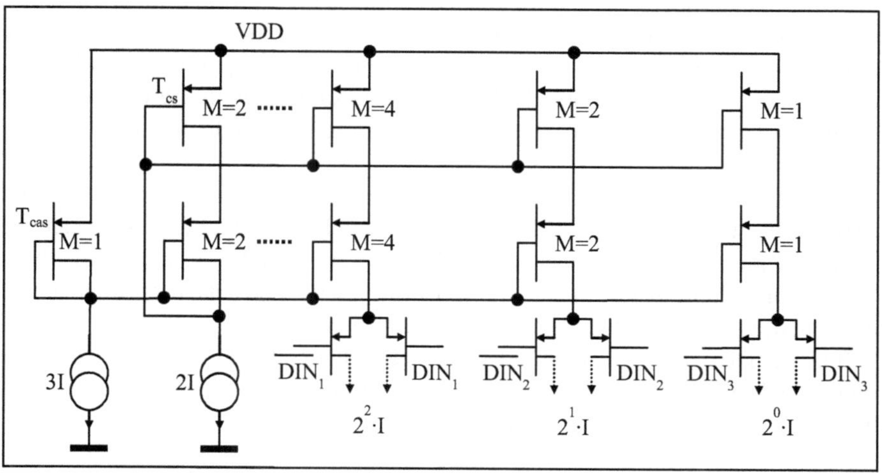

Fig. 7.5 Circuit with current mirror and switches of a current steering DAC

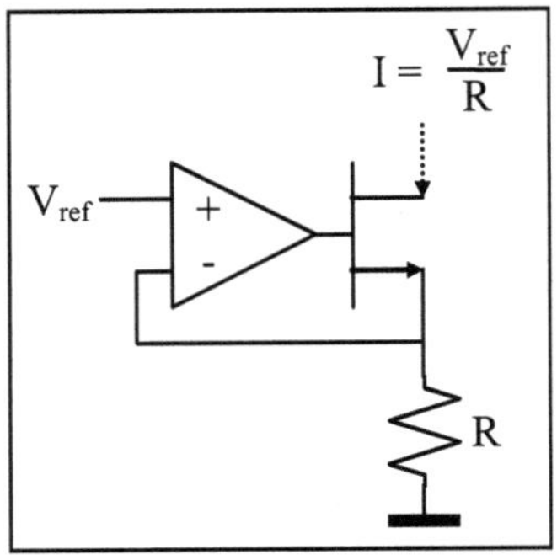

Fig. 7.6 Possible circuit for a current source

difficult to integrate particularly if good matching is required. The currents from the LSB DAC can be scaled with current mirrors into the MSB DAC.

The current source can be designed such as in Fig. 7.6. The amplifier will regulate the voltage at the resistor to the reference voltage V_{ref}, so that the current I equals the reference voltage divided by the resistance R. If the resistor matches the output resistors R_{op} and R_{on} of the DAC, then the output voltage is independent of the absolute resistor value but only on the resistor matching. Consequently, the gain error will improve dramatically.

The DC accuracy of the current steering DAC is limited, but the currents can be steered very fast. High conversion rates up to 2.4 GSPS are common today. Table 7.3 illustrates the performance parameters on the example DAC5670 from Texas Instruments [4]. The DAC5670 is today one of the fastest commercially available DACs. The thermal noise is integrated over a wide bandwidth and will limit the SNR. The high bandwidth will furthermore require high currents and cause high power dissipation.

In the layout, care must be taken that the current sources will match well [5]. Due to the high switching frequencies, care must be taken that the parasitic capacitance at the nodes V_{op} and V_{on} remain as low as possible. With output resistors of 500 Ω and parasitic capacitors around 2 pF, which for example might be caused by the bondpads, the time constant is already 1 ns resulting in a −3 dB frequency of only 160 MHz.

The current steering DACs are naturally used in high speed applications and particularly in communication. Examples are satellite communication, wideband communication, cellular base stations, cable modems or radar systems. Other high speed applications, where the current steering DAC finds its use, are medical ultrasound, video or digital TV or wave form generators.

If higher DC accuracy is required at less speed, then other DAC architectures are more suited. A DAC with extremely good differential linearity and monotonicity is the string DAC, which is discussed next.

Table 7.3 Exemplary specifications of a current steering DAC (DAC5670 from Texas Instruments [4])

Parameter	Min	Typ	Max	Unit
Accuracy				
Resolution	14			Bit
Integral linearity error	−7.5	±1.5	7.5	LSB
Differential linearity error	−0.98	±0.8	1.75	LSB
Monotonicity	14			Bit
Gain error	−6	±1.6	6	%FSR
Gain drift		75		ppm FSR/°C
Full-scale output current			30	mA
Offset error	−0.45	±0.09	0.45	%FSR
Offset drift		75		ppm FSR/°C
Dynamic performance				
SFDR ($f_{out} = 100$ MHz)	47	55		dBc
SFDR ($f_{out} = 500$ MHz)	44	50		dBc
SNR (fout = 100 MHz)	63	70		dBc
SNR (fout = 500 MHz)	53	60		dBc
Output noise voltage f = 0.1 Hz to 10 Hz		2		μV_{pp}
Power supply				
AVDD	3	3.3	3.6	V
I_{AVDD}		560	650	mA
Power dissipation		1800	2350	mW
PSRR		0.4	1.3	%FSR/V

7.3 String DAC

The string DAC operates similarly as the flash ADC, which was already described in Chap. 1. The various output voltages of the DAC are generated with a resistive divider. Switches are then connecting the potential, which is selected by the digital input code, to the output. Figure 7.7 is illustrating the principle.

Note that the output resistance of the resistive divider is code dependent. The output impedance is set by the parallel resistance of the sum of resistors from the selected node to the positive reference REF_p and the sum or resistors from the selected node to the negative reference potential REF_n. In the example of Fig. 7.7, the resistance to REF_p is 4.5·R and the resistance to REF_n is 3.5·R. The output impedance R_{out} is then

$$R_{out} = \frac{1}{\frac{1}{3.5 \cdot R} + \frac{1}{4.5 \cdot R}} = 1.97 \cdot R \tag{7.4}$$

If a string DAC with n-bit resolution and unity resistance R is used, then the maximum resistance is reached at mid-scale, where approximately 2^{n-1} resistors

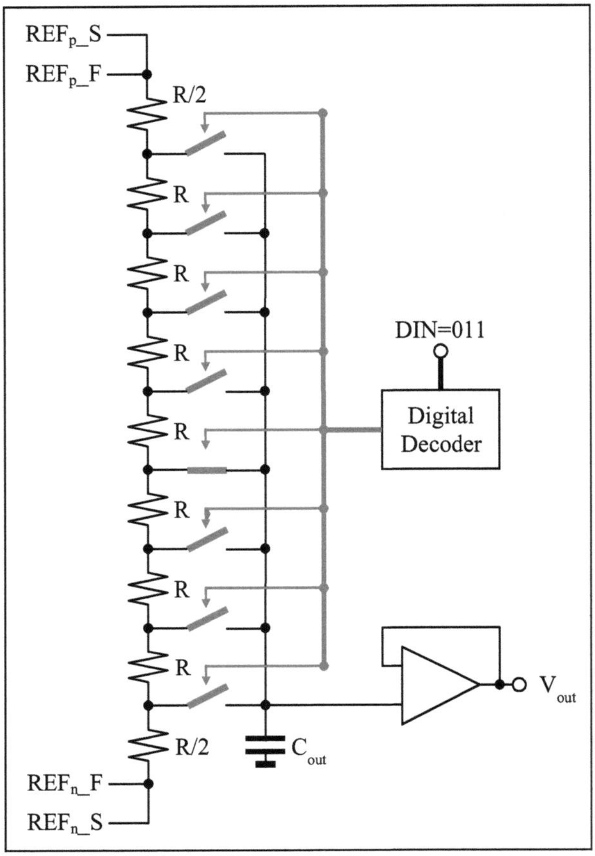

Fig. 7.7 Block diagram of a string DAC

connect from the tap to REF_p and the same number to REF_n. The output resistance is then

$$R_{out} = \frac{1}{\frac{1}{2^{n-1} \cdot R} + \frac{1}{2^{n-1} \cdot R}} = 2^{n-2} \cdot R \tag{7.5}$$

The code dependent output resistance will cause a code dependent thermal noise. Optionally, a capacitor C_{out} can be added, which would change the thermal noise of the resistors into a kT/C noise that would be independent of the resistor value. The noise would then be equal for each output voltage, which is selected by the digital input code.

The code dependent output resistance will furthermore cause a code dependent settling time with the time constant

$$\tau = R_{out} \cdot C_{out} \tag{7.6}$$

This code dependent time constant will cause harmonic distortion. Most string DACs therefore minimize the output capacitance to the parasitic switch and wire capacitance and buffer the node with an output amplifier. The amplifier's bandwidth will then limit the noise of the DAC.

Figure 7.8 is showing a more advanced switching scheme, where the switches are cascaded. The idea is to organize the resistors in blocks that can easily be addressed with MSB bits only. In the example below, the second column from the left is selected with $D_1D_2 = 10$. Thus only four switches are connected to the amplifier instead of 16, the parasitic capacitance at this node can be lowered and the conversion rate will increase.

As another advantage the digital decoding simplifies dramatically. In the example above, it can be seen that all switches, which are addressed with a digital input word with the LSBs being $D_3D_4 = 01$, can be closed in parallel. The second switch, which is addressed with D_1D_2, will do the final selection. In this way, the switches that are within one complete row can be selected with a simple gate that has only two inputs. In Fig. 7.7, each switch must be controlled separately, with a digital gate that requires four inputs.

Figure 7.7 also shows force (REF_p_F, REF_n_F) and sense lines (REF_{p}_S, REF_n_S) for the reference voltage. The reference input path is associated with resistance from the ESD cell and particularly from the chip internal wiring that is causing gain and offset errors as the parasitic resistors affect the resistive divider. This is even more critical for R-2R DACs, which will be explained in the next chapter. Parasitic wire resistance in R-2R DACs can cause essential linearity errors. A sense path is added, which can be used as feedback path for operational amplifiers as shown in Fig. 7.9. The negative amplifier input is forced to *REF* by

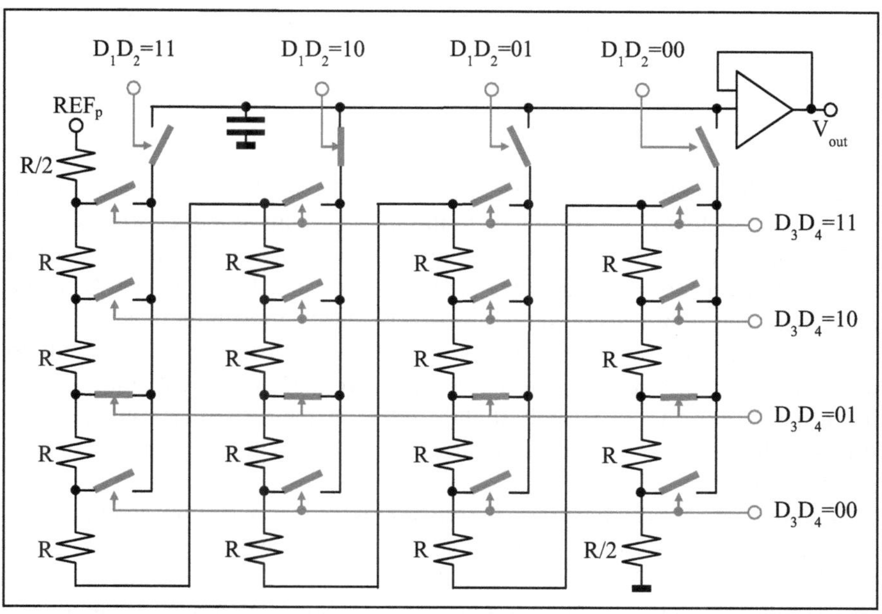

Fig. 7.8 String DAC with cascaded switching scheme reducing digital decoding logic

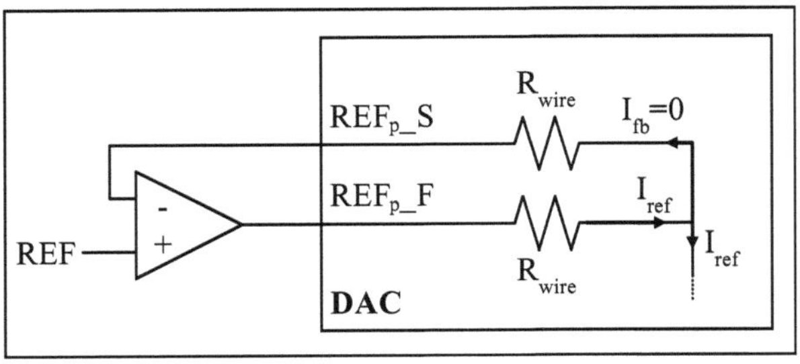

Fig. 7.9 Connectivity of force and sense lines to avoid offset caused by parasitic wire resistors R_{wire}

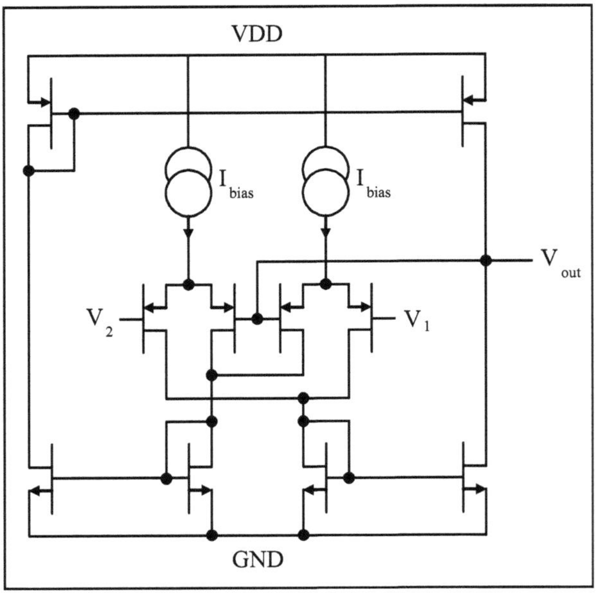

Fig. 7.10 Amplifier with dual input stage for interpolation of voltages V_1 and V_2 used in string DACs

the gain of the amplifier. As the feedback path of the amplifier is basically current free the same potential will be forced to the starting point of the resistive divider.

The current I_{ref} is flowing through each resistor, which is causing the voltage drop $I_{ref} \cdot R$. Resistors within an integrated circuit are matching up to 0.1 % so that the voltage step from resistor to resistor is very constant resulting in an extremely good DNL. Also the output amplifier has a linear transfer function, which will not affect the DNL. Unfortunately, resistors have a voltage coefficient associated, so that resistors close to REF_p will slightly differ in their resistance compared to resistors close to REF_n. This difference is adding up to an essential integral non-linearity.

Similar to the flash converter, the circuit overhead is getting essential with increasing resolution n as $2^n + 1$ resistors and 2^n switches will be required. This would limit the resolution of the string DAC. Also, the thermal noise increases with the resistance. Fortunately, there is a workaround. Some additional resolution can be gained by the operational amplifier as illustrated in Fig. 7.10.

The output of an amplifier in buffer configuration will regulate the output voltage to be equal to the positive input voltage. If two identical amplifiers in almost identical operating points are operating in parallel with the first one having an input voltage of V_1 and the second one an input voltage of V_2, then the output voltage V_{out} has to end up in the middle.

$$V_{out} = \frac{V_1 + V_2}{2} \tag{7.7}$$

This is also true, if two identical input stages connect to the same output stage as shown in Fig. 7.10. Note that each input stage also requires its own current source. Otherwise, the input stage with the lower input potential will draw more current than the input stage with the higher input potential assuming PCH input transistors. The two stages would not be in a similar operating point anymore.

This interpolating effect can be utilized for the string DAC. The MSBs will select a resistor. Now, the upper and lower nodes of the resistor are forwarded to the amplifier. If the LSB of the digital input DIN_{LSB} = ‘0’, then both inputs V_1 and V_2 connect to the resistor node with the lower potential. The output voltage follows the potential. With the LSB being one, then V_2 is connecting to the higher resistor node, so that the output voltage will settle in the middle of the higher resistor node and the lower resistor node. This can be seen in Fig. 7.11.

Note that the various input stages are not really increasing the amplifier layout. For noise reasons, the input transistors of the amplifier require a certain width anyhow. For good matching, the positive and negative transistors would be placed with a certain multiplier and would be routed interleaved. A simplified version without dummy transistors and with limited multiplier is illustrated in Fig. 7.12. The interpolated version would basically just change the routing, which is shown in Fig. 7.13. The two according schematics are shown in Fig. 7.14.

The interpolation is not limited to two differential stages. Four or five bits can easily be achieved by this technique. Note that the interpolation is also strictly monotonic. This enables the design of 16 bit string DACs such as the DAC8550

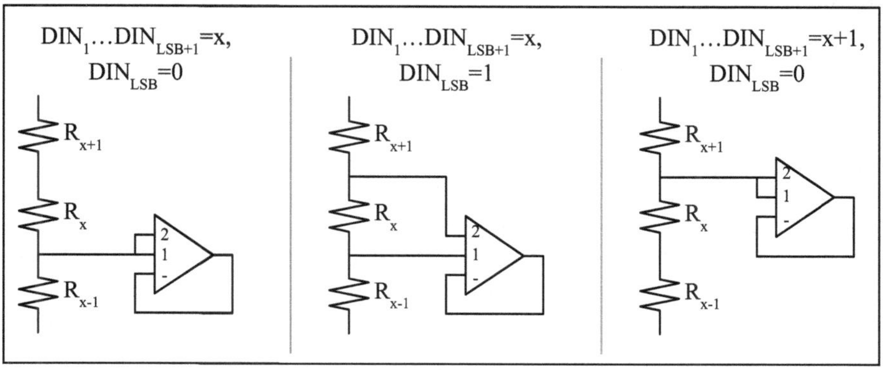

Fig. 7.11 Connecting an interpolating amplifier in a string DAC [6]

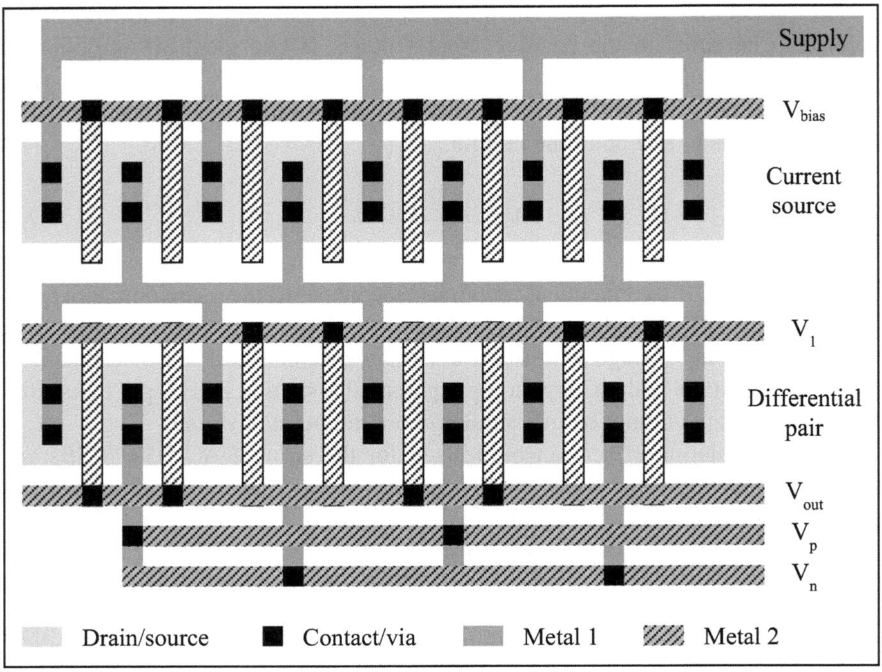

Fig. 7.12 Exemplary layout of a normal differential pair with fingered transistors

from Texas Instruments [7]. A section of the datasheet with the most interesting parameters is shown in Table 7.4.

Note the extremely good DNL performance. The INL or relative accuracy is extremely well for string DACs and indicate the usage of a semiconductor technology with ultra-low resistive voltage coefficient. INL values with ±64 LSB for 16 bit DACs are common. The AC performance is only so well, as the output signal is filtered with a 20 kHz low-pass filter.

The good linearity is surely also a result from optimized layout techniques [8, 9]. Similar to the CDAC in SAR ADCs, the resistors should be placed symmetrically in the layout to avoid mismatch due to a derivation on a chip such as the thickness of a resistor. If resistors with such a derivation are not placed symmetrically, then the derivation would strongly affect the INL. Also care must be taken that the wiring between the resistors always has identical resistance and that the wiring including the contact is not an essential part of a unity resistance.

The performance of string DACs is interesting for closed loop servo drives, where the relative accuracy is less important. String DACs can be very cost effective and therefore are used in applications such as process control, data acquisition systems, PC peripherals or portable instrumentation, if the INL performance is sufficient. If better INL accuracy is required, then R-2R DACs are the better choice.

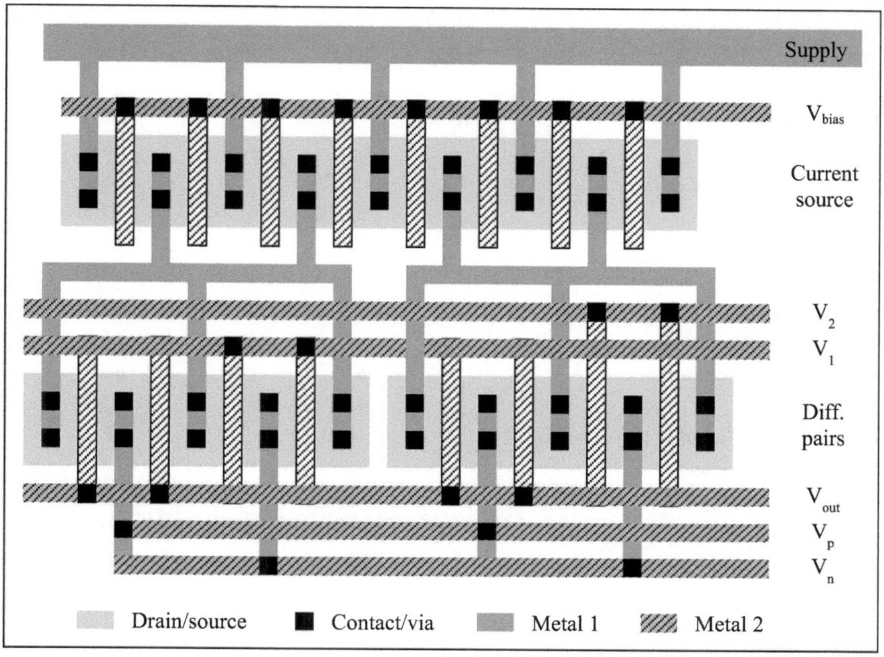

Fig. 7.13 Exemplary layout of a interpolating differential pair

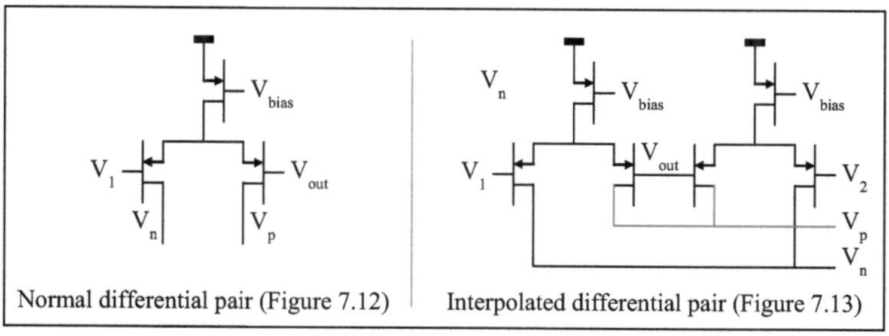

Fig. 7.14 Schematic of normal and interpolated differential pair according to Figs. 7.12 and 7.13

7.4 R-2R DAC

The resistive divider can also be configured in an R-2R configuration as shown in Fig. 7.15. The resistors within the arrows below the circuitry sum up to the indicated resistance R or $2R$. It can be seen that the output resistance is constant with the value R. This way, the time constant at the at the output node V_{out} is code independent, so that the circuitry qualifies for good harmonic distortion [10].

The analysis of the circuitry in Fig. 7.15 can be simplified by replacing the switches with voltage sources V_1, V_2, V_3 and V_4 as shown in Fig. 7.16 and then by

Table 7.4 Exemplary specifications of a string DAC (DAC8550 from Texas Instruments [7])

Parameter	Min	Typ	Max	Unit
Accuracy				
Resolution	16			Bit
Relative accuracy		±3	±8	LSB
Differential linearity error		±0.25	±1	LSB
Gain error		±0.02	±0.2	%FSR
Gain temperature coefficient		±1		ppm FSR/°C
Offset error		±2	±12	mV
Offset drift		±5		μV/°C
Output characteristics				
Output voltage settling time		8	10	μs
Slew rate		1.8		V/μs
Capacitive load stability ($R_L = 2$ kΩ)		1000		pF
Dynamic performance				
SNR ($f_{out} = 1$ kHz)		95		dB
THD ($f_{out} = 1$ kHz)		−85		dB
SFDR (fout = 1 kHz)		87		dB
SINAD (fout = 1 kHz)		84		dB
Power supply				
AVDD	2.7		5.5	V
I_{AVDD}		160	250	μA
Power dissipation (5.5 V)		0.88	1.375	mW
PSRR		0.75		mV/V

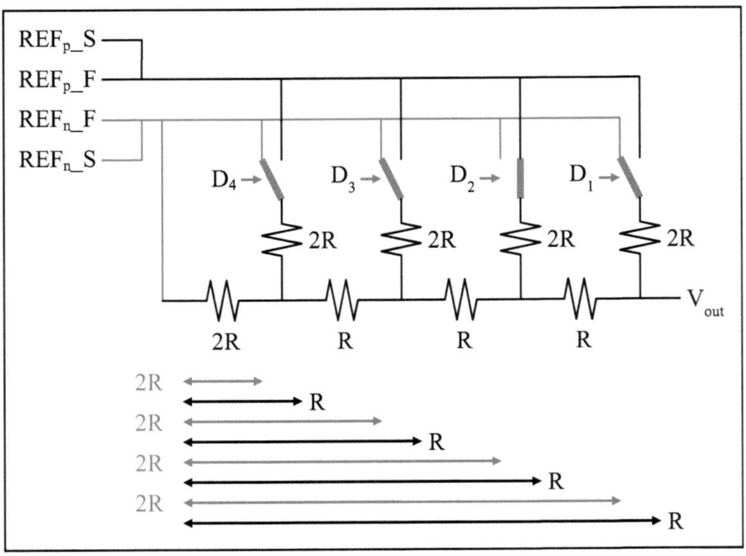

Fig. 7.15 Block diagram of a 4 bit R-2R DAC

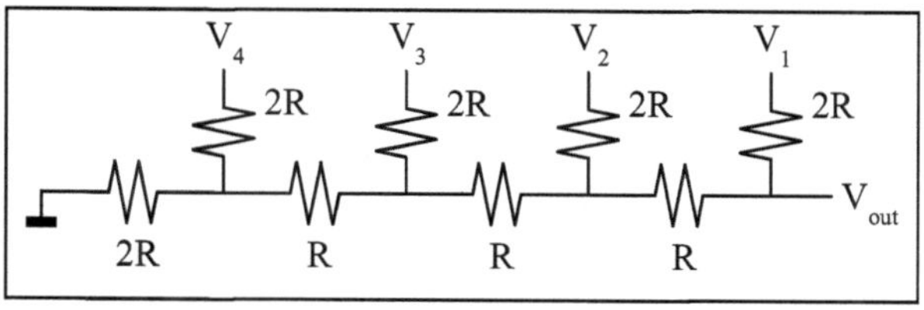

Fig. 7.16 Equivalent circuitry for the R-2R DAC in Fig. 7.16

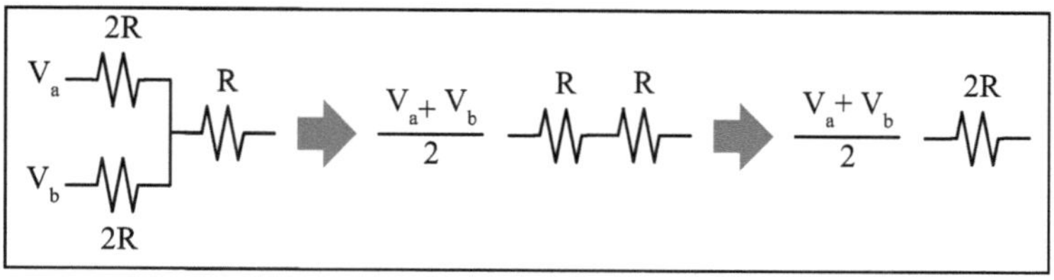

Fig. 7.17 Equivalent circuit of the partial resistor network inside a R-2R DAC

replacing the circuitry piecewise with equivalent voltages and resistors based on Fig. 7.17.

With the two basic transformations from Figs. 7.16 and 7.17, the analysis can start from the left. Figure 7.18 shows the various steps.

Based on Fig. 7.18, V_4 will act as the LSB and V_1 as the MSB. Consequently, the resistor connecting to V_1 represents 2^{n-1} LSBs with n being the resolution of the DAC, which in this case is 4 bit. Any resistive mismatch will directly lead to DNL errors. This can easily be seen in step 4 of Fig. 7.18. If the resistor to V_1 has a value of $2R+\Delta R$ and consequently a mismatch of ΔR and if all LSBs are connected to ground while V_1 connects to the reference voltage REF (code 1000) then the resistive network operates as pure resistive divider that follows Eq. (7.8).

$$V_{out}(1000) = REF \cdot \frac{2R}{\Delta R + 4R} \tag{7.8}$$

If all LSBs for the same resistor network are all connected to REF and the MSB to ground (code 0111), then the output voltage is

$$V_{out}(0111) = REF \cdot \left(1 - \frac{1}{8}\right) \cdot \frac{\Delta R + 2R}{\Delta R + 4R} \tag{7.9}$$

The difference in output voltage of the two adjacent codes 1000 and 0111 should certainly be REF/16. However, due to the mismatch it can be calculated to

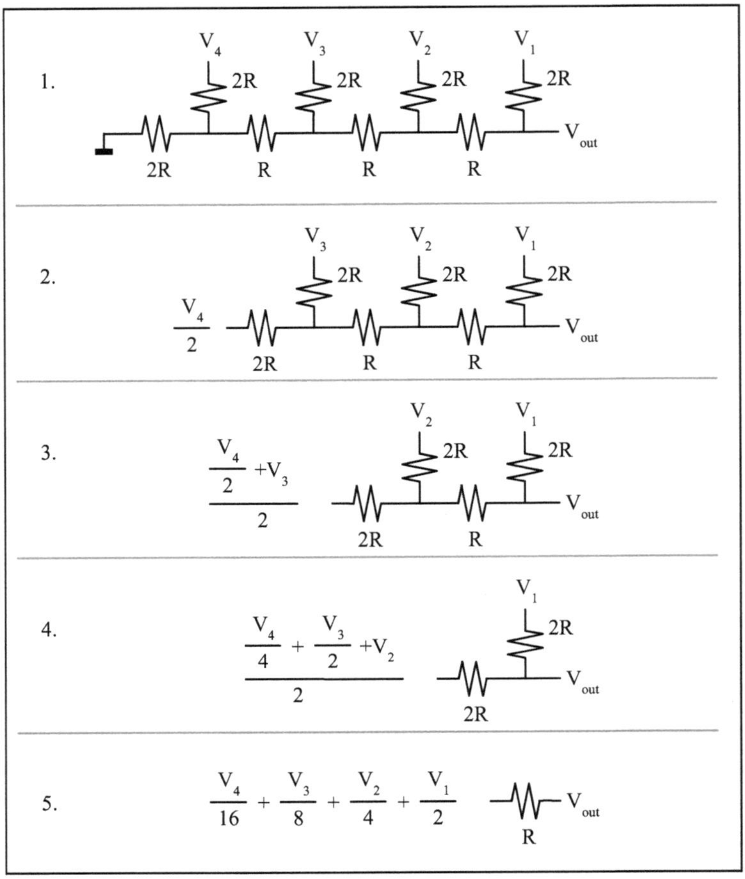

Fig. 7.18 Circuit analysis of a 4 bit R-2R DAC

$$\Delta V_{out} = REF \cdot \frac{2R}{\Delta R + 4R} - REF \cdot \left(1 - \frac{1}{8}\right) \cdot \frac{\Delta R + 2R}{\Delta R + 4R}$$

$$\Delta V_{out} = \frac{REF}{\Delta R + 4R} \cdot \left[2R - \left(1 - \frac{1}{8}\right) \cdot (\Delta R + 2R)\right]$$

$$\Delta V_{out} = \frac{REF}{\Delta R + 4R} \cdot \left[2R - \Delta R - 2R + \frac{1}{8}\frac{\Delta R + 4R}{2} + \frac{1}{8}\frac{\Delta R}{2}\right]$$

$$\Delta V_{out} = \frac{REF}{16} - REF \cdot \frac{15}{16} \cdot \frac{\Delta R}{\Delta R + 4R} \tag{7.10}$$

Based on Eq. (7.10), ΔR will directly generate a significant error in the step size and therefore in the DNL. If resistors match to 0.1 % accuracy, then ΔR in the denominator can be neglected. The term 15/16 in Eq. (7.10) is caused by the resistor $2R$, which is permanently connected to ground. It is expressing the more general term of $1 - 2^{-n}$ with n being the resolution again. For large resolution, this term converges to 1. The DNL can then be simplified to

$$DNL(1000) = FSR \cdot \frac{\Delta R}{4R} \tag{7.11}$$

As a consequence, if the resistors only match to 0.1 %, then trimming is required for DACs with resolutions above 10 bit. This is particularly true for resistors, which are representing the MSBs. This is very similar to the SAR ADC. The resistors used to be adjusted by laser trimming. Alternatively, resistors might also be connected or disconnected with transistors and in-package trim. The parasitic switch resistance must be considered similarly as the switch capacitance for the SAR ADC. For good matching, the resistors require large area.

Step 5 in Fig. 7.18 shows that the output resistance is constant with the value R. This resistor will directly influence the output settling time together with any capacitance at the output. The output resistance also generates thermal noise that was already discussed previously.

$$n_{DAC} = \sqrt{4kT \cdot R \cdot \Delta f} \tag{7.12}$$

A small resistance comes with the disadvantage of a high reference current. Unfortunately, this reference current is code dependent similarly to the SAR ADC again. Any resistance in the reference path will not only generate a gain error, but also a significant INL. Most high resolution R-2R DACs therefore provide force and sense lines, so that the parasitic resistance can be suppressed by the gain of a driving amplifier.

As another disadvantage, the voltages at all resistors will change when switching the input code from 0111 to 1000. This will cause a significant distortion, which is expressed with the parameter 'glitch energy'. Even if R-2R DACs can settle within 1us, the THD is often specified at much lower update rates to reduce the effects of the glitch energy.

The currently best R-2R DAC on the market is the AD5791 from Analog Devices [11]. Its parameters are illustrated in Table 7.5. It achieves a DNL of ±1 LSB even at a resolution of 20 bit. The output resistance R is typically 3.4 kΩ, so that one LSB is expressed by the resistance R_{LSB} of only

$$R_{\text{LSB}} = \frac{2 \cdot R}{2^{19}} = 6.5 \ \text{m}\Omega \tag{7.13}$$

This value illustrates that not only the resistors need to match well, but also the switches and the wiring of the various resistors.

Similar to SAR ADCs, the power dissipation often neglects the dissipation from the reference voltage even if the power dissipation of the reference path is dominating. A particular example is specifying a current consumption of 2 µA [12]. The exemplary part specifies the reference voltage to ±10 V and reference input

Table 7.5 Exemplary specifications of a R-2R DAC (AD5791 from Analog Devices [11])

Parameter	Min	Typ	Max	Unit
Accuracy				
Resolution	20			Bit
Relative accuracy		±0.25	±1.5	LSB
Differential linearity error		±0.5	±1	LSB
Gain error		±0.1	±7	LSB
Gain temperature coefficient		±0.02		ppm FSR/°C
Offset error		±0.1	±7	LSB
Offset drift		±0.04		ppm FSR/°C
Output characteristics				
Output voltage range	−10		10	V
Output voltage settling time		1		µs
Slew rate		50		V/µs
Output noise spectral density		7.5		nV/sqrt(Hz)
Output voltage noise		1.1		µVpp
Midscale glitch impulse		3.1		nV-s
DC output impedance		3.4		kΩ
SFDR ($f_{out} = 1$ kHz, $f_{sample} = 10$ kHz)		100		dB
THD ($f_{out} = 1$ kHz, $f_{sample} = 10$ kHz)		97		dB

resistance to 7.5 kΩ, which is suggesting a 2.7 mA current in the reference path or a power dissipation of larger than 50 mW.

The R-2R DACs can particularly be found in applications such as test equipment. In regulation loops for industrial or medical instrumentation, it is used as accurate analog feedback to an input voltage. In PLC equipment (programmable logic control), R-2R DACs will find use in analog I/O cards. It can further be used to control fans with DC motors or to control the energy of diodes in optical networks.

If even better THD and noise performance is required, then the delta-sigma DAC can be used, which is explained next.

7.5 Delta-Sigma DAC

It was mentioned above that the glitch energy is causing distortion, which particularly in audio applications is undesirable. The delta-sigma DACs are used in such applications. These DACs enabled the CD player. IEEE actually announced the delta-sigma DACs from Burr-Brown to an IEEE milestone [13].

The idea is to bring the digital input word into the bit stream format, which is already known from the delta-sigma ADCs. While ADCs filter the bitstream digitally to receive a high resolution digital output word, the DACs filter the bitstream in the analog domain and receive a high resolution analog output signal. The scheme is shown in Fig. 7.19. This is reducing the analog circuitry to a linear filter so that a very linear transfer function can be achieved.

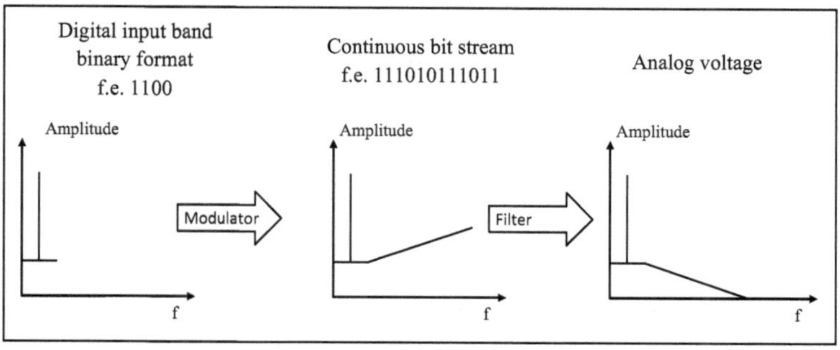

Fig. 7.19 Signal flow inside a delta-sigma DAC

The digital input word, which is provided with small data rate, has a limited bandwidth of half the data rate based on the Nyquist theorem. The noise floor is defined by the limited resolution of the digital input word (for example 24 bits) and in first order can be assumed to be white noise, which is equally distributed. A digital delta-sigma modulator will bring the data format in a continuous running bitstream, which is known from the delta-sigma ADC, at higher data rate. The bitstream format is advantageous as it can be transferred to an analog signal by pure filtering. The conversion from the binary format to the bitsteam format will cause some quantization error, which will be shaped again, so that noise only occurs at high frequencies and not in the interesting signal frequencies. The bitstream will then be filtered with an analog filter to suppress the quantization noise and to receive the analog output signal.

The input of the filter is connected to a positive reference potential, if the bitstream is '1' and to a negative reference potential, if the bitstream is '0'. The filter will basically apply the reference voltage scaled by the digital input at its output. This generates requirements for the reference voltage and to the analog filter that are very similar to the continuous time delta-sigma modulator.

A significant error source is the distortion of the switching process between the positive and the negative reference voltage potential. If the reference load changes, the reference potential will first follow the load change due to the limited bandwidth and settle back to an accuracy, which is defined by the load rejection of the reference voltage.

The switching frequency is significantly higher at midscale compared to full-scale output voltages. Therefore the switching process is code dependent and will not only generate noise but also distortion. Charge injection is adding to this effect. The filter should therefore be realized fully differential to keep the reference load constant and to have equal charge injection in the positive and negative filter path.

The average input voltage will also be affected by the clock jitter. The reference voltage will be applied for too long or too short based on the jitter. Normally, the jitter is basically generating noise. However, the current consumption of the DAC might be code dependent due to the bitstream nature. Internal wire resistance or external parasitic inductance in the supply path will generate a modulation on the internal supply voltage. The modulated supply will affect the clock frequency, as the internal clock generator has limited power supply rejection and the delay of gates in the clock tree is strongly supply voltage dependent. For best clock stability a crystal is often used to set the clock frequency.

Both, the reference settling and the clock jitter effect, can be suppressed by using a switched-capacitor filter approach, which samples the reference voltage in a first clock phase and then transfers the charge to an integrator in the second clock phase. The output of the integrator is also the filter output.

The reference potential still shows distortion, when the sample capacitor is coupled to the reference for charging the capacitor. However, the reference is settled back to its nominal value by the time the charge on the capacitor is frozen. This suppresses the reference settling error. Independent of the clock jitter, an equal amount of charge is transferred to the filter during one clock cycle. The code dependency of the clock jitter is therefore eliminated.

The integrator output will show voltage steps, which are a function of the capacitor ratio between the integrator capacitor C_i and the sample capacitor C_s.

$$\Delta V_{out} = REF \cdot \frac{C_s}{C_i} \tag{7.14}$$

Such voltage steps are undesirable and are not applicable for continuous-time filters. Commercial devices therefore often show a combination of switched-capacitor filters and continuous-time filters as shown in Fig. 7.20. The integration process will also generate distortion at the input of the continuous time filter, but the voltage steps are significantly lower in amplitude.

Figure 7.20 shows the typical block diagram for a delta-sigma DAC. The digital input has a slow data rate. A sample rate of 192 kHz is typical in audio applications. Inputs with a minimum wire count such as I^2C or SPI are typically used. The input, which has binary format, is then applied to the digital delta-sigma modulator with an order a. The output of the modulator is a digital bitstream with significantly higher data rate. The higher the modulator data rate is compared to the digital input rate, the lower is the quantization noise in the signal band, which is generated by the modulator.

The bitstream will then control, if the sampled reference charge is integrated to the positive or negative path of the switched capacitor filter that has the order b. The

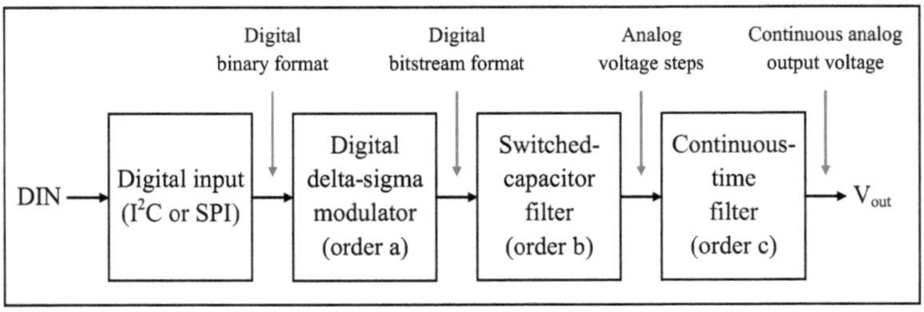

Fig. 7.20 Typical block diagram of a delta-sigma DAC

output finally connects to the continuous time filter with order c. Similar to the delta-sigma ADC, the total order $(b+c)$ of the filter should be higher than the order of the modulator a by at least one. If the order is even higher, then the additional filter effect is minor, but delay, noise and non-linearity from the analog filter circuitry are added. The analog filter design is a wide field and will not be discussed in this book. A large variety of literature is available [14, 15].

The 1/f-noise will play an essential role, as the output bandwidth is very limited. Offset cancellation schemes such as mentioned in Chap. 4 could be added into the switched-capacitor filter. This would suppress the 1/f-noise at least for the switched-capacitor components. The 1/f-noise can be reduced by using large transistors for the differential input pairs of the amplifiers used in the filters.

There are no special layout requirements, which were not discussed previously in Chaps. 2 and 4. The supply of the digital circuitry should be separated from the analog supply to avoid coupling of noise and distortion through the power supply. Care must be taken to avoid clock jitter. The switched capacitor filter should clearly separate the digital circuitry and the analog circuitry to avoid coupling from digital signals into the analog circuitry. The switching process can also generate noise and distortion in the analog supply. Care must be taken that these supply ripples do not affect the analog outputs. The capacitors and resistors in the filters should match well to minimize gain and offset errors, which are actually less important for audio applications.

Table 7.6 illustrates the interesting specifications of an audio DAC. It can be seen, that gain and offset are specified widely as they do not matter in AC

Table 7.6 Exemplary performance specifications of a ΔΣ DAC (PCM1794 from Texas Instruments [16])

Parameter	Min	Typ	Max	Unit
Accuracy				
Resolution	24			Bit
Gain error	−6	±2	6	% FSR
Gain mismatch (channel-to-channel)	−3	±0.5	3	% FSR
Bipolar zero error	−3	±0.5	3	% FSR
Dynamic performance				
SNR ($2V_{rms}$)	123	127		dB
THD + N		0.0004	0.0008	%
Dynamic range	123	127		dB
Channel separation	120	123		dB
Sample frequency (f_s)	10		200	kHz
Power supply				
AVDD	4.75	5	5.25	V
DVDD	3	3.3	3.6	V
I_{AVDD}		37		mA
I_{DVDD}		45		mA
Power dissipation (5.5 V)		335		mW

applications. The noise is typically 127 dB, which is 30 dB better than the highest performance R-2R DAC [11]. Note that the SNR is even better in modes with higher output amplitude. The THD is limited by the voltage coefficient of capacitors and resistors. It is slightly better if compared to R-2R DACs, which suffer from high glitch energy.

The power consumption is unexpectedly high. A significant amount is generated by the digital circuitry and could be reduced by moving to a smaller process technology. The analog current consumption is partly required to keep the thermal noise low.

The flatness of the filters in the pass-band is particularly important for audio applications and is therefore specified as shown in Table 7.7. The filters will also generate a time delay similar to the digital filters, which is in the millisecond range. However, such a delay is not important for applications such as audio.

Delta-sigma DACs can also be used in other applications. The DAC1220 from Texas Instruments [17] is a commercial example. Its specifications are illustrated in Table 7.8.

Such DACs are particularly interesting for test and measurement applications. In the industrial environment, this could include calibration routines in production. For such applications, the DC specifications and particularly the drift specifications of gain and offset are essential.

It is visible that all commercial product examples were either from Analog Devices or Texas Instruments. They are dominating the market of high-performance converters today. Unfortunately, the high-performance data converter market is very limited. Further interesting manufacturers are Linear Technologies, Maxim Integrated and soon eesy-ic.

Table 7.7 Exemplary filter specifications of a ΔΣ DAC (PCM1794 from Texas Instruments [16])

Parameter	Min	Typ	Max	Unit
Filter 1 characteristics				
Passband (0.00001 dB)			0.454	f_s
Passband (−3 dB)			0.46	f_s
Passband ripple			±0.00001	dB
Stop band	0.546			f_s
Stop band attenuation	−130			dB
Delay time		55		$1/f_s$
Filter 2 characteristics				
Passband (0.04 dB)			0.254	f_s
Passband (−3 dB)			0.46	f_s
Passband ripple			±0.001	dB
Stop band	0.732			f_s
Stop band attenuation	−100			dB
Delay time		18		$1/f_s$

Table 7.8 Exemplary specifications of a delta-sigma (DAC1220 from Texas Instruments [17])

Parameter	Min	Typ	Max	Unit
Accuracy				
Monotonicity	20			Bit
Linearity error			±15	ppm FSR
Gain error			±150	ppm FSR
Gain error drift		2		ppm FSR/°C
Unipolar offset error			±60	ppm FSR
Unipolar offset drift		1		ppm/°C
Output characteristics				
Output voltage settling time (120 ppm)		1.8	2	ms
Output voltage settling time (30 ppm)		15		ms
Output noise voltage (0.1 Hz to 10 Hz)		1		μV_{rms}
Capacitive load stability ($R_L = 2\ k\Omega$)		1000		pF
Power supply				
AVDD	4.75		5.25	V
I_{AVDD}		460		μA
I_{DVDD}		140		μA
Power dissipation (5.5 V)		2.5	3.5	mW

References

1. Product datasheet of the DAC9881, Texas Instruments, 2008, URL: http://www.ti.com/lit/ds/symlink/dac9881.pdf, status of 3/11/2015.
2. Georgi Radulov, Patrick Quinn, Hans Hegt, Arthur H.M. van Roermund, *Smart and Flexible Digital-to-Analog Converters*, Springer, New York/Dordrecht/Heidelberg/London, 2011.
3. Martin Clara, *High-performance D/A converters*, Springer, New York/Dordrecht/Heidelberg/London, 2013.
4. Product Datasheet of the DAC5670, Texas Instruments, 2013, URL: http://www.ti.com/lit/ds/symlink/dac5670.pdf, status of 3/11/2015.
5. Yongjian Tang, Hans Hegt, Arthur van Roermund, *Dynamic-Mismatch Mapping for Digitally-Assisted DACs*, Springer, New York/Dordrecht/Heidelberg/London, 2013.
6. Abdullah Yilmaz, *LSB interpolation circuit and method for segmented digital-to-analog converter*, US Patent 6246351, Burr-Brown, 2001, URL: http://www.google.com/patents/US6246351, status of 3/11/2015.
7. Product datasheet of the DAC8550, Texas Instruments, 2012, URL: http://www.ti.com/lit/ds/symlink/dac8550.pdf, status of 3/11/2015.
8. Abdullah Yilmaz, *Resistor string integrated circuit and method for reduced linearity error*, US Patent 6496133, Texas Instruments, 2002, URL: http://www.google.com.ar/patents/US6496133, status of 3/11/2015.
9. Abdullah Yilmaz, Jerry Doorenbos, *Integral nonlinearity error correction for interpolating string DAC*, US Patent 6970122, Texas Instruments, 2005, URL: http://www.google.com.ar/patents/US6970122, status of 3/11/2015.
10. K.A. Efstathiou, D.S. Karadimas, *An R-2R Ladder-Based Architecture for High Linearity DACs*, in the proceedings of the IEEE Conference on Instrumentation and Measurement Technology (IMTC2007), pp. 1-5, 2007.

11. Product datasheet of the AD5791, Analog Devices, 2011, URL: http://www.analog.com/static/imported-files/data_sheets/AD5791.pdf, status of 3/11/2015.
12. Product Datasheet of the DAC8871, Texas Instruments, 2008, URL: http://www.ti.com/lit/ds/symlink/dac8871.pdf, status of 3/11/2015.
13. IEEE Global History Network, *Milestones: 16-bit Monolithic DAC*, IEEE, 2009, URL: http://www.ieeeghn.org/wiki/index.php/Milestones:16-bit_Monolithic_DAC,_1981, status of 3/11/2015.
14. M. Shibata, A. Yasuda, K. Sato, T. Soga, *A cascaded delta-sigma DAC with an analog FIR filter reducing mismatch-effects*, in the proceedings of the 48th Midwest Symposium on Circuits and Systems, pp. 1263-1266, 2005.
15. S. Tao, R. Schreier, F. Hudson, *Mismatch shaping for a current-mode multibit delta-sigma DAC*, in the IEEE journal of Solid-State Circuits, Volume 34, Issue 3, pp. 331-338, 1999.
16. Product Datasheet of the PCM1794, Texas Instruments, 2006, URL: http://www.ti.com/lit/ds/symlink/pcm1794.pdf, status of 3/11/2015.
17. Product Datasheet of the DAC1220, Texas Instruments, 2009, URL: http://www.ti.com/lit/ds/symlink/dac1220.pdf, status of 3/11/2015.

Index

A
Absolute temperature, 151
Accumulator, 217
Accuracy, 156
AC measurement, 292
AC specifications, 11
All-pass CDAC, 129
Amplifier, 315
Analog filter, 323
Analog market, 1
Analog section, 278
Analog-to-digital converters (ADC), 2
Aperture jitter, 24
Auto-zero functionality, 94
Auto-zero stage, 87

B
Bandwidth, 93, 105, 237, 268
Best-fit, 11
Bipolar, 210
Bipolar transistors, 150
Bipolar zero, 9
Bit, 322
Bit stream, 32
Boltzmann constant, 150
Buried N-layer, 114

C
Calibration, 60, 216, 285, 309
Capability study, 299
Capacitive DAC (CDAC), 56, 149, 177
Cascaded current source, 309
Cascading modulators, 213
CDAC. *See* Capacitive DAC (CDAC)
Charge injection, 52, 161, 323
Chip Select, 23
Clock jitter, 24, 210, 239, 323
CMOS, 20
CMRR. *See* Common-mode rejection ratio (CMRR)
Coarse ADC, 175
Coarse CDAC, 132
Code change glitch, 307
Code of interest, 287
Coherent sampling, 15
Common-mode boost, 167
Common-mode current, 248
Common-mode input voltage, 271
Common-mode rejection, 56, 66, 109, 272, 275
Common-mode rejection ratio (CMRR), 69
Common-mode shift, 176
Common-mode voltage, 6, 53, 147, 166, 170
Comparator, 84, 168, 199, 207
Compensate, 253
Compensation capacitor, 276
Consumer market, 1
Contactor, 281
Continuity test, 295
Continuous time, 209
Convergence, 149
Conversion rate, 156
Correlated double sampling, 230
Cross-coupled latch, 88
Current consumption, 24
Current input, 248
Current measurement, 39
Current source, 248
Current steering DAC, 308

F. Ohnhäuser, *Analog-Digital Converters for Industrial Applications Including an Introduction to Digital-Analog Converters*, DOI 10.1007/978-3-662-47020-6

D

DAC. *See* Digital-to-analog converters (DAC)
Data rate, 209
Datasheet limits, 296
DC specifications, 5, 281
Decoupling, 114, 192, 197
Decoupling capacitor, 279, 294
Deep power-down, 166
Delay, 207
Delta-sigma ADC, 31
Delta-sigma converter, 25
Delta-sigma DAC, 322
Device-under-test (DUT), 281
Die picture, 179
Die size, 192, 201
Differential non-linearity (DNL), 10, 64, 83, 144, 155, 182, 207, 267, 277, 283, 305, 308, 314, 321
Differential pairs, 87, 90
Digital circuitry, 192
Digital filter, 31
Digital functionality, 294
Digital gate density, 194
Digital ground currents, 278
Digital interface, 20
Digital output, 145
Digital sections, 278
Digital-to-analog converters (DAC), 2, 107, 149, 285, 305
Digital trim, 116, 179, 187
Digital trimming, 195
Discrete Fourier transformation (DFT), 15
Distortion, 52, 90, 137, 323
DNL. *See* Differential non-linearity (DNL)
Driver, 243
Dual CDAC, 159, 174, 195
  architecture, 130
Dual gate oxide, 179
Dual integrated circuits, 164
Dummy switches, 233
Dummy transistors, 224
DUT. *See* Device-under-test (DUT)
Dynamic element matching, 73
Dynamic errors, 164
  correction, 112, 140, 143, 157, 159, 176, 195

E

Effective bandwidth, 228
Effective number of bits (ENOB), 18, 36
Electrical fuse, 83
Electrical motor drives, 38
e-meter, 44
Encoder, 39
ENOB. *See* Effective number of bits (ENOB)
Error correction, 149
ESD-cell, 295
Exponential function, 84
Exponential settling, 227, 269

F

Fast Fourier transformation, 15
Feedback, 220, 247
Feedback-loop, 207
Final test limits, 296
Fine CDAC, 198
Flash converter, 4
Flicker noise, 94, 108
Flipflop (FF), 88
Floating, 132
Floor plan, 73, 93
1/f-noise, 306, 325
Force, 314
Fourier transformation, 13
Frequency –3 dB, 85
Full-scale error, 216
Full-scale input voltage, 146
Full-scale range, 8
Fully differential, 6, 126, 144, 167, 174, 210, 323
Fully differential latch, 88
Fuses, 81

G

Gain, 5–10, 65, 120, 149, 313, 321
Gain-bandwidth-product, 259
Gain drift, 243
Gain errors, 227, 272
Gain trim, 143
Gate-source voltage, 170
Gaussian distribution, 164
Glitch energy, 321
Ground, 5
Ground currents, 278
Ground planes, 278
Group delay, 35
Guardbands, 296, 299

H

Hall effect, 39
Handler, 281
Handler interface board (HIB), 281

Higher order systems, 211
High impedance, 73, 136
High-pass transfer function, 208
High voltage, 138
Histogram, 19, 156, 181, 185, 282
Hits per code, 156, 282
Hold switch, 52, 166, 171, 195

I
I2C, 23
Industrial application, 2
INL. *See* Integral non-linearity (INL)
In-package, 61, 76
Input range, 65
Input switch, 52
Input voltage, 267
Input voltage range, 8, 74
Instrumentation amplifiers, 275
Integral linearity, 54
Integral non-linearity (INL), 11, 182, 207, 277, 305, 308, 314, 321
  trim, 179
Integrator, 288
  leakage, 233
Interpolating amplifier, 315
Interpolation, 314
Isolated PWELLs, 114

J
Joint Electron Tube Engineering Council (JEDEC), 20

L
Laser cutting, 61
Laser links, 75
Layout, 90, 154, 179, 202, 233, 310, 325
Leakage, 171
Least-significant bits (LSB), 3, 74
Life time, 130
Linear voltage coefficient, 145
Linearity, 214, 227
Load rejection, 227, 323
Lookup table, 217
Low impedance node, 77
Low-pass, 208
Low-voltage, 200
LSB. *See* Least-significant bits (LSB)

M
Main frame, 281
Manual actuator, 281
Master fuse, 81
Matching, 59, 64, 70, 272, 309
Memory, 81, 144, 194
Mismatch, 139
Missing code, 11
Molding compound, 61, 75
Most significant bit (MSB), 58
Multi-bit feedback, 249
Multi-bit quantizer, 213

N
NAP mode, 166
Negative full-scale, 3, 215, 308
Noise, 74, 93, 105, 156, 181, 306, 321
Noise density, 95, 105, 228, 256, 270
Noise-shaping, 31, 229
Noise transfer function, 210
Non-linearity, 244
Non-monotonic, 306
Non-overlapping clocks, 223
Nyquist theorem, 15, 40

O
Offsets, 5–10, 68, 90, 159, 313
  cancellation, 87
  error, 272, 285
One-Time-Programmable ROM (OTPs), 81
On-resistance, 129, 241
Operational transconductance amplifier (OTA), 88
Oscillation, 113, 192, 269
Output resistance, 311, 317
Overdrives, 84
Oversampling, 31, 73, 111, 177, 207
Oversampling ratio, 209

P
Parallel converter, 4
Parallel interface, 21
Parameters, 296
Partial ramp, 164
PCB. *See* Printed circuit board (PCB)
Pedestal DAC, 286
Performance parameters, 3
Phase margin, 269
Pipeline ADCs, 27
Pipeline converter, 25
Point symmetrical, 72
Positive full-scale (PFS), 3, 308
Power consumption, 24, 165, 181, 192, 237, 294
Power dissipation, 201

Power relays, 43
Power supply, 113
  rejection, 284
Power transistor, 81
Printed circuit board (PCB), 279
Probe interface board (PIB), 281
Probe-needles, 281
Prober, 281
Process, 192, 193
Production test, 156, 281
Pseudo differential architecture, 70
Pseudo differential inputs, 6, 174
Pulse responds, 35

Q
Quadratic voltage coefficient, 145
Quad site, 164
Quality flow, 302
Quantization, 13, 17, 31
  noise, 31, 213, 323, 324
Quantizer, 207
Quiescent current, 87

R
Rail-to-rail amplifiers, 271
Reference, 5, 104, 140, 147
Reference voltage, 275
Register, 81
Relative accuracy, 306
Reliability, 130
Repeatability, 288
Resistive feedback, 241
Resolution, 3, 5
Resolver, 40
rms-noise, 19
R-2R, 317

S
Sample-and-hold, 6
Sample capacitance, 74
Sample capacitors, 52, 145
Sampled signal, 14
Sample rate, 14
Sampling, 120, 166
  frequency, 14
  time, 52, 268
SAR. *See* Successive approximation register (SAR)
Saturation voltage, 170
Scale, 142
Scale-down capacitor, 59, 78
Scale-down trim, 79, 140, 179, 195
Sense lines, 314
Serial port interface (SPI), 23
Servo loop, 156, 287
Settling, 104, 130, 227
Settling time, 35, 84, 307, 312
Shunt resistors, 39
Signal amplitude, 120
Signal to noise and distortion (SINAD), 18
Signal-to-noise ratios (SNR), 77, 132, 207, 250, 251
Signal transfer function, 210
SINC filter, 34
Single integrated circuit, 164
Single-ended, 5, 224
Single supply voltage, 271
SLEEP mode, 166
Slewing phase, 269
Slew rate, 307
Slope converters, 26
Smart meters, 42
Snapshot, 29, 111
SNR. *See* Signal-to-noise ratios (SNR)
Soft turn on, 196
Specification, 305
Specific density, 98
Spurious-free-dynamic-range (SFDR), 19
Stability, 212
String DAC, 107, 153, 311
Successive approximation register (SAR), 29, 149, 293
  ADC, 52
  converter, 25
Supply voltage, 130, 192, 323
Supporting points, 214
Switched-capacitor, 31, 84, 255, 324

T
Temperature coefficient, 97, 272
Temperature range, 296
Test head, 281
Test machine, 281
Test time, 156
THD. *See* Total-harmonic-distortion (THD)
Thermal energy, 97
Thermal noise, 94
Thermal noise density, 306
Thin film resistor, 99
Time constant, 84, 105, 312, 317
Time-discrete, 211
Total geometrical sum, 110

Total-harmonic-distortion (THD), 18, 293
Transfer function, 3, 140, 181, 185, 187, 207, 214
Transition, 3
Trim, 60
Trim times, 164

U
Unipolar input voltage range, 210
Unipolar zero, 9
Unique charge, 151
Unity capacitors, 70
Unstable, 269

V
Voltage coefficient, 54, 107, 145, 314, 326
Voltage controlled current sources, 151
Voltmeter, 286

W
Wafer level, 281
Window functions, 16

Z
Zero-power sampling, 116, 192, 195

MIX
Papier aus verantwortungsvollen Quellen
Paper from responsible sources
FSC® C105338

If you have any concerns about our products,
you can contact us on
ProductSafety@springernature.com

In case Publisher is established outside the EU,
the EU authorized representative is:
Springer Nature Customer Service Center GmbH
Europaplatz 3, 69115 Heidelberg, Germany

Printed by Libri Plureos GmbH
in Hamburg, Germany